MS324 Waves, diffusion and variational principles

Block III

Variational principles

This publication forms part of an Open University course. Details of this and other Open University courses can be obtained from the Student Registration and Enquiry Service, The Open University, PO Box 197, Milton Keynes MK7 6BJ, United Kingdom: tel. +44 (0)845 300 6090, email general-enquiries@open.ac.uk

Alternatively, you may visit the Open University website at http://www.open.ac.uk where you can learn more about the wide range of courses and packs offered at all levels by The Open University.

To purchase a selection of Open University course materials visit http://www.ouw.co.uk, or contact Open University Worldwide, Walton Hall, Milton Keynes MK7 6AA, United Kingdom, for a brochure: tel. +44 (0)1908 858793, fax +44 (0)1908 858787, email ouw-customer-services@open.ac.uk

The Open University, Walton Hall, Milton Keynes, MK7 6AA.

First published 2005. Second edition 2010.

Copyright © 2005, 2010 The Open University

All rights reserved. No part of this publication may be reproduced, stored in a retrieval system, transmitted or utilised in any form or by any means, electronic, mechanical, photocopying, recording or otherwise, without written permission from the publisher or a licence from the Copyright Licensing Agency Ltd. Details of such licences (for reprographic reproduction) may be obtained from the Copyright Licensing Agency Ltd, Saffron House, 6–10 Kirby Street, London EC1N 8TS; website http://www.cla.co.uk.

Open University course materials may also be made available in electronic formats for use by students of the University. All rights, including copyright and related rights and database rights, in electronic course materials and their contents are owned by or licensed to The Open University, or otherwise used by The Open University as permitted by applicable law.

In using electronic course materials and their contents you agree that your use will be solely for the purposes of following an Open University course of study or otherwise as licensed by The Open University or its assigns.

Except as permitted above you undertake not to copy, store in any medium (including electronic storage or use in a website), distribute, transmit or retransmit, broadcast, modify or show in public such electronic materials in whole or in part without the prior written consent of The Open University or in accordance with the Copyright, Designs and Patents Act 1988.

Edited, designed and typeset by The Open University, using the Open University TeX System.

Printed in the United Kingdom by Cambrian Printer Limited, Aberystwyth.

ISBN 978 0 7492 5272 4

2.1

The paper used in this publication contains pulp sourced from forests independently certified to the Forest Stewardship Council (FSC) principles and criteria. Chain of custody certification allows the pulp from these forests to be tracked to the end use. (see www.fsc-uk.org)

Contents

CHAPTER 1   THE CALCULUS OF VARIATIONS   5

1.1 Introduction   5
1.2 The shortest distance between two points   7
    1.2.1 The distance between two points on a given curve   7
    1.2.2 The stationary distance   10
    1.2.3 Geodesics   13
    1.2.4 The shortest path   14
1.3 Two generalisations   15
    1.3.1 Functionals depending only upon $y'(x)$ 15
    1.3.2 Functionals depending upon x and $y'(x)$ 17
1.4 Notation   18
1.5 Some examples of functionals   20
    1.5.1 The brachistochrone   21
    1.5.2 Minimal surface of revolution   22
    1.5.3 A problem in navigation   22
    1.5.4 The isoperimetric problem   22
    1.5.5 Fermat's principle   23
    1.5.6 The catenary   25
    1.5.7 Coordinate-free formulation of Newton's equations   26
1.6 Summary and Outcomes   28
1.7 Further Exercises   28
1.8 Harder Exercises   29
Solutions to Exercises in Chapter 1   32

CHAPTER 2   THE EULER–LAGRANGE EQUATION   41

2.1 Introduction   41
2.2 Preliminary remarks   42
    2.2.1 Relation to differential calculus   42
    2.2.2 The differentiation of a functional   43
2.3 The fundamental lemma of the calculus of variations   47
2.4 The Euler–Lagrange equation   48
2.5 Minimal surface of revolution   51
    2.5.1 Derivation of the functional   52
    2.5.2 Application of the Euler–Lagrange equation   53
    2.5.3 The solution in a special case   53
    2.5.4 Summary of Section 2.5   58
2.6 Soap films   60
2.7 The brachistochrone   62
    2.7.1 The cycloid   62
    2.7.2 Formulation of the problem   65
    2.7.3 A solution   66
2.8 Further Exercises   69

2.9	Harder Exercises	72
	Solutions to Exercises in Chapter 2	76

CHAPTER 3 FURTHER DEVELOPMENTS OF THE THEORY   95

3.1	Introduction	95
3.2	Invariance of the Euler–Lagrange equation	96
	3.2.1 Changing the independent variable	97
	3.2.2 Changing both the dependent and independent variables	99
3.3	Functionals containing many dependent variables	103
	3.3.1 Two dependent variables	104
	3.3.2 Many dependent variables	107
3.4	Changing dependent variables	109
3.5	Symmetries (Optional)	112
	3.5.1 Invariance under translations	112
	3.5.2 Noether's theorem	116
3.6	Further Exercises	123
3.7	Harder Exercises	126
	Solutions to Exercises in Chapter 3	127

CHAPTER 4 NEWTONIAN DYNAMICS   143

4.1	Introduction	143
	4.1.1 On nomenclature	144
4.2	Newton's laws	146
	4.2.1 Particle moving on a parabolic wire	146
	4.2.2 The simple pendulum	149
	4.2.3 The double pendulum	152
4.3	A discussion of Newton's laws	155
	4.3.1 Internal and external forces	158
	4.3.2 The centre of mass	160
	4.3.3 Generalised coordinates	162
	4.3.4 Kinetic and potential energy	166
4.4	Lagrange's equations and Hamilton's principle	174
4.5	Applications of Lagrange's equations	181
	4.5.1 The simple pendulum	182
	4.5.2 The double pendulum	186
	4.5.3 A simple pendulum with moving support	188
	4.5.4 Transverse vibrations of a taut string (Optional)	190
4.6	Further Exercises	195
4.7	Harder Exercises	197
	Solutions to Exercises in Chapter 4	199

INDEX   217

CHAPTER 1
The calculus of variations

1.1 Introduction

In this final block of MS324 we introduce two related mathematical entities, which are of use in many areas of applied mathematics and are of immense importance in formulating the laws of theoretical physics. These are the *functional* and *variational principles*. The theory of these entities is called *the calculus of variations*.

A *functional* is a generalisation of a function. A real function of a single real variable maps an interval of the real line to real numbers; for instance, the function $f(x) = 1/(1+x^2)$ maps the whole real line to the interval $(0,1]$; the function $\ln x$ maps the positive real axis to the whole real line. Similarly, a real function of n real variables maps a domain of $\mathbb{R}^n$ into the real numbers.

A functional maps a given class of functions to real numbers. A simple example of a functional is

$$S[y] = \int_0^1 dx\, y'(x)^2, \quad y(0) = 0,\ y(1) = 1, \qquad (1.1)$$

which associates a real number with any real function, $y(x)$, satisfying the boundary conditions and for which the integral exists. Here $y'(x) = dy/dx$, and throughout this block we use the notation $f'(x)$ to denote the derivative of $f(x)$ with respect to its argument when $f(x)$ is a function of a single variable. The square bracket notation $S[y]$ is used to emphasise the fact that the functional depends upon the choice of function used to evaluate the integral. Later in this chapter we shall see that a wide variety of problems can be described in terms of functionals.

Real functions of n real variables can have various properties; for instance they can be continuous, they may be differentiable or they can have stationary points and local maxima and minima. Functionals share many of these properties. In particular, the notion of a stationary point of a function has an important analogy in the theory of functionals and this gives rise to the idea of a *variational principle*, which arises when the solution to a problem is given by a function that makes a given functional stationary. Variational principles are very common and important in the natural sciences.

A simple example of a variational principle is that of finding the shortest distance between two points. Suppose that the two points lie in a plane, with one point at the origin, O, and the other at point A with coordinates $(1,1)$, and that $y(x)$ represents a smooth curve passing through O and A. The distance between O and A along this curve is given by the functional

$$S[y] = \int_0^1 dx\, \sqrt{1 + y'(x)^2}, \quad y(0) = 0,\ y(1) = 1. \qquad (1.2)$$

This result is proved in Section 1.2, where we also prove the physically obvious result that the shortest distance is given by the straight line joining O to A. This may seem like using a sledgehammer to crack a walnut, but the methods introduced are important parts of the calculus of variations and can be used to solve more general and difficult problems.

Variational principles are important for three principal reasons. First, many problems are naturally formulated in terms of a functional and an associated variational principle. For instance, the shape of a closed curve of given length that encloses the largest area is defined by a variational principle. When the closed curve is on a flat surface, this becomes a traditional problem that was first posed and solved by Zenodorus (c. 180 BC), but for curves on other surfaces the solution is not so easy. Other examples are the shape assumed by a hanging chain, which is determined by minimising the functional for the energy, and the path taken by a ray of light, which minimises the time of passage. These, and other, variational principles will be described later in this chapter.

Second, most equations of mathematical physics can be derived from variational principles. This is important partly because it suggests a unifying theme in our description of nature and partly because such formulations are independent of any particular coordinate system, so making the essential mathematical structure of the equations more transparent and easier to understand. We explore some of these aspects in Chapter 4, where we reformulate Newton's equationsof motion as a variational principle.

Finally, variational principles provide powerful computational tools; this use is not considered here.

The structure of this block is as follows. In this introductory chapter we first consider the particular variational principle defining the shortest distance between two points in a plane. This simple example is used to introduce many of the important ideas associated with functionals and also the notation required to deal with the general theory. Chapter 1 ends with a discussion of some of the problems that can be formulated in terms of variational principles.

Chapter 2 contains the main theory, where we deal with a general class of variational principles described by the Euler–Lagrange equation. We develop and generalise the ideas introduced here, in order to obtain necessary conditions for a function to make this class of functionals stationary.

Chapter 3 is in two parts. In the first part we develop a minor generalisation of the theory described in Chapter 2, which is required later in order to reformulate Newton's equations as a variational principle. In the second part we consider how some general properties of functionals can impose constraints upon the behaviour of the stationary functions. For instance, this theory shows how momentum conservation in a mechanical system is caused by the invariance of the associated functional under translations in space; similarly, energy conservation is a consequence of the invariance of the associated functional under translations in time.

In Chapter 4 we discuss the application of Newton's equations to simple mechanical systems and show how this task is often made easier using Lagrange's formulation. We also show that, under certain widely applicable circumstances, Newton's three laws can be expressed as a simpler variational principle, called Hamilton's principle.

Exercise 1.1

Determine the value of the functional

$$S[y] = \int_0^1 dx\, y'(x)^2, \quad y(0) = 0,\ y(1) = 1,$$

for the following functions:

(a) $y(x) = x$

(b) $y(x) = x^a$ where $a > \frac{1}{2}$

(c) $y(x) = \sin(\pi x / 2)$

1.2 The shortest distance between two points

The distance between two points on a given surface can be expressed in terms of a functional depending on the path between the two points and the properties of the surface. The simplest surface is a two-dimensional plane, and we deal with this case here.

In this section two significant results are derived. First, we derive the functional for the distance between two points in a plane along a given smooth curve. Second, we show that the stationary path is a straight line.

It is well known that the shortest path between two points in a plane is the straight line joining them; indeed, it is Euclid's definition of a straight line. However, it is almost always easiest to understand a new idea by applying it to a simple familiar problem, so here we introduce the ideas of the calculus of variations by finding the equation of this line. The algebra may seem over-complicated for this simple problem, but far more complicated problems may be solved with very little extra effort.

See Book I of Euclid's *Elements*, Definition 4.

1.2.1 The distance between two points on a given curve

First, we need an expression for the length of a curve between two given points P_a and P_b in a plane, having the Cartesian coordinates (a, A) and (b, B), respectively. If the curve is represented by $y = f(x)$ with $a \leq x \leq b$, then, since it passes through P_a and P_b, we must have $f(a) = A$ and $f(b) = B$, as shown in Figure 1.1. We also assume that $f(x)$ is differentiable for $a \leq x \leq b$.

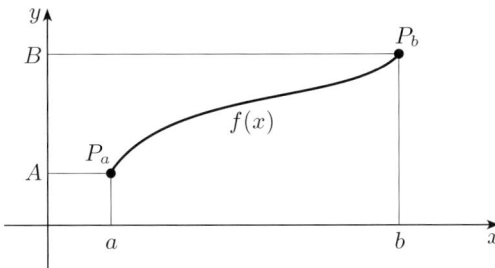

Figure 1.1 The graph of an arbitrary curve passing through P_a and P_b, with coordinates (a, A) and (b, B), respectively.

We assume that the curve joining P_a and P_b is described by a single-valued function $f(x)$, i.e. for each x in the interval $a \leq x \leq b$, $f(x)$ has one and only one value. This means that curves such as those shown in Figures 1.2 and 1.3 are not considered, because they are not the graphs of single-valued functions.

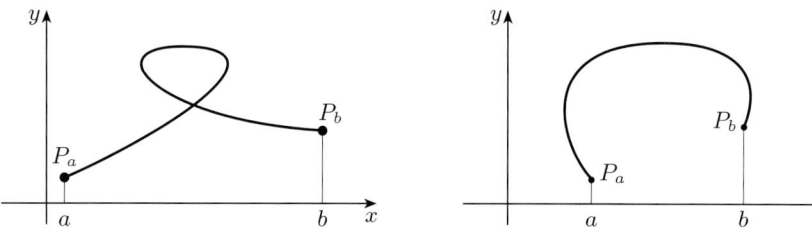

Figure 1.2   Figure 1.3

Divide the interval between a and b into N equal-length segments (x_{k-1}, x_k), $k = 1, 2, \ldots, N$, where $x_0 = a$ and $x_N = b$, so

$$x_{k+1} = x_k + \delta x \quad \text{and} \quad N\delta x = b - a, \tag{1.3}$$

as shown in Figure 1.4.

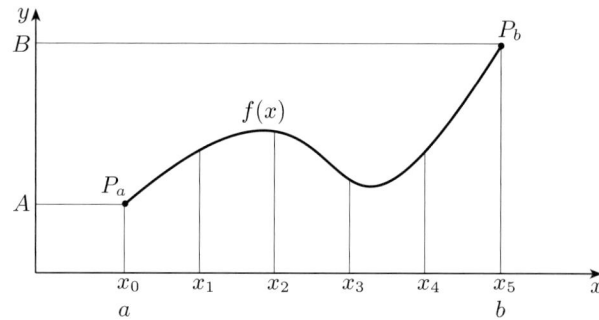

Figure 1.4 Diagram showing the subdivision of the interval $[a, b]$ into N equal intervals, where $x_{k+1} - x_k = \delta x$, $k = 0, 1, \ldots, N-1$; in this example $N = 5$.

The curve is now approximated by a continuous curve C_N comprising N straight-line segments, the kth segment joining the points (x_{k-1}, y_{k-1}) and (x_k, y_k), where $y_k = f(x_k)$. The length of the original curve is *defined* to be the limit as $N \to \infty$ of the length of C_N. Thus we need to determine the length of each segment of C_N, to add these together and then to take the limit as $N \to \infty$.

A more rigorous definition of the length allows for the increments along the x-axis to be different lengths.

First, we find the distance along the straight line between (x_k, y_k) and (x_{k+1}, y_{k+1}). Since $y_k = f(x_k)$, $k = 0, 1, \ldots, N$, are the points defining C_N the relation between successive values of y is obtained using Taylor's expansion,

$$\begin{aligned} y_{k+1} &= f(x_k + \delta x) \\ &= y_k + \delta x f'(x_k) + O(\delta x^2), \quad y_k = f(x_k). \end{aligned} \tag{1.4}$$

The length of the required segment, denoted by δs_k, is given by Pythagoras' theorem, as illustrated in Figure 1.5.

1.2 The shortest distance between two points

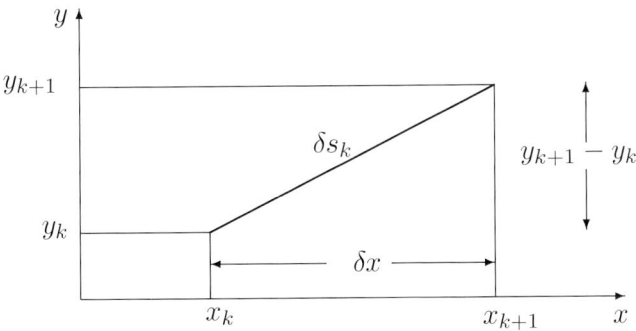

Figure 1.5

The length δs_k is determined by the end points:
$$\delta s_k^2 = (x_{k+1} - x_k)^2 + (y_{k+1} - y_k)^2,$$
$$= \delta x^2 + \left(f'(x_k)\delta x + O(\delta x^2)\right)^2 = \left(1 + f'(x_k)^2\right)\delta x^2 + O(\delta x^3), \quad (1.5)$$

where we have used equation (1.4) to write $y_{k+1} - y_k = f'(x_k)\delta x + O(\delta x^2)$. On taking the square root of each side of this equation, we obtain an expression for δs_k

$$\delta s_k = \delta x \sqrt{1 + f'(x_k)^2 + O(\delta x)}$$
$$= \delta x \sqrt{1 + f'(x_k)^2} \sqrt{1 + \frac{O(\delta x)}{1 + f'(x_k)^2}}, \quad (1.6)$$

and hence, using the binomial expansion for the right-hand side,

$$\delta s_k = \delta x \sqrt{1 + f'(x_k)^2} + O(\delta x^2). \quad (1.7)$$

The approximate distance from (a, A) to (b, B) along the curve $y = f(x)$ is given by the sum of all the segment lengths:

$$S[f] \simeq \sum_{k=0}^{N-1} \delta s_k = \sum_{k=0}^{N-1} \delta x \sqrt{1 + f'(x_k)^2} + O(N\delta x^2). \quad (1.8)$$

Since $N\delta x = b - a$, as $\delta x \to 0$ this sum becomes the integral

$$S[f] = \int_a^b dx \sqrt{1 + f'(x)^2}, \quad f(a) = A, \ f(b) = B. \quad (1.9)$$

The distance between the two points P_a and P_b, along the curve $y = f(x)$, is given by the above expression: for each suitable real function, $f(x)$, the integral has a numerical value denoted by $S[f]$, the functional for the distance. For instance, if P_a is at the origin and $P_b = (1, 0)$ the distance from P_a to P_b along the straight line joining these points is 1, whereas along the semi-circle of which P_aP_b is a diameter, the distance is $\pi/2$.

The functional (1.9) defines the distance between the points (a, A) and (b, B) along a given curve passing through these points. In the next subsection we consider how the length of the path varies with $f(x)$ and we shall define the notion of a stationary function; we also show that the straight line through the end points is a stationary function. This may seem an elaborate method of deriving the fourth definition given in Euclid's *Elements*, particularly since we have already used this definition to derive said functional. However, the method of deriving this functional and finding its stationary paths can easily be generalised to more difficult problems; for instance, it may be used to derive functionals for the shortest distance between two points on curved surfaces. Some examples of such problems are discussed in Exercises 1.26 and 1.27.

Exercise 1.2

If P_a is at the origin and $P_b = (b, 0)$, where $b > 0$, show the following.

(a) If $y = f(x)$ is the straight line joining P_a to P_b, then the integral (1.9) has the value b.

(b) If $y = f(x)$ is a semi-circle with diameter $P_a P_b$, so that $f(x)^2 = bx - x^2$, show that

$$S[f] = \frac{b}{2} \int_0^b dx \, \frac{1}{\sqrt{x(b-x)}}.$$

By putting $x = b \sin^2 \phi$, show that $S = \pi b / 2$.

1.2.2 The stationary distance

We require the equation defining the functions for which $S[y]$ is stationary, that is the stationary paths. However, as in ordinary calculus, the idea of a stationary point of a function is preceded by the notion of the rate of change of functions. Similarly, in the calculus of variations we first need to consider the change of $S[y]$ with changes in y; then we can define the idea of a stationary path and, hopefully, find these paths. These ideas are introduced here, and developed in Chapter 2, using the analogy of the stationary points of functions of many real variables.

You will recall from the discussion about stationary points of a function of n variables (Block 0, Subsection 2.2.7) that we define a stationary point to be one at which the values of the function at all neighbouring points are 'almost' the same as at the stationary point. To be precise, if $G(\boldsymbol{x})$ is a suitably behaved function of n real variables, $\boldsymbol{x} = (x_1, x_2, \ldots, x_n)$, we compare values of G at $\boldsymbol{x}$ and the nearby point $\boldsymbol{x} + \epsilon \boldsymbol{\xi}$, where $|\epsilon| \ll 1$ and $|\boldsymbol{\xi}| = 1$,

$$G(\boldsymbol{x} + \epsilon \boldsymbol{\xi}) - G(\boldsymbol{x}) = \epsilon \sum_{k=1}^{n} \frac{\partial G}{\partial x_k} \xi_k + O(\epsilon^2). \tag{1.10}$$

In real analysis bold characters are often used to denote sets of variables; this notation will be used frequently in this block, particularly in Chapter 3.

Definitions of the symbols $\ll$ and $\gg$ are given in the Handbook.

A stationary point is *defined* to be one for which the term $O(\epsilon)$ is zero for *all* $\boldsymbol{\xi}$. This gives the familiar conditions for a point to be stationary, namely $\partial G / \partial x_k = 0$ for $k = 1, 2, \ldots, N$.

For a functional we proceed in the same way. That is, we choose two adjacent paths joining P_a to P_b and compare the values of S along these paths. If one is represented by a differentiable function $y(x)$, adjacent paths may be represented by $y(x) + \epsilon g(x)$, where ϵ is a real variable and $g(x)$ is another differentiable function. Since both paths pass through P_a and P_b, we require $y(a) = A$, $y(b) = B$ and $g(a) = g(b) = 0$; otherwise $g(x)$ is arbitrary. The difference

$$\delta S = S[y + \epsilon g] - S[y] \tag{1.11}$$

may be considered as a function of the real variable ϵ, for arbitrary $y(x)$ and $g(x)$ and for small values of $|\epsilon|$. When $\epsilon = 0$, $\delta S = 0$, and for small $|\epsilon|$ we expect δS to be proportional to ϵ; this is true in general, as seen in equation (1.17) below.

However, there may be some paths for which δS is proportional to ϵ^2, rather than ϵ. These paths are special, and we *define* them to be the *stationary paths*, *stationary curves* or *stationary functions*. Thus, a *necessary* condition for a path $y(x)$ to be a stationary path is that

$$S[y + \epsilon g] - S[y] = O(\epsilon^2) \tag{1.12}$$

1.2 The shortest distance between two points

for *any* sufficiently well behaved function $g(x)$ satisfying $g(a) = g(b) = 0$. The equation for the stationary function $y(x)$ is obtained by examining this difference more carefully.

The distances along these two adjacent curves are

$$S[y] = \int_a^b dx \, \sqrt{1 + y'(x)^2} \tag{1.13}$$

and

$$S[y + \epsilon g] = \int_a^b dx \, \sqrt{1 + [y'(x) + \epsilon g'(x)]^2}. \tag{1.14}$$

We proceed by expanding the integrand of $S[y + \epsilon g]$ in powers of ϵ, retaining only the terms proportional to ϵ. One way of making this expansion is to consider the integrand as a function of ϵ and to use the Taylor series to expand in powers of ϵ; in this manipulation x is treated as a constant, and hence so are $y'(x)$ and $g'(x)$. Thus

$$\sqrt{1 + (y' + \epsilon g')^2} = \sqrt{1 + y'^2} + \epsilon \left[\frac{d}{d\epsilon}\sqrt{1 + (y' + \epsilon g')^2}\right]_{\epsilon=0} + O(\epsilon^2)$$

$$= \sqrt{1 + y'^2} + \epsilon \left[\frac{(y' + \epsilon g')g'}{\sqrt{1 + (y' + \epsilon g')^2}}\right]_{\epsilon=0} + O(\epsilon^2)$$

$$= \sqrt{1 + y'(x)^2} + \epsilon \frac{y'(x)g'(x)}{\sqrt{1 + y'(x)^2}} + O(\epsilon^2). \tag{1.15}$$

Substituting this expansion into the integral gives

$$S[y + \epsilon g] = \int_a^b dx \left(\sqrt{1 + y'(x)^2} + \frac{\epsilon \, y'(x)g'(x)}{\sqrt{1 + y'(x)^2}}\right) + O(\epsilon^2)$$

$$= S[y] + \epsilon \int_a^b dx \, \frac{y'(x)}{\sqrt{1 + y'(x)^2}} g'(x) + O(\epsilon^2). \tag{1.16}$$

The difference between the two distances is therefore

$$S[y + \epsilon g] - S[y] = \epsilon \int_a^b dx \, \frac{y'(x)}{\sqrt{1 + y'(x)^2}} g'(x) + O(\epsilon^2). \tag{1.17}$$

This difference depends upon both $y(x)$ and $g(x)$, just as for functions of n real variables the difference $G(\boldsymbol{x} + \epsilon \boldsymbol{\xi}) - G(\boldsymbol{x})$, equation (1.10), depends upon both $\boldsymbol{x}$ and $\boldsymbol{\xi}$, the equivalents of $y(x)$ and $g(x)$ respectively.

If $S[y]$ is stationary it follows, by definition, that

$$\int_a^b dx \, \frac{y'(x)}{\sqrt{1 + y'(x)^2}} g'(x) = 0, \tag{1.18}$$

and that this equation must hold for *all* functions $g(x)$ for which $g(a) = g(b) = 0$ and $g'(x)$ is continuous; we shall see in the next chapter that these properties are sufficient to determine $y(x)$ uniquely. Here, however, we simply show that if

$$\frac{y'(x)}{\sqrt{1 + y'(x)^2}} = \alpha = \text{constant}, \tag{1.19}$$

then the integral in equation (1.18) is zero for all $g(x)$. Thus, if equation (1.19) is true then equation (1.18) becomes

$$\int_a^b dx \, \alpha g'(x) = \alpha[g(b) - g(a)] = 0 \quad \text{since } g(a) = g(b) = 0. \tag{1.20}$$

In Section 2.3 we shall show that condition (1.19) is *necessary* as well as sufficient for equation (1.18) to hold.

Rearranging equation (1.19) gives $y'(x) = m$, where m is a constant. Integration now gives the general solution,

$$y(x) = mx + c, \tag{1.21}$$

for another constant c: this is the equation of a straight line, as expected. The constants m and c are determined by the conditions that the straight line passes through P_a and P_b:

$$y(x) = \frac{B-A}{b-a}x + \frac{Ab-Ba}{b-a}. \tag{1.22}$$

This analysis shows that the functional $S[y]$ defined in equation (1.9) is *stationary* along the straight line joining P_a to P_b. We have *not* shown that this gives a minimum distance: this is done in Exercise 1.7.

Exercise 1.3

Show that the equation of the straight line, $y = mx + c$, passing through the distinct points (a, A) and (b, B) is given by equation (1.22). Why does this equation fail if $b = a$ and $A \neq B$?

Exercise 1.4

Use the stationary path defined by equation (1.22) in the functional (1.9) to show that along this path the distance between P_a and P_b is $S = \sqrt{(b-a)^2 + (B-A)^2}$.

Exercise 1.5

Use the method described in Subsection 1.2.2 on the functional

$$S[y] = \int_0^1 dx\, \sqrt{1 + y'(x)}, \quad y(0) = 0,\ y(1) = B > -1,$$

to show that on a stationary path $y'(x) = $ constant. Hence show that the stationary function is the straight line $y(x) = Bx$ and also that the value of the functional on this line is $S[y] = \sqrt{1 + B}$.

Exercise 1.6

Consider the functional

$$S[y] = \int_1^2 dx\, xy'^2, \quad y(1) = 0,\ y(2) = 1.$$

Show that

$$S[y + \epsilon g] - S[y] = 2\epsilon \int_1^2 dx\, xy'g' + O(\epsilon^2)$$

and, employing the same analysis as used to derive equation (1.22), show that, if $g(1) = g(2) = 0$, the stationary path satisfies $xy'(x) = $ constant. Integrate this equation to show that the stationary function is $y(x) = \ln x / \ln 2$.

1.2.3 Geodesics

Problems involving shortest distances on surfaces other than a plane illustrate other features of variational problems. For example, if we replace the plane by the surface of a sphere then the shortest distance between two points on the surface can be shown to be the length of an arc of the great circle joining the two points, i.e. the circle created by the intersection of the spherical surface and the plane passing through the two points and the centre of the sphere. Now, for most points, there are two stationary paths corresponding to the long and the short arcs of the great circle. However, if the points are at opposite ends of a diameter, there are infinitely many shortest paths. This example suggests that solutions to variational problems can be complicated.

In general, the stationary paths between two points on a surface are called geodesics. For a plane surface the only geodesics are straight lines; for a sphere, most pairs of points are joined by just two geodesics, i.e. both segments of the great circle through the points. For other surfaces there may be several stationary paths; an example of the consequences of such complications is described next.

Gravitational lensing

In Einstein's general theory of relativity, the path taken by light from a source to an observer is along a geodesic on a surface in a four-dimensional space. In this theory gravitational forces are represented by distortions to this surface. The theory therefore predicts that light is 'bent' by gravitational forces, a prediction that was first observed in 1919 by Eddington's measurements of the positions of stars during a total solar eclipse. These observations provided the first direct confirmation of Einstein's general theory of relativity.

Albert Einstein, 1879–1955.

Arthur Stanley Eddington, 1882–1944.

The departure from a straight-line path depends upon the mass of the body between the source and observer. If it is sufficiently massive then two or more images may be seen, as illustrated schematically in Figure 1.6.

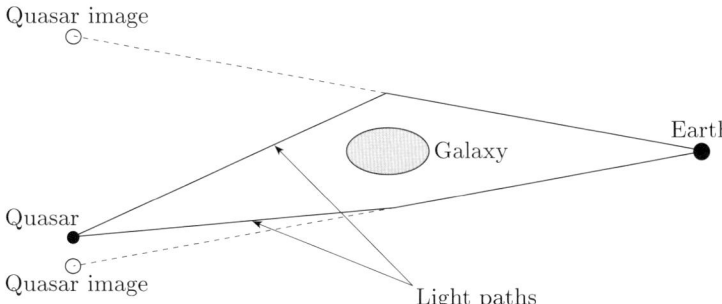

Figure 1.6 Diagram showing, schematically, how an intervening galaxy can sufficiently distort a path of light from a bright object, such as a quasar, to provide the two stationary paths and hence the two images shown. Many examples of such multiple images, and more complicated but similar optical effects, have now been observed. Usually there are more than two stationary paths.

1.2.4 The shortest path

In this subsection we show that the straight line given by equation (1.22) gives a minimum distance. This involves the expansion of $S[y + \epsilon g]$ to $O(\epsilon^2)$. The general theory that describes the behaviour of this second-order term is difficult, and it is not part of this course; the material of this subsection is therefore not assessed.

First, recall that for functions of n variables the nature of a stationary point is determined by making a second-order Taylor expansion about a point $\boldsymbol{x} = \boldsymbol{a}$,

$$G(\boldsymbol{a} + \epsilon \boldsymbol{\xi}) = G(\boldsymbol{a}) + \epsilon \sum_{k=1}^{n} \frac{\partial G}{\partial x_k} \xi_k + \tfrac{1}{2}\epsilon^2 \sum_{k=1}^{n} \sum_{j=1}^{n} \frac{\partial^2 G}{\partial x_k \partial x_j} \xi_k \xi_j + \cdots, \quad (1.23)$$

where all derivatives are evaluated at $\boldsymbol{x} = \boldsymbol{a}$. This is a generalisation of the three-dimensional Taylor expansion given in Block 0 (Subsection 2.2.7). If $G(\boldsymbol{x})$ is stationary at $\boldsymbol{x} = \boldsymbol{a}$ then all first derivatives are zero. The nature of the stationary point is usually determined by the behaviour of the second-order term. For a stationary point to be a minimum it is necessary for the quadratic term to be strictly positive for all $\boldsymbol{\xi}$, i.e.

$$\sum_{k=1}^{n} \sum_{j=1}^{n} \frac{\partial^2 G}{\partial x_k \partial x_j} \xi_k \xi_j > 0 \quad \text{for all } \xi_k, \xi_j, \; k, j = 1, 2, \ldots, n, \quad (1.24)$$

provided $|\boldsymbol{\xi}| = 1$. The stationary point is a maximum if this quadratic form is strictly negative. Usually for large n it is difficult to determine whether these inequalities are satisfied, although there are well-defined tests. The relevant test for $n = 2$ is given in Chapter 2 of Block 0.

For a functional we proceed in the same way: the nature of a stationary path is determined by the second-order expansion. If $S[y]$ is stationary then, by definition,

$$S[y + \epsilon g] - S[y] = K[y, g]\epsilon^2 + O(\epsilon^3) \quad (1.25)$$

for some quantity $K[y, g]$, depending upon both y and g; examples of this expansion are found in Exercises 1.7 and 1.8. Then $S[y]$ is a minimum if $K[y, g] > 0$ for all $g(x)$, and a maximum if $K[y, g] < 0$ for all $g(x)$. Normally it is difficult to establish these inequalities, and the necessary general theory is not part of this course. For the functional defined by equation (1.9), however, the proof is straightforward; the following exercise guides you through it.

Exercise 1.7

(a) Using either the binomial expansion or a Taylor expansion show that

$$\sqrt{1 + (\alpha + \epsilon\beta)^2} = \sqrt{1 + \alpha^2} + \frac{\alpha\beta\epsilon}{\sqrt{1 + \alpha^2}} + \frac{\beta^2 \epsilon^2}{2(1 + \alpha^2)^{3/2}} + O(\epsilon^3).$$

(b) Use this result to show that if $y(x)$ is the straight line defined in equation (1.22) and $S[y]$ is the functional (1.9), then

$$S[y + \epsilon g] - S[y] = \frac{\epsilon^2}{2(1 + m^2)^{3/2}} \int_a^b dx\, g'(x)^2, \quad m = \frac{B - A}{b - a}.$$

Deduce that the straight line is a minimum distance between P_a and P_b.

Exercise 1.8

In this exercise we consider the functional defined in Exercise 1.5 in more detail.

By expanding the integrand of $S[y + \epsilon g]$ to second order in ϵ, show that if $y(x)$ is the stationary path, then

$$S[y + \epsilon g] = S[y] - \frac{\epsilon^2}{8(1+B)^{3/2}} \int_0^1 dx\, g'(x)^2 + O(\epsilon^3), \quad B > -1.$$

Hence deduce that the path $y(x) = Bx$, $B > -1$, $0 \leq x \leq 1$, is a maximum of this functional.

1.3 Two generalisations

In this section we extend the preceding analysis to deal with more general functionals. First, we consider functionals having integrands of the form $F(y')$, where F is any suitably well-behaved, real function of a single real variable. By setting $F = \sqrt{1 + y'^2}$, we regain the example treated in Subsection 1.2.2. Second, we generalise this by considering integrands of the form $F(x, y')$, which depend explicitly upon x and y'; an example is $F = \sqrt{x^2 + y'^2}$.

1.3.1 Functionals depending only upon $y'(x)$

The functional (1.9) (page 9), depends only upon the derivative of the unknown function. Although this is a special case, it is worth considering it in more detail in order to develop the notation we need.

If $F(z)$ is a function of z, which we need to be differentiable, then a general functional of the form (1.9) is

$$S[y] = \int_a^b dx\, F(y'), \quad y(a) = A, \quad y(b) = B, \tag{1.26}$$

where $F(y')$ simply means that in $F(z)$ all occurrences of z are replaced by $y'(x)$. Thus in the previous example

$$F(z) = \sqrt{1 + z^2} \quad \text{so} \quad F(y') = \sqrt{1 + y'(x)^2}. \tag{1.27}$$

Note that the symbols $F(y')$ and $F(y'(x))$ denote the same function.

The difference between the functional evaluated along $y(x)$ and the adjacent paths $y(x) + \epsilon g(x)$, where $|\epsilon| \ll 1$ and $g(a) = g(b) = 0$, is

$$S[y + \epsilon g] - S[y] = \int_a^b dx\, \left[F(y' + \epsilon g') - F(y') \right]. \tag{1.28}$$

Now we need to express $F(y' + \epsilon g')$ as a series in ϵ; because $F(z)$ is differentiable, Taylor's theorem gives

$$F(z + \epsilon u) = F(z) + \epsilon u \frac{d}{dz} F(z) + O(\epsilon^2). \tag{1.29}$$

The expansion of $F(y' + \epsilon g')$ is obtained from this simply by replacing z by $y'(x)$ and u by $g'(x)$, which gives

$$F(y' + \epsilon g') - F(y') = \epsilon g'(x) \frac{d}{dy'} F(y') + O(\epsilon^2), \tag{1.30}$$

where the notation dF/dy' means

$$\frac{d}{dy'}F(y') = \frac{d}{dz}F(z)\bigg|_{z=y'(x)}. \tag{1.31}$$

For instance, if $F(z) = \sqrt{1+z^2}$ then

$$\frac{dF}{dz} = \frac{z}{\sqrt{1+z^2}} \quad \text{and} \quad \frac{dF}{dy'} = \frac{y'(x)}{\sqrt{1+y'(x)^2}}. \tag{1.32}$$

Exercise 1.9

Find the expressions for dF/dy' for the following cases of $F(y')$:

(a) $F(y') = (1 + y'^2)^{1/4}$

(b) $F(y') = \sin y'$

(c) $F(y') = \exp(y')$

Substituting the difference (1.30) into the equation (1.28) gives

$$S[y + \epsilon g] - S[y] = \epsilon \int_a^b dx\, g'(x)\frac{d}{dy'}F(y') + O(\epsilon^2). \tag{1.33}$$

By definition the functional $S[y]$ is stationary if the term $O(\epsilon)$ is zero for *all* suitable functions $g(x)$. We now give a sufficient condition, deferring until Chapter 2 the proof that it is also necessary. In this analysis it is important to remember that $F(z)$ is a given function and that $y(x)$ is an unknown function that we need to find. Observe that if

$$\frac{d}{dy'}F(y') = \alpha = \text{constant}, \tag{1.34}$$

then equation (1.33) becomes

$$\begin{aligned} S[y + \epsilon g] - S[y] &= \epsilon\alpha \int_a^b dx\, g'(x) + O(\epsilon^2) \\ &= \epsilon\alpha(g(b) - g(a)) + O(\epsilon^2) \\ &= O(\epsilon^2). \end{aligned} \tag{1.35} \quad \text{Since } g(a) = g(b) = 0.$$

In general, equation (1.34) is true only if $y'(x)$ is also constant, so

$$y(x) = mx + c. \tag{1.36}$$

Therefore

$$y(x) = \frac{B-A}{b-a}x + \frac{Ab - Ba}{b-a}, \tag{1.37}$$

using the boundary conditions $y(a) = A$ and $y(b) = B$.

This is the same solution as given in equation (1.22). Thus, for this type of functional, the stationary function is independent of the form of the integrand although the nature of the stationary function is not; see for instance Exercise 1.22 (page 29). The exception is when $F(z)$ is linear, in which case the value of $S[y]$ depends only upon the end points and not the values of $y(x)$ in between; then there is no stationary path, as shown in the next exercise.

1.3 Two generalisations

Exercise 1.10

If $F(z) = Cz + D$, where C and D are constants, show that equation (1.34) does *not* imply that $y'(x) = $ constant.

In this case show that the value of the functional $S[y] = \int_a^b dx\, F(y')$ is independent of the chosen path.

Note that this is the only example where the equation $dF/dy' = $ constant does not imply that $y'(x) = $ constant.

1.3.2 Functionals depending upon x and $y'(x)$

Now consider the slightly more general functional

$$S[y] = \int_a^b dx\, F(x, y'), \quad y(a) = A, \quad y(b) = B, \tag{1.38}$$

where the integrand $F(x, y')$ depends explicitly upon the two variables x and y', which are treated as independent variables. The difference in the value of the functional along adjacent paths is

$$S[y + \epsilon g] - S[y] = \int_a^b dx\, \left[F(x, y' + \epsilon g') - F(x, y')\right]. \tag{1.39}$$

In this example $F(x, z)$ is a function of two variables and we require the expansion

$$F(x, z + \epsilon u) = F(x, z) + \epsilon u \frac{\partial F}{\partial z} + O(\epsilon^2), \tag{1.40}$$

where Taylor's series for functions of two variables is used. Comparing this with the expression preceding equation (1.30) we see that the only difference is that the total derivative with respect to y' has been replaced by a partial derivative. As before, on replacing z by $y'(x)$ and u by $g'(x)$, equation (1.39) becomes

$$S[y + \epsilon g] - S[y] = \epsilon \int_a^b dx\, g'(x) \frac{\partial}{\partial y'} F(x, y') + O(\epsilon^2), \tag{1.41}$$

where

$$\frac{\partial}{\partial y'} F(x, y') = \frac{\partial}{\partial z} F(x, z) \bigg|_{z = y'}. \tag{1.42}$$

If $y(x)$ is the stationary path, it is necessary that

$$\int_a^b dx\, g'(x) \frac{\partial}{\partial y'} F(x, y') = 0 \quad \text{for all } g(x). \tag{1.43}$$

As before a sufficient condition for this is that $\partial F/\partial y' = $ constant, which gives a first-order differential equation for $y(x)$,

$$\frac{\partial}{\partial y'} F(x, y') = c, \quad y(a) = A, \quad y(b) = B, \tag{1.44}$$

where c is a constant. This is the equivalent of equation (1.34), but now the explicit presence of x in the equation means that $y'(x) = $ constant is not a solution.

Exercise 1.11

Consider the functional

$$S[y] = \int_0^1 dx\, \sqrt{1+x+y'^2}, \quad y(0) = A > 0, \quad y(1) = B > A.$$

Show that the function $y(x)$, defined by the relation

$$y'(x) = c\sqrt{1+x+y'(x)^2},$$

where c is a constant, makes $S[y]$ stationary. By expressing $y'(x)$ in terms of x, solve this equation to show that

$$y(x) = A + \frac{(B-A)}{(2^{3/2}-1)}\left((1+x)^{3/2} - 1\right).$$

1.4 Notation

In the previous sections we used the notation $F(y')$ to denote a function of the derivative of $y(x)$ and proceeded to treat y' as an independent variable, so that the expression dF/dy' had the meaning defined in equation (1.31). This notation and its generalisation are very important in subsequent analysis; it is therefore essential that you are familiar with them and can use them.

Consider a function $F(x, u, v)$ of three variables, for instance $F = x\sqrt{u^2 + v^2}$, and assume that all necessary partial derivatives of $F(x, u, v)$ exist. If $y(x)$ is a function of x we may form a function of x by replacing u by $y(x)$ and v by $y'(x)$, thus

$$F(x, u, v) \quad \text{becomes} \quad F(x, y, y').$$

Although this is a function of x, it is more often convenient to consider it as a function of three independent variables (x, y, y'). The first partial derivatives with respect to y and y' are simply

$$\frac{\partial}{\partial y} F(x, y, y') = \left.\frac{\partial}{\partial u} F(x, u, v)\right|_{u=y, v=y'} \tag{1.45}$$

and

$$\frac{\partial}{\partial y'} F(x, y, y') = \left.\frac{\partial}{\partial v} F(x, u, v)\right|_{u=y, v=y'}. \tag{1.46}$$

Because y depends upon x, we may also form the total derivative of $F(x, y, y')$ with respect to x using the chain rule

$$\frac{dF}{dx} = \frac{\partial F}{\partial x} + \frac{\partial F}{\partial y}\frac{dy}{dx} + \frac{\partial F}{\partial y'}\frac{dy'}{dx}$$

$$= \frac{\partial F}{\partial x} + \frac{\partial F}{\partial y}y'(x) + \frac{\partial F}{\partial y'}y''(x). \tag{1.47}$$

In the particular case $F(x, u, v) = x\sqrt{u^2 + v^2}$, with $u = y$ and $v = y'$, these rules give

$$\frac{\partial F}{\partial x} = \sqrt{y^2 + y'^2}, \quad \frac{\partial F}{\partial y} = \frac{xy}{\sqrt{y^2 + y'^2}}, \quad \frac{\partial F}{\partial y'} = \frac{xy'}{\sqrt{y^2 + y'^2}}. \tag{1.48}$$

1.4 Notation

Similarly, the second-order derivatives are

$$\frac{\partial^2 F}{\partial y^2} = \frac{\partial^2 F}{\partial u^2}\bigg|_{u=y,v=y'}, \quad \frac{\partial^2 F}{\partial y'^2} = \frac{\partial^2 F}{\partial v^2}\bigg|_{u=y,v=y'}$$

and $\quad \dfrac{\partial^2 F}{\partial y \partial y'} = \dfrac{\partial^2 F}{\partial u \partial v}\bigg|_{u=y,v=y'}.$ (1.49)

Because you must be familiar with and be able to use this notation we suggest that, before proceeding, you attempt as many of the following exercises as time permits.

Exercise 1.12

For the functions

$$F_1(x,y) = \sin(x+y), \quad F_2(x,y) = \cos(x+y^2) \quad \text{and} \quad F_3(x,y) = \exp(xy),$$

where $y = y(x)$, find $\partial F_k/\partial x$, $\partial F_k/\partial y$ and dF_k/dx, for $k = 1, 2, 3$. Also, show that

$$\frac{d}{dx}\left(\frac{\partial F_1}{\partial y'}\right) \neq \frac{\partial}{\partial y'}\left(\frac{dF_1}{dx}\right).$$

Exercise 1.13

For the function $F = \sqrt{x^2 + y'^2}$, find

$$\frac{\partial F}{\partial x}, \quad \frac{\partial F}{\partial y}, \quad \frac{\partial F}{\partial y'}, \quad \frac{dF}{dx}.$$

Also, show that

$$\frac{d}{dx}\left(\frac{\partial F}{\partial y'}\right) = \frac{\partial}{\partial y'}\left(\frac{dF}{dx}\right).$$

Exercise 1.14

For the function $F = \sqrt{x^2 + yy'^2}$, find

$$\frac{\partial F}{\partial x}, \quad \frac{\partial F}{\partial y}, \quad \frac{\partial F}{\partial y'} \quad \text{and} \quad \frac{dF}{dx}.$$

Exercise 1.15

Show that for an arbitrary differentiable function $F(x, y, y')$

$$\frac{d}{dx}\left(\frac{\partial F}{\partial y'}\right) = \frac{\partial^2 F}{\partial y'^2}y'' + \frac{\partial^2 F}{\partial y \partial y'}y' + \frac{\partial^2 F}{\partial x \partial y'}.$$

Hence show that

$$\frac{d}{dx}\left(\frac{\partial F}{\partial y'}\right) \neq \frac{\partial}{\partial y'}\left(\frac{dF}{dx}\right),$$

with equality only if F does not depend explicitly upon y.

Exercise 1.16

Use the first identity found in Exercise 1.15 to show that the equation

$$\frac{d}{dx}\left(\frac{\partial F}{\partial y'}\right) - \frac{\partial F}{\partial y} = 0$$

is equivalent to the second-order differential equation

$$\frac{\partial^2 F}{\partial y'^2}y'' + \frac{\partial^2 F}{\partial y \partial y'}y' + \frac{\partial^2 F}{\partial x \partial y'} - \frac{\partial F}{\partial y} = 0.$$

Note: the first equation will later be seen as crucial to the general theory described in Chapter 2. The fact that it is a second-order differential equation means that unique solutions can be obtained only if two initial or two boundary conditions are given. Also the fact that the coefficient of $y''(x)$ is $\partial^2 F/\partial y'^2$ is very important in the general theory of the existence of solutions of this type of equation.

Exercise 1.17

(a) If $F(y, y') = y\sqrt{1 + y'^2}$, find

$$\frac{\partial F}{\partial y}, \quad \frac{\partial F}{\partial y'}, \quad \frac{\partial^2 F}{\partial y'^2}, \quad \frac{\partial^2 F}{\partial y \partial y'}$$

and show that

$$\frac{d}{dx}\left(\frac{\partial F}{\partial y'}\right) - \frac{\partial F}{\partial y} = \left(1 + y'^2\right)^{-3/2}\left(y^2 \frac{d}{dx}\left(\frac{y'}{y}\right) - 1\right).$$

This is a hard exercise which you are advised to do for revision.

(b) By solving the equation $y^2(y'/y)' = 1$ show that a non-zero solution of

$$\frac{d}{dx}\left(\frac{\partial F}{\partial y'}\right) - \frac{\partial F}{\partial y} = 0 \quad \text{is} \quad y = \frac{1}{A}\cosh(Ax + B),$$

for some constants A and B.

[Hint: let y be the independent variable and define a new variable z by the equation $yz(y) = dy/dx$ to obtain an expression for dy/dx that can be integrated.]

1.5 Some examples of functionals

In this section we describe a variety of problems that can be formulated in terms of functionals, with solutions that are stationary paths of these functionals. This list is provided because it is likely that you will not be familiar with these descriptions and will be unaware of the wide variety of problems for which variational principles are useful, and sometimes essential. You should not spend long on this section if time is short; in this case you should aim at obtaining a rough overview of the examples. Indeed, you may move directly to Chapter 2 and return to this section at a later date, if necessary.

In each of the following subsections, a different problem is described and the relevant functional is written down; some of these are derived later. In compiling this list one aim has been to describe a reasonably wide range of applications: if you are unfamiliar with the underlying physical ideas behind any of these examples, do not worry because they are not an assessed part of the course. Another aim is to show that there are subtly different types of variational problems, for instance the isoperimetric and catenary problems, described on pages 22 and 25, respectively. These variants are not considered further in this course, but you should know of their existence.

1.5.1 The brachistochrone

Given two points $P_a = (a, A)$ and $P_b = (b, B)$ in the same vertical plane, as illustrated in the diagram below, we require the shape of the smooth wire joining P_a to P_b such that a bead sliding on the wire under gravity, with no friction, and starting at P_a with a given speed shall reach P_b in the shortest possible time.

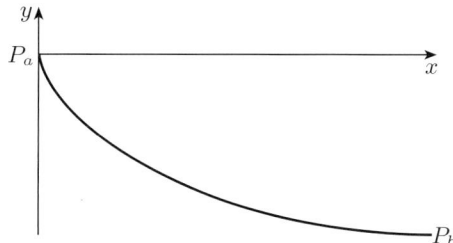

Figure 1.7 The curved line joining P_a to P_b is a segment of a cycloid, as shown in Section 2.7. In this diagram the axes are chosen to give $a = A = 0$.

The name given to this curve is the *brachistochrone*, which comes from the Greek *brachystos*, meaning shortest, and *chronos*, meaning time.

If the y-axis is vertical, it can be shown that the time taken along the curve $y(x)$ is

$$T[y] = \int_a^b dx \sqrt{\frac{1 + y'(x)^2}{C - 2gy(x)}}, \quad y(a) = A, \quad y(b) = B, \qquad (1.50)$$

where g is the acceleration due to gravity and C a constant depending upon the end points and the initial energy of the particle. This expression is derived in Section 2.7.

This problem was first considered by Galileo in his 1638 work *Two new sciences*, but lacking the necessary mathematical methods he concluded, erroneously, that the solution is the arc of a circle passing vertically through P_a.

Galileo Galilei, 1564–1642.

It was Bernoulli, however, who made the problem famous when in June 1696 he challenged the mathematical world to solve it. He followed his statement of the problem by a paragraph reassuring readers that the problem was very useful in mechanics, that it is not the straight line through P_a and P_b and that the curve is well known to geometers. He also stated that he would show that this is so at the end of the year, provided no one else had.

Johann Bernoulli, 1654–1705.

In December 1696 Bernoulli extended the time limit to Easter 1697, though by this time he was in possession of Leibniz's solution, sent in a letter dated 16 June 1696, Leibniz having received notification of the problem on 9 June. Newton also solved the problem quickly, apparently on the day of receipt, and published his solution anonymously.

Gottfried Wilhelm Leibniz, 1646–1716.

Further details of this history and these solutions may be found in Chapter 1 of H. H. Goldstine, *A history of the calculus of variations from the 17th through the 19th century* (Springer-Verlag, 1980).

The curve giving this shortest time is a segment of a *cycloid*, which is the curve traced out by a point fixed on the circumference of a vertical circle rolling, without slipping, along a straight line. The parametric equations of the cycloid shown in Figure 1.7 are

$$x = a(\theta - \sin\theta), \quad y = -a(1 - \cos\theta), \qquad (1.51)$$

where a is the radius of the circle: these equations are derived in Subsection 2.7.1.

1.5.2 Minimal surface of revolution

Here the problem is to find a curve $y(x)$ passing through two given points $P_a = (a, A)$ and $P_b = (b, B)$, with $A \geq 0$ and $B > 0$, as shown in Figure 1.8, such that when rotated about the x-axis the area of the curved surface formed is a minimum.

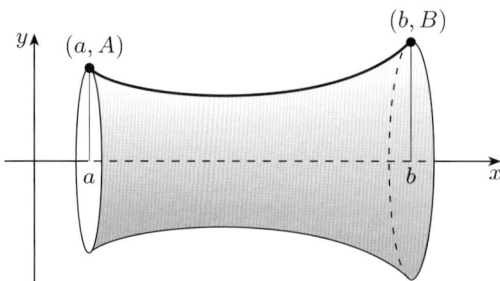

Figure 1.8 Diagram showing the cylindrical shape produced when a curve $y(x)$, joining (a, A) to (b, B), is rotated about the x-axis.

The area of this surface is shown in Section 2.5 to be

$$S[y] = 2\pi \int_a^b dx\, y(x) \sqrt{1 + y'(x)^2}, \tag{1.52}$$

and we shall see that this problem has solutions that can be expressed in terms of differentiable functions only for certain combinations of A, B and $b - a$.

1.5.3 A problem in navigation

Given a river with straight, parallel banks a distance a apart and a boat that can travel with constant speed c in still water, the problem is to cross the river in the shortest time, starting and landing at given points.

If the y-axis is chosen to be the left bank, the starting point to be the origin and the water is assumed to be moving parallel to the banks with speed $v(x)$, a known function of the distance from the left-hand bank, then the time of passage along the path $y(x)$ is, assuming $c > \max(v(x))$,

$$T[y] = \int_0^a dx\, \frac{\sqrt{c^2(1 + y'^2) - v(x)^2} - v(x)y'}{c^2 - v(x)^2}, \tag{1.53}$$

with boundary conditions $y(0) = 0$ and $y(a) = A$, where the final destination is a distance A along the right-hand bank. The derivation of this result is set in Exercise 1.28, one of the harder exercises at the end of this chapter.

1.5.4 The isoperimetric problem

Among all curves, represented by functions with continuous derivatives, that join the two points P_a and P_b in the upper half plane and have given length L, determine the one that encompasses the largest area, $A[y]$, shown in Figure 1.9.

1.5 Some examples of functionals

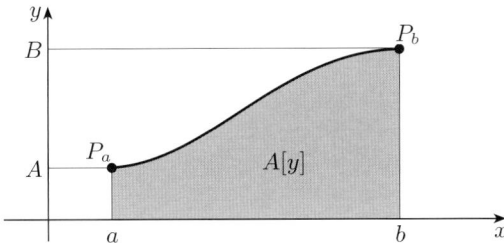

Figure 1.9 Diagram showing the area $A[y]$ under a curve of given length joining P_a to P_b.

This is a classic problem discussed by Pappus of Alexandria in about AD 300. Pappus showed that of two regular polygons having equal perimeters the one with the greater number of sides has the greater area. In the same book he demonstrates that for a given perimeter the circle has a greater area than does any regular polygon. This work seems to follow closely the earlier work of Zenodorus (c. 180 BC): extant fragments of his work include a proposition that of all solid figures, the surface areas of which are equal, the sphere has the greatest volume.

This intuitively obvious result is enshrined in the myth describing the foundation of the Phoenician city of Carthage in 814 BC: this is that Dido, also known as Elissa, having fled from Tyre after her brother, King Pygmalion, had killed her husband, was granted by the Libyans as much land as an oxhide could cover. By cutting the hide into thin strips, she was able to claim far more ground than anticipated: this early display of deviousness was considered by the Greeks and Romans a Punic trait. As with all foundation myths there is no trace of evidence for its veracity.

Returning to Figure 1.9, a modern analytic treatment of the problem requires a differentiable function $y(x)$ satisfying $y(a) = A$, $y(b) = B$, such that the area,

$$A[y] = \int_a^b dx\, y(x) \tag{1.54}$$

is largest when the length of the curve,

$$L = \int_a^b dx\, \sqrt{1 + y'(x)^2}, \tag{1.55}$$

is given. It transpires that a circular arc is the solution.

This problem differs from the first three because an additional constraint – the length of the curve – is imposed. In this course we do not consider this type of problem, although another such example is described in Section 1.5.6.

1.5.5 Fermat's principle

Light, and other forms of electromagnetic radiation, is a wave phenomenon. However, in many common circumstances light may be considered to travel along lines joining the source to the observer: these lines are called rays and are often straight lines. This is why most shadows have distinct edges and why eclipses of the Sun are so spectacular.

In a vacuum, and normally in air, these rays are straight lines and the speed of light is $c \simeq 2.9 \times 10^{10}\,\mathrm{cm\,s^{-1}}$, independent of its colour. In other uniform media, for example water, the rays also travel in straight lines, but the speed is different: if the speed of light in a uniform medium is c_m then the refractive index is defined to be the ratio $n = c/c_m$. The refractive index usually depends on the wavelength: thus for water it is 1.333 for red light (wavelength $6.5 \times 10^{-5}\,\mathrm{cm}$) and 1.343 for blue light (wavelength $7.5 \times 10^{-5}\,\mathrm{cm}$); this difference in the refractive index is one cause of rainbows. In non-uniform media, in which the refractive index depends upon the position, light rays follow curved paths. Mirages are one consequence of a position-dependent refractive index, where light passes through layers of air of differing densities.

A simple example of the ray description of light is the reflection of light in a plane mirror, as shown in Figure 1.10. In the diagram the source is S and the light ray is reflected from the mirror at R to the observer at O. The plane of the mirror is perpendicular to the page and it is assumed that the plane SRO is in the page.

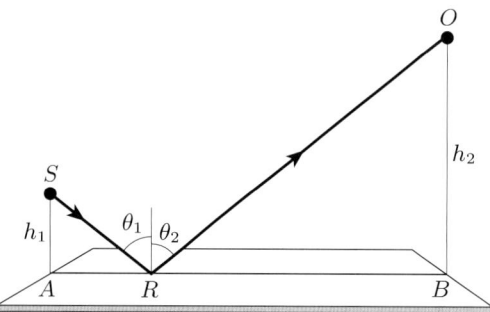

Figure 1.10 Diagram showing light travelling from a source S to an observer at O, via a reflection at R. The angles of incidence and reflection are defined to be θ_1 and θ_2, respectively.

Light travels in straight lines and is reflected from the mirror at a point R as shown in the diagram, but without further information the position of R is unknown. Observations, however, show that the angle of incidence, θ_1, and the angle of reflection, θ_2, are equal. This law of reflection was known to Euclid (c. 300 BC) and Aristotle (384–322 BC), but it was Hero of Alexandria (c. 125 BC) who showed by geometric argument that the equality of the angles of incidence and reflection is a consequence of the Aristotelian principle that nature does nothing the hard way. That is, if light is to travel from the source S to the observer O via a reflection in the mirror then it travels along the shortest path.

This result was generalised by the French mathematician Fermat into what is now known as *Fermat's principle*, which states that the path taken by light rays is that which minimises the *time* of passage. Pierre de Fermat, (1601–1665).

Fermat's original statement was that light travelling between two points seeks a path such that the number of waves is equal, as a first approximation, to that in a neighbouring path. This formulation has the form of a variational principle, which is remarkable because Fermat announced this result in 1658, before the calculus of either Newton or Leibniz was developed.

For the mirror, because the speed along SR and RO is the same, this just means that the distance along SR plus RO is a minimum. If $AB = d$ and

$AR = x$ the total distance travelled by the light ray depends only upon x and is

$$f(x) = \sqrt{x^2 + h_1^2} + \sqrt{(d-x)^2 + h_2^2}. \tag{1.56}$$

This function has a minimum when $\theta_1 = \theta_2$, that is when the angle of incidence, θ_1, equals the angle of reflection, θ_2. This result is derived in Exercise 1.20.

In general, for light moving in the Oxy-plane, in a medium with refractive index $n(x, y)$, with the source at the origin and observer at (a, A) the time of passage, T, along an arbitrary path $y(x)$ joining these points is

$$T[y] = \frac{1}{c} \int_0^a dx\, n(x,y)\sqrt{1+y'^2}, \quad y(0) = 0, \quad y(a) = A. \tag{1.57}$$

This follows because the time taken to travel along an element of length δs is $n(x,y)\delta s/c$ and $\delta s = \sqrt{1+y'(x)^2}\,\delta x$. If the refractive index, $n(x,y)$, is constant, this integral reduces to the integral (1.9) and the path of a ray is a straight line, as would be expected.

Fermat's principle can be used to show that for light reflected at a mirror the angle of incidence equals the angle of reflection. For light crossing the boundary between two media it gives Snell's law,

Willebrord Snell, 1591-1626.

$$\frac{\sin \alpha_1}{\sin \alpha_2} = \frac{c_1}{c_2}, \tag{1.58}$$

where α_1 and α_2 are the angles between the ray and the normal to the boundary and c_k, $k = 1, 2$, are the speeds of light in the media, as shown in Figure 1.11. For example, in water the speed of light is approximately $c_2 = c_1/1.3$, where c_1 is the speed of light in air, so $1.3 \sin \alpha_2 = \sin \alpha_1$.

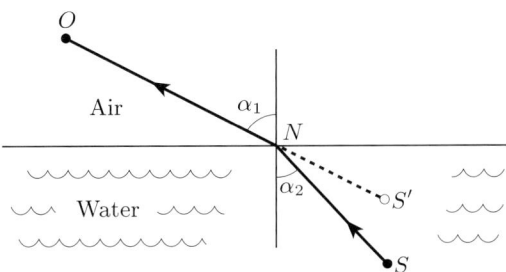

Figure 1.11 Diagram showing the refraction of light at the surface of water. The angles of incidence and refraction are defined to be α_2 and α_1 respectively; these are connected by Snell's law.

In this figure the observer at O sees an object S in a pond and the light ray from S to O travels along the two straight lines SN and NO, but the observer perceives the object to be at S', on the straight line ON. This explains, for instance, why a stick put partly into water appears bent.

1.5.6 The catenary

A catenary is the shape assumed by an inextensible chain of uniform density hanging between supports at both ends. In Figure 1.12 we show an example of such a curve when the points of support, $(-a, A)$ and (a, A), are at the same height.

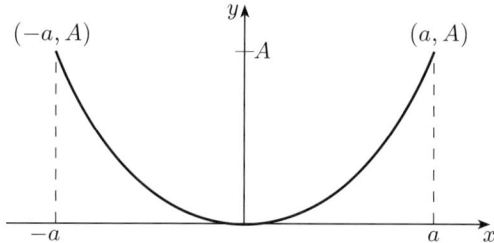

Figure 1.12 The catenary formed by a uniform chain hanging between two points at the same height.

If the lowest point of the chain is taken as the origin, the catenary equation is

$$y = c\left(\cosh\left(\frac{x}{c}\right) - 1\right), \tag{1.59}$$

for some constant c determined by the length of the chain and the value of a.

If the shape of the chain is described by the differentiable function $y(x)$ it can be shown (see Exercise 1.25) that the potential energy E of the chain is proportional to the functional

$$S[y] = \int_{-a}^{a} dx\, y\sqrt{1 + y'^2}. \tag{1.60}$$

The function, y, that minimises this functional, subject to the length of the curve $L = \int_{-a}^{a} dx\, \sqrt{1 + y'^2}$ remaining constant, is the shape assumed by the hanging chain.

1.5.7 Coordinate-free formulation of Newton's equations

Newton's second law for an isolated system of N interacting particles relates the acceleration of the kth particle to the force acting upon it from all the other particles,

$$m_k \frac{d^2 \boldsymbol{r}_k}{dt^2} = \boldsymbol{F}_k(\boldsymbol{r}_1, \boldsymbol{r}_2, \ldots, \boldsymbol{r}_N), \quad k = 1, 2, \ldots, N, \tag{1.61}$$

where m_k is the mass of the kth particle at $\boldsymbol{r}_k$. This law of motion accurately describes a significant portion of the physical world, from the motion of large molecules to the motion of galaxies. However, when formulated in this manner a number of difficulties arise, some practical and some theoretical, the latter concerning the underlying structure of this system of nonlinear equations and their solutions.

More general systems, affected by external forces, can also be described by Newton's equations and we use one such system to illustrate the practical difficulties frequently encountered when using Newton's laws as originally formulated. This system is the simple pendulum, constrained to move in a vertical plane, as shown in the following diagram. Here the mass at P is attached to the point O, taken to be the origin, by a light rod of length l, swinging freely, that is without friction, about O in the plane of the paper.

1.5 Some examples of functionals

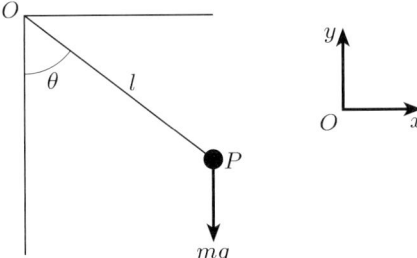

Figure 1.13 Diagram showing the geometry of the simple pendulum.

If the y-axis points vertically upwards, the position of the mass at P has the coordinates $\boldsymbol{r} = l(\sin\theta, -\cos\theta)$, where the angle θ is defined in the figure. The external gravitational force acting on the mass is $\boldsymbol{F} = (0, -mg)$. In addition there is the tension of magnitude T in the rod, responsible for ensuring that the distance OP is constant. Because the length OP is fixed the coordinates of P depend upon only one parameter, θ, but Newton's equations of motion involve the two coordinates of $\boldsymbol{r}$ and the two forces $\boldsymbol{F}$ and T.

Often we are not interested in the tension because it does not affect the motion, until it exceeds the breaking strain of the rod. Thus, ideally, we require a more 'natural' formulation of the equation of motion for θ which does not involve T. Also, observe that the vector velocity and acceleration of P in Cartesian coordinates are

$$\dot{\boldsymbol{r}} = l(\dot\theta \cos\theta, \dot\theta \sin\theta) \qquad (1.62)$$

and

$$\ddot{\boldsymbol{r}} = l\left(\ddot\theta \cos\theta - \dot\theta^2 \sin\theta, \ddot\theta \sin\theta + \dot\theta^2 \cos\theta\right). \qquad (1.63)$$

The acceleration depends on $\ddot\theta$, $\dot\theta$ and θ in a fairly complicated manner; in more complex problems the equivalent algebra is more complicated and adds additional and unnecessary difficulties to the formulation of the equations of motion (see Chapter 4).

In this example the tension T is referred to as a constraining force – because it constrains the distance OP to be constant. Another example of a constraining force is the reaction between a curved surface and a particle moving smoothly on it: in this case the constraint stops the particle falling through the surface.

Constraining forces are clearly of physical importance, but it is often helpful to use a formulation of Newton's equations that avoids their use. By using a variational principle, not only are these forces avoided but so are the difficulties encountered when expressing $\ddot{\boldsymbol{r}}$ in non-Cartesian coordinates. Furthermore, the variational formulation highlights the underlying mathematical structure of Newton's equations, which allows sophisticated methods to be used to help solve the equations. It also provides guides to the structure of the equations needed when Newton's laws are not valid, as for example when speeds approach that of the speed of light (the relativistic limit), or when the momentum of particles become very small, that is atomic and nuclear systems, when quantum mechanics is needed. In this course, in particular in Chapter 4, we lay the foundations for the development of some of these ideas.

1.6 Summary and Outcomes

These few examples provide some idea of the significance of variational principles. In summary, they are important for three distinct reasons:
- A variational principle is often the easiest or the only method of formulating a problem.
- A variational formulation provides a coordinate free method of expressing the laws of dynamics, allowing powerful analytic techniques to be used in ordinary Newtonian dynamics. They also pave the way for the formulation of dynamical laws describing motion of objects moving at speeds close to that of light (special relativity) and particles interacting through gravitational forces (general relativity) and the laws of the microscopic world (quantum mechanics).
- Often conventional boundary-value problems may be reformulated in terms of a variational principle; this provides a powerful tool for approximating solutions. (This theory is not covered in this course.)

After studying this chapter you should:
- have an idea of what a functional is and some appreciation of their applications;
- be able to determine the conditions required of a differentiable path that makes functionals of the form

$$S[y] = \int_a^b dx\, F(y') \quad \text{or} \quad S[y] = \int_a^b dx\, F(x, y') \tag{1.64}$$

stationary;
- be able to perform elementary calculations on the partial derivatives of functions of three variables.

1.7 Further Exercises

Exercise 1.18

Functionals do not need to have the particular form considered in this chapter. The following expressions, where $a(x)$ and $b(x)$ are prescribed functions, also map functions to real numbers.

(a) $D[y] = y'(1)$

(b) $K[y] = \int_0^1 dx\, a(x) \left[y(x) + y(1)y'(x) \right]$

(c) $L[y] = \left[xy(x)y'(x) \right]_0^1 + \int_0^1 dx\, \left[a(x)y'(x) + b(x)y(x) \right]$

Find the values of these functionals for each of the functions $y(x) = x^2$ and $y(x) = \cos \pi x$ when $a(x) = x$ and $b(x) = 1$.

Exercise 1.19

If $F(y, y') = \exp(ay + y')$, where a is a constant, show by direct calculation that

$$\frac{d}{dx}\left(\frac{\partial F}{\partial y'}\right) \neq \frac{\partial}{\partial y'}\left(\frac{dF}{dx}\right) \quad \text{except if } a = 0.$$

Exercise 1.20

Show that the function

$$f(x) = \sqrt{x^2 + h_1^2} + \sqrt{(d-x)^2 + h_2^2},$$

where h_1, h_2 are defined in Figure 1.10 (page 24) and x and d denote the lengths AR and AB respectively, is stationary when $\theta_1 = \theta_2$ where

$$\sin\theta_1 = \frac{x}{\sqrt{x^2 + h_1^2}}, \quad \sin\theta_2 = \frac{d-x}{\sqrt{(d-x)^2 + h_2^2}}.$$

Show that at this stationary value $f(x)$ has a minimum.

Exercise 1.21

Show that the functionals

$$S_1[y] = \int_a^b dx\,(1+xy')\,y' \quad \text{and} \quad S_2[y] = \int_a^b dx\,xy'^{\,2},$$

where $b > a > 0$, $y(b) = B$ and $y(a) = A$, are both stationary on the same curve, namely

$$y(x) = A + (B-A)\frac{\ln(x/a)}{\ln(b/a)}.$$

Explain why the same function makes both functionals stationary.

Exercise 1.22

Show that the functional

$$S[y] = \int_0^1 dx\,\left(1 + y'(x)^2\right)^{1/4}, \quad y(0) = 0, \quad y(1) = B,$$

is stationary for the straight line $y(x) = Bx$.

In addition, show that this straight line gives a minimum value of the functional only if $B < \sqrt{2}$, while for $B > \sqrt{2}$ it gives a maximum.

1.8 Harder Exercises

Exercise 1.23

Consider the functional

$$S[y] = \int_0^1 dx\,y'\sqrt{1+y'}, \quad y(0) = 0, \quad y(1) = B > -1.$$

(a) Show that the stationary function is the straight line $y(x) = Bx$ and that the value of the functional on this line is $S[y] = B\sqrt{1+B}$.

(b) Use the binomial expansion to expand the integrand of $S[y+\epsilon g]$ to second order in ϵ to show that if $y(x) = Bx$,

$$S[y+\epsilon g] = S[y] + \frac{(4+3B)\epsilon^2}{8(1+B)^{3/2}} \int_0^1 dx\,g'(x)^2, \quad B > -1.$$

Deduce that on this path the functional has a minimum.

Exercise 1.24

In this exercise the theory developed in Subsection 1.3.1 is extended. The function $F(z)$ has a continuous, non-constant second derivative and the functional S is defined by the integral

$$S[y] = \int_a^b dx\, F(y').$$

(a) Show that

$$S[y+\epsilon g] - S[y] = \epsilon \int_a^b dx\, \frac{dF}{dy'}g'(x) + \frac{1}{2}\epsilon^2 \int_a^b dx\, \frac{d^2 F}{dy'^2}g'(x)^2 + O(\epsilon^3),$$

where $g(a) = g(b) = 0$.

(b) Show that if $y(x)$ is chosen to make dF/dy' constant then the functional is stationary.

(c) Determine the conditions that make the functional either a maximum or a minimum.

Exercise 1.25

A uniform, flexible, inextensible chain of length L is suspended between two supports having the coordinates $(-a, A)$ and (a, A), with the y-axis pointing vertically upwards. Show that, if the shape assumed by the chain is described by the differentiable function $y(x)$, then its length and potential energy are given respectively by

$$L[y] = \int_{-a}^{a} dx\, \sqrt{1+y'^2} \quad \text{and} \quad E[y] = g\rho \int_{-a}^{a} dx\, y\sqrt{1+y'^2},$$

where ρ is the line density of the chain and g the acceleration due to gravity.

Exercise 1.26

This question is about the shortest distance between two points on the surface of a right-circular cylinder, so is a generalisation of the theory developed in Section 1.2.1.

(a) If the cylinder axis coincides with the z-axis we may use cylindrical polar coordinates (ρ, ϕ, z) to label points on the cylindrical surface, where ρ is the cylinder radius. Show that the Cartesian coordinates of a point (x, y) in the plane are given by $x = \rho\cos\phi$, $y = \rho\sin\phi$ and hence that the distance between two adjacent points on the cylinder, (ρ, ϕ, z) and $(\rho, \phi + \delta\phi, z + \delta z)$ is, to first-order, given by

$$\delta s^2 = \rho^2 \delta\phi^2 + \delta z^2.$$

(b) A curve on the surface may be defined by prescribing z as a function of ϕ. Show that the length of a curve from $\phi = \phi_1$ to $\phi = \phi_2$ is

$$L[z] = \int_{\phi_1}^{\phi_2} d\phi\, \sqrt{\rho^2 + z'(\phi)^2}.$$

(c) Deduce that the shortest distance on the cylinder between the two points $(\rho, 0, 0)$ and (ρ, α, ζ) is along the curve $z = \zeta\phi/\alpha$.

Exercise 1.27

An inverted cone has its apex at the origin and axis along the z-axis. Let α be the angle between this axis and the sides of the cone, and define a point on the conical surface by the coordinates (ρ, ϕ), where ρ is the perpendicular distance to the z-axis and ϕ is the polar angle measured from the x-axis, as in Exercise 1.26.

Show that the distance on the cone between adjacent points (ρ, ϕ) and $(\rho + \delta\rho, \phi + \delta\phi)$ is, to first-order,

$$\delta s^2 = \rho^2 \delta\phi^2 + \frac{\delta\rho^2}{\sin^2 \alpha}.$$

Hence show that if $\rho(\phi)$, $\phi_1 \leq \rho \leq \phi_2$, is a curve on the conical surface then its length is

$$L[\rho] = \int_{\phi_1}^{\phi_2} d\phi \sqrt{\rho^2 + \frac{\rho'^2}{\sin^2 \alpha}}.$$

Exercise 1.28

A straight river of uniform width a flows with velocity $(0, v(x))$, where the axes are chosen so the left-hand bank is the y-axis and where $v(x) > 0$. A boat can travel with constant speed $c > \max(v(x))$ relative to still water. If the starting and landing points are chosen to be the origin and (a, A), respectively, show that the path giving the shortest time of crossing is given by minimising the functional

$$T[y] = \int_0^a dx\, \frac{\sqrt{c^2(1+y'(x)^2) - v(x)^2} - v(x)y'(x)}{c^2 - v(x)^2}, \quad y(0) = 0, \quad y(a) = A.$$

Hint: the derivation of this result requires use of the identity

$$1 + \sqrt{X} = \left(1 + \sqrt{X}\right)\left(\frac{1 - \sqrt{X}}{1 - \sqrt{X}}\right) = \frac{1 - X}{1 - \sqrt{X}}.$$

Solutions to Exercises in Chapter 1

Solution 1.1

(a) If $y = x$, then $y' = 1$ and the functional becomes $S = \int_0^1 dx = 1$.

(b) If $y = x^a$, then $y' = ax^{a-1}$ and the functional becomes
$$S = a^2 \int_0^1 dx\, x^{2(a-1)} = \frac{a^2}{2a-1}, \quad a > \tfrac{1}{2}.$$

Note that the integral does not exist if $a \leq 1/2$ because the integrand tends to zero too rapidly at the origin.

(c) If $y = \sin(\pi x/2)$, then $y' = (\pi/2)\cos(\pi x/2)$ and the functional becomes
$$S = \frac{\pi^2}{4}\int_0^1 dx\, \cos\left(\frac{\pi}{2}x\right)^2 = \frac{\pi^2}{8}\int_0^1 dx\,(1+\cos\pi x) = \frac{\pi^2}{8}.$$

Solution 1.2

(a) On this straight line $y = 0$ so the value of the functional is $S = \int_0^b dx = b$.

(b) If $f^2 = bx - x^2$ then $2ff' = b - 2x$ and the functional is
$$S[f] = \int_0^b dx\, \sqrt{1 + \frac{(b-2x)^2}{4(bx-x^2)}} = \frac{b}{2}\int_0^b dx\, \frac{1}{\sqrt{x(b-x)}}.$$

Setting $x = b\sin^2\phi$, so $dx/d\phi = 2b\sin\phi\cos\phi$ and $x(b-x) = (b\sin\phi\cos\phi)^2$, gives $S[f] = b\int_0^{\pi/2} d\phi = \pi b/2$, as is obvious without this analysis.

Solution 1.3

The line passes through (a, A) and (b, B) so the constants m and c are determined from the two equations,

$A = am + c \quad \text{and} \quad B = bm + c.$

Subtracting these gives $B - A = m(b - a)$, giving the required expression for m. Substituting this into the first equation gives
$$c = A - \frac{B-A}{b-a}a = \frac{Ab - Ba}{b-a}.$$

If $a = b$ and $A \neq B$ the two points are distinct and there is a straight line joining them. This line *cannot* be represented by an equation of the form $y = mx + c$, however, because on this line x is constant.

In this case it is necessary to use parametric equation for the line, which is
$$x = (b-a)t + a, \quad y = (B-A)t + A, \quad \text{where } 0 \leq t \leq 1,$$

and which is valid for all distinct points P_a and P_b. You are not expected to provide this part of the solution.

Solution 1.4

Since $y' = (B-A)/(b-a)$ the value of the functional (1.9) is
$$S[y] = \int_a^b dx\, \sqrt{1 + \frac{(B-A)^2}{(b-a)^2}} = \frac{\sqrt{(b-a)^2 + (B-A)^2}}{b-a}\int_a^b dx$$
$$= \sqrt{(b-a)^2 + (B-A)^2}.$$

Solution 1.5

In order to find the stationary function we need to compute the difference $\delta S = S[y + \epsilon g] - S[y]$ only to $O(\epsilon)$. However, because we will require the second-order term in Exercise 1.8, here we evaluate the difference to $O(\epsilon^2)$. The difference is

$$\delta S = \int_0^1 dx \left(\sqrt{1 + y'(x) + \epsilon g'(x)} - \sqrt{1 + y'(x)} \right),$$

where $g(0) = g(1) = 0$. But

$$\sqrt{1 + y'(x) + \epsilon g'(x)} = \sqrt{1 + y'(x)} \left(1 + \frac{\epsilon g'(x)}{1 + y'(x)} \right)^{1/2}$$

$$= \sqrt{1 + y'(x)} \left(1 + \frac{\epsilon g'(x)}{2(1 + y'(x))} - \frac{\epsilon^2}{8} \left(\frac{g'(x)}{1 + y'(x)} \right)^2 + \cdots \right),$$

where we have used the binomial expansion $(1 + z)^{1/2} = 1 + \frac{1}{2}z - \frac{1}{8}z^2 + \cdots$, which is equivalent to using the Taylor series for $(1 + z)^{1/2}$.

Thus to first order in ϵ we obtain

$$\delta S = \frac{\epsilon}{2} \int_0^1 dx \, \frac{g'(x)}{\sqrt{1 + y'(x)}} + O(\epsilon^2).$$

The functional is stationary if the first-order term is zero for all $g(x)$, otherwise δS would change sign with ϵ. Using the result quoted in the text (after equation (1.19)) this gives $\sqrt{1 + y'(x)} = $ constant, that is $y'(x) = $ constant and $y(x) = \alpha x + \beta$. The boundary conditions, $y(0) = 0$ and $y(1) = B$, then give $y = Bx$ for the stationary path. With this value for $y(x)$, the integrand is real if $B > -1$ and has the value $S = \sqrt{1 + B}$.

Solution 1.6

The difference between $S[y + \epsilon g]$ and $S[y]$ is

$$S[y + \epsilon g] - S[y] = \int_1^2 dx \, x \left((y' + \epsilon g')^2 - y'^2 \right) = 2\epsilon \int_1^2 dx \, x y' g' + O(\epsilon^2).$$

If $xy' = a = $ constant, this difference is $O(\epsilon^2)$ and $y(x)$ is a stationary path.

This differential equation is separable and becomes

$$\frac{dy}{dx} = \frac{a}{x} \quad \text{with general solution} \quad y = b + a \ln x.$$

But $y(1) = 0$, so $b = 0$ and $y(2) = 1$, so $a = 1/\ln 2$, hence the required solution.

Solution 1.7

(a) The required expansion is given by first writing the square root as

$$\sqrt{1 + \alpha^2 + 2\epsilon\alpha\beta + \epsilon^2\beta^2} = \sqrt{1 + \alpha^2} \left(1 + \frac{2\epsilon\alpha\beta}{1 + \alpha^2} + \frac{\epsilon^2\beta^2}{1 + \alpha^2} \right)^{1/2}.$$

Now use the binomial expansion $(1 + z)^{1/2} = 1 + \frac{1}{2}z - \frac{1}{8}z^2 + \cdots$ to give

$$\sqrt{1 + \frac{2\epsilon\alpha\beta}{1 + \alpha^2} + \frac{\epsilon^2\beta^2}{1 + \alpha^2}} = 1 + \frac{1}{2}\left(\frac{2\epsilon\alpha\beta}{1 + \alpha^2} + \frac{\epsilon^2\beta^2}{1 + \alpha^2}\right) - \frac{1}{8}\left(\frac{2\epsilon\alpha\beta}{1 + \alpha^2} + \frac{\epsilon^2\beta^2}{1 + \alpha^2}\right)^2 + \cdots$$

$$= 1 + \frac{\epsilon\alpha\beta}{1 + \alpha^2} + \epsilon^2 \left(\frac{\beta^2}{2(1 + \alpha^2)} - \frac{\alpha^2\beta^2}{2(1 + \alpha^2)^2} \right) + O(\epsilon^3).$$

But

$$\frac{1}{1 + \alpha^2} - \frac{\alpha^2}{(1 + \alpha^2)^2} = \frac{1}{(1 + \alpha^2)^2},$$

so the final result is
$$\sqrt{1+(\alpha+\epsilon\beta)^2} = \sqrt{1+\alpha^2}\left(1+\frac{\epsilon\alpha\beta}{1+\alpha^2}+\frac{\epsilon^2\beta^2}{2(1+\alpha^2)^2}+O(\epsilon^3)\right)$$
$$= \sqrt{1+\alpha^2}+\frac{\epsilon\alpha\beta}{\sqrt{1+\alpha^2}}+\frac{\epsilon^2\beta^2}{2(1+\alpha^2)^{3/2}}.$$

(b) With $\alpha = y'(x)$ and $\beta = g'(x)$ we see, using the argument described in the text, that the term $O(\epsilon)$ in the expansion of $S[y+\epsilon g]-S[y]$ is zero if $y'(x)=$ constant, hence the line defined by equation (1.22) makes the functional stationary. With this choice of $y(x)$, $\alpha = m$ and the second term in the expansion derived in part (a) is zero and the quoted result follows. The second-order term in the expansion of the functional is positive for $\epsilon \neq 0$ and all $g(x)$, so the functional has a minimum along this line.

Solution 1.8

It was shown in the solution of Exercise 1.5 (page 33) that to second order in ϵ,
$$S[y+\epsilon g]-S[y] = \frac{\epsilon}{2}\int_0^1 dx\,\frac{g'(x)}{\sqrt{1+y'(x)}}-\frac{\epsilon^2}{8}\int_0^1 dx\,\frac{g'(x)^2}{(1+y'(x))^{3/2}}+O(\epsilon^3).$$

With $y=Bx$ the first-order term on the right-hand side is zero by definition, hence
$$\delta S = -\frac{\epsilon^2}{8(1+B)^{3/2}}\int_0^1 dx\,g'(x)^2 < 0,\quad B>-1.$$

This term is always negative, so for sufficiently small $|\epsilon|$ we have $S[y_s+\epsilon g]<S[y_s]$, where $y_s(x)=Bx$ is the stationary path, which is therefore a local maximum.

Solution 1.9

(a) If $F(y')=(1+y'^2)^{1/4}$, then
$$\frac{dF}{dy'} = \frac{y'}{2(1+y'^2)^{3/4}}.$$

(b) If $F(y')=\sin y'$, then $dF/dy'=\cos y'$.

(c) Since $\frac{d}{dz}(e^z)=e^z$ we have $dF/dy'=F$.

Solution 1.10

If $F(z)=Cz+D$, $F'(z)=C$ and equation (1.34) gives $C=\alpha$. In this case the functional becomes
$$S[y] = \int_a^b dx\,(Cy'(x)+D) = C\,[y(b)-y(a)]+D(b-a).$$

This depends only upon C, D and the boundaries a and b: the value of the functional is therefore independent of the chosen path.

Solution 1.11

In this example $F(x,v)=\sqrt{1+x+v^2}$ and equation (1.44) becomes
$$v = c\sqrt{1+x+v^2},$$
where $v=y'(x)$. Squaring and rearranging this equation gives
$$\left(\frac{dy}{dx}\right)^2 = a^2(1+x),\quad a^2 = \frac{c^2}{1-c^2}.$$

The square root of this equation gives $dy/dx = \pm a\sqrt{1+x}$. Since $x>0$, $y'(x)\neq 0$; also $y(1)>y(0)$, so $y'(x)>0$ and we need the positive sign. Now separate variables to write the solution in the form
$$\int_A^y dv = a\int_0^x du\,\sqrt{1+u},$$

which includes the boundary condition at $x = 0$. Integrating this gives the solution in the form
$$y(x) - A = \tfrac{2}{3}a\left((1+x)^{3/2} - 1\right).$$

The value of a is obtained from the boundary condition $y(1) = B$, i.e.
$$\tfrac{2}{3}a = \frac{B-A}{2^{3/2}-1} \quad \text{and hence} \quad y(x) = A + \frac{(B-A)}{(2^{3/2}-1)}\left((1+x)^{3/2}-1\right).$$

Solution 1.12

We have
$$\frac{\partial F_1}{\partial x} = \frac{\partial}{\partial x}\sin(x+y) = \cos(x+y) \quad \text{and} \quad \frac{\partial F_1}{\partial y} = \frac{\partial}{\partial y}\sin(x+y) = \cos(x+y).$$

Note that for this function $\partial F_1/\partial x = \partial F_1/\partial y$. Also,
$$\frac{dF_1}{dx} = \frac{\partial F_1}{\partial x} + \frac{\partial F_1}{\partial y}y' = (1+y')\cos(x+y).$$

For F_2 we have
$$\frac{\partial F_2}{\partial x} = -\sin(x+y^2), \quad \frac{\partial F_2}{\partial y} = -2y\sin(x+y^2)$$

and
$$\frac{dF_2}{dx} = \frac{\partial F_2}{\partial x} + y'\frac{\partial F_2}{\partial y} = -(1+2yy')\sin(x+y^2).$$

For F_3 we have
$$\frac{\partial F_3}{\partial x} = yF_3, \quad \frac{\partial F_3}{\partial y} = xF_3 \quad \text{and} \quad \frac{dF_3}{dx} = (y+xy')F_3 = (xy)'F_3.$$

For the last part, we note that the left-hand side is zero since $\partial F_1/\partial y' = 0$. But
$$\frac{dF_1}{dx} = (1+y')\cos(x+y), \quad \text{so} \quad \frac{\partial}{\partial y'}\left(\frac{dF_1}{dx}\right) = \cos(x+y).$$

Solution 1.13

The partial derivatives are
$$\frac{\partial F}{\partial x} = \frac{x}{\sqrt{x^2+y'^2}}, \quad \frac{\partial F}{\partial y} = 0, \quad \frac{\partial F}{\partial y'} = \frac{y'}{\sqrt{x^2+y'^2}},$$

and the total derivative is
$$\frac{dF}{dx} = \frac{\partial F}{\partial x} + \frac{\partial F}{\partial y}y' + \frac{\partial F}{\partial y'}y'' = \frac{x}{\sqrt{x^2+y'^2}} + \frac{y'y''}{\sqrt{x^2+y'^2}}.$$

Hence
$$\frac{d}{dx}\left(\frac{\partial F}{\partial y'}\right) = \frac{y''}{\sqrt{x^2+y'^2}} - \frac{xy'}{(x^2+y'^2)^{3/2}} - \frac{y'^2 y''}{(x^2+y'^2)^{3/2}}$$

and, using the above expression for dF/dx
$$\frac{\partial}{\partial y'}\left(\frac{dF}{dx}\right) = -\frac{xy'}{(x^2+y'^2)^{3/2}} + \frac{y''}{\sqrt{x^2+y'^2}} - \frac{y'^2 y''}{(x^2+y'^2)^{3/2}}.$$

Solution 1.14

The partial derivatives are
$$\frac{\partial F}{\partial x} = \frac{x}{\sqrt{x^2+yy'^2}}, \quad \frac{\partial F}{\partial y} = \frac{y'^2}{2\sqrt{x^2+yy'^2}}, \quad \frac{\partial F}{\partial y'} = \frac{yy'}{\sqrt{x^2+yy'^2}}.$$

The total derivative is
$$\frac{dF}{dx} = \frac{\partial F}{\partial x} + \frac{\partial F}{\partial y}y' + \frac{\partial F}{\partial y'}y'' = \frac{x}{\sqrt{x^2+yy'^2}} + \frac{y'^3}{2\sqrt{x^2+yy'^2}} + \frac{yy'y''}{\sqrt{x^2+yy'^2}}.$$

Solution 1.15

The chain rule applied to a function $G(x, y(x), y'(x))$ has the form

$$\frac{dG}{dx} = \frac{\partial G}{\partial y'}\frac{dy'}{dx} + \frac{\partial G}{\partial y}\frac{dy}{dx} + \frac{\partial G}{\partial x}.$$

In this example, where $G = \partial F/\partial y'$, this expression becomes

$$\frac{d}{dx}\left(\frac{\partial F}{\partial y'}\right) = \frac{\partial}{\partial y'}\left(\frac{\partial F}{\partial y'}\right)\frac{dy'}{dx} + \frac{\partial}{\partial y}\left(\frac{\partial F}{\partial y'}\right)\frac{dy}{dx} + \frac{\partial}{\partial x}\left(\frac{\partial F}{\partial y'}\right)$$

$$= \frac{\partial^2 F}{\partial y'^2}y'' + \frac{\partial^2 F}{\partial y'\partial y}y' + \frac{\partial^2 F}{\partial x \partial y'},$$

which gives the required expression and is the left-hand side of the inequality.

The right-hand side of the inequality is

$$\frac{\partial}{\partial y'}\left(\frac{dF}{dx}\right) = \frac{\partial}{\partial y'}\left(\frac{\partial F}{\partial x} + \frac{\partial F}{\partial y}y' + \frac{\partial F}{\partial y'}y''\right)$$

$$= \frac{\partial^2 F}{\partial x \partial y'} + \frac{\partial F}{\partial y} + \frac{\partial^2 F}{\partial y \partial y'}y' + \frac{\partial^2 F}{\partial y'^2}y'',$$

which differs from the left-hand side by the term $\partial F/\partial y$. Thus if F is independent of y, the derivatives are equal.

Solution 1.16

Subtract the term $\partial F/\partial y$ to obtain the required result.

Solution 1.17

(a) Direct differentiation gives

$$\frac{\partial F}{\partial y} = \sqrt{1+y'^2}, \quad \frac{\partial F}{\partial y'} = \frac{yy'}{\sqrt{1+y'^2}}.$$

Differentiating the second expression gives

$$\frac{\partial^2 F}{\partial y'^2} = \frac{y}{\sqrt{1+y'^2}} - \frac{yy'^2}{(1+y'^2)^{3/2}} = \frac{y}{(1+y'^2)^{3/2}} \quad \text{and} \quad \frac{\partial^2 F}{\partial y \partial y'} = \frac{y'}{\sqrt{1+y'^2}}.$$

Using the expression derived in Exercise 1.16, namely

$$z = \frac{d}{dx}\left(\frac{\partial F}{\partial y'}\right) - \frac{\partial F}{\partial y} = y''\frac{\partial^2 F}{\partial y'^2} + y'\frac{\partial^2 F}{\partial y \partial y'} - \frac{\partial F}{\partial y} = 0, \quad \text{since} \quad \frac{\partial^2 F}{\partial x \partial y'} = 0,$$

we obtain

$$z = \frac{yy''}{(1+y'^2)^{3/2}} + \frac{y'^2}{(1+y'^2)^{1/2}} - (1+y'^2)^{1/2}$$

$$= \frac{1}{(1+y'^2)^{3/2}}\left(yy'' + (1+y'^2)y'^2 - (1+y'^2)^2\right)$$

$$= \frac{1}{(1+y'^2)^{3/2}}\left(yy'' - y'^2 - 1\right).$$

But

$$\frac{d}{dx}\left(\frac{y'}{y}\right) = \frac{y''}{y} - \frac{y'^2}{y^2}$$

giving

$$yy'' - y'^2 = y^2\frac{d}{dx}\left(\frac{y'}{y}\right), \quad \text{if } y \neq 0,$$

hence

$$\frac{d}{dx}\left(\frac{\partial F}{\partial y'}\right) - \frac{\partial F}{\partial y} = \frac{1}{(1+y'^2)^{3/2}}\left(y^2\frac{d}{dx}\left(\frac{y'}{y}\right) - 1\right).$$

(b) If the left-hand side is zero we have
$$y^2 \frac{d}{dx}\left(\frac{y'}{y}\right) = 1 \quad \text{or} \quad y^2 y' \frac{d}{dy}\left(\frac{y'}{y}\right) = 1.$$

Now define $z = y'/y$ and consider z to be a function of y, so in the following $z' = dz/dy$. Now put the second equation in the form $y^3 z\, z'(y) = 1$, which can be integrated directly to give

Note that this is possible because x may be considered a function of y, so y'/y can be expressed in terms of y.

$$\frac{1}{2}z^2 = \frac{C^2}{2} - \frac{1}{2y^2},$$

for some constant C. Hence, since $z = y'/y$,

$$\frac{dy}{dx} = \sqrt{(Cy)^2 - 1},$$

giving

$$\int \frac{dy}{\sqrt{(Cy)^2 - 1}} = x + D.$$

Finally, set $Cy = \cosh\phi$ to give $\phi = C(x+D)$, i.e. $y = (1/C)\cosh(C(x+D))$, which is the required solution if $C = A$ and $CD = B$.

Solution 1.18

(a) If $y(x) = x^2$, then $y' = 2x$ and $D[y] = 2$.
If $y(x) = \cos\pi x$, then $y' = -\pi\sin\pi x$ and $D[y] = 0$.

(b) If $a(x) = x$, then:

if $y(x) = x^2$, $\quad K[y] = \int_0^1 dx\, x(x^2 + 2x) = \frac{11}{12}$;

if $y(x) = \cos\pi x$, $\quad K[y] = \int_0^1 dx\, x \cos\pi x = -\frac{2}{\pi^2}$.

(c) If $a(x) = x$ and $b(x) = 1$, then:

if $y(x) = x^2$, $\quad L[y] = \left[2x^4\right]_0^1 + \int_0^1 dx\,(3x^2) = 3$;

if $y(x) = \cos\pi x$, $\quad L[y] = \left[-\frac{\pi}{2}x\sin 2\pi x\right]_0^1 + \int_0^1 dx\,(-\pi x \sin\pi x + \cos\pi x) = -1$.

Solution 1.19

Consider the left-hand side, $\partial F/\partial y' = F$ and $dF/dx = (ay' + y'')F$. Thus

$$\frac{d}{dx}\left(\frac{\partial F}{\partial y'}\right) = (ay' + y'')F \quad \text{and} \quad \frac{\partial}{\partial y'}\left(\frac{dF}{dx}\right) = aF + (ay' + y'')F.$$

Solution 1.20

The derivative of $f(x)$ is $f'(x) = x/\sqrt{x^2 + h_1^2} - (d-x)/\sqrt{(d-x)^2 + h_2^2}$. Since

$$\sin\theta_1 = \frac{AR}{SR} = \frac{x}{\sqrt{x^2 + h_1^2}} \quad \text{and} \quad \sin\theta_2 = \frac{RB}{RO} = \frac{d-x}{\sqrt{(d-x)^2 + h_2^2}},$$

where the distances are defined in Figure 1.10 (page 24), we see that the distance travelled by the light is stationary when $\sin\theta_1 = \sin\theta_2$, that is $\theta_1 = \theta_2$. Further, since

$$f''(x) = \frac{h_1^2}{(x^2 + h_1^2)^{3/2}} + \frac{h_2^2}{((d-x)^2 + h_2^2)^{3/2}} > 0,$$

the stationary point is a minimum.

Solution 1.21

Observe that
$$S_1[y] = S_2[y] + \int_a^b dx\, y'(x) = S_2[y] + B - A,$$

that is, the values of the two functionals differ by a constant, independent of the path. Hence the stationary paths of the two functionals are the same.

Consider the difference $\delta = S_2[y + \epsilon g] - S_2[y]$, where $g(a) = g(b) = 0$:
$$\delta = 2\epsilon \int_a^b dx\, xy'(x)g'(x) + O(\epsilon^2)$$

so that $\delta = O(\epsilon^2)$ if $xy'(x) = c$, where c is a constant. Integrating this equation gives $y(x) = d + c\ln(x/a)$, where d is another constant. The boundary conditions now give
$$A = d \quad \text{and} \quad B = d + c\ln(b/a)$$

and hence
$$y(x) = A + (B - A)\frac{\ln(x/a)}{\ln(b/a)}.$$

Solution 1.22

In this example $F(y') = (1 + y'^2)^{1/4}$, and is a function of y' only, so the stationary path is a straight line, $y = mx + c$, with the constants m and c being chosen to satisfy the boundary conditions. These give $y = Bx$. Also,
$$\frac{\partial F}{\partial y'} = \frac{y'}{2(1+y'^2)^{3/4}}, \quad \text{and} \quad \frac{\partial^2 F}{\partial y'^2} = \frac{2 - y'^2}{4(1+y'^2)^{7/4}}.$$

Using the result derived in Exercise 1.24, and the fact that $y' = B$ on the stationary path, we see that
$$S[y + \epsilon g] - S[y] = \frac{(2 - B^2)\epsilon^2}{8(1+B^2)^{7/4}} \int_0^1 dx\, g'(x)^2 + O(\epsilon^3).$$

Thus if $B < \sqrt{2}$ the difference is positive for all $g(x)$ and ϵ, if sufficiently small, so the functional is a minimum along the line $f(x) = Bx$. For $B > \sqrt{2}$ the difference is negative and the functional is a maximum. If $B = \sqrt{2}$ the nature of the stationary path can be determined only by expanding to higher order in ϵ.

Solution 1.23

(a) We need the difference $\delta = S[y + \epsilon g] - S[y]$, where $g(0) = g(1) = 0$, otherwise $g(x)$ is an arbitrary continuous function. Now, using the binomial expansion
$$\sqrt{1 + \alpha + \epsilon\beta} = \sqrt{1 + \alpha}\left(1 + \frac{\epsilon\beta}{2(1+\alpha)} - \frac{\epsilon^2\beta^2}{8(1+\alpha)^2} + O(\epsilon^3)\right),$$

so
$$(\alpha + \epsilon\beta)\sqrt{1 + \alpha + \epsilon\beta} = \alpha\sqrt{1+\alpha}\left(1 + \frac{\epsilon\beta}{2(1+\alpha)} - \frac{\epsilon^2\beta^2}{8(1+\alpha)^2} + \cdots\right)$$
$$+ \epsilon\beta\sqrt{1+\alpha}\left(1 + \frac{\epsilon\beta}{2(1+\alpha)} + \cdots\right)$$
$$= \alpha\sqrt{1+\alpha} + \frac{\epsilon\beta(2 + 3\alpha)}{2\sqrt{1+\alpha}} + \frac{\epsilon^2\beta^2(4 + 3\alpha)}{8(1+\alpha)^{3/2}} + \cdots.$$

Now substitute $\alpha = y'$ and $\beta = g'$ to obtain
$$\delta = \epsilon \int_0^1 dx\, \frac{2 + 3y'}{2\sqrt{1+y'}}g'(x) + \frac{\epsilon^2}{8}\int_0^1 dx\, \frac{4 + 3y'}{(1+y')^{3/2}}g'(x)^2 + O(\epsilon^3).$$

If $y(x)$ is a stationary path of S, then the term $O(\epsilon)$ is zero. Since $g(0) = g(1) = 0$ it follows, as in the text, that $y'(x) = $ constant is a possible solution. Since $y(0) = 0$ and $y(1) = B$ this gives $y(x) = Bx$ and $S[y] = B\sqrt{1+B}$.

Alternatively, using equation (1.34) (page 16), with $F(y') = y'\sqrt{1+y'}$, we see that the stationary path is given by $F'(y') =$ constant and hence $y' =$ constant, that is $y = mx + c$: since $y(0) = 0$ and $y(1) = B$ this gives $y(x) = Bx$.

(b) On substituting Bx for $y(x)$ we see that
$$\delta = \frac{\epsilon^2(4+3B)}{8(1+B)^{3/2}} \int_0^1 dx\, g'(x)^2 + O(\epsilon^3).$$

Then, provided $B > -1$, δ is positive and the functional is a minimum on the stationary path.

Solution 1.24

(a) Consider the difference $\delta = S[y + \epsilon g] - S[y]$ where $g(a) = g(b) = 0$, so we need the Taylor expansion
$$F(y' + \epsilon g') = F(y') + \epsilon g'\frac{dF}{dy'} + \frac{1}{2}\epsilon^2 g'^2 \frac{d^2F}{dy'^2} + \cdots.$$

Hence
$$\delta = \epsilon \int_a^b dx\, \frac{dF}{dy'}g'(x) + \frac{1}{2}\epsilon^2 \int_a^b dx\, \frac{d^2F}{dy'^2}g'(x)^2 + O(\epsilon^3).$$

(b) If $dF/dy' =$ constant then $\delta = O(\epsilon^2)$ so $S[y]$ is stationary. Further, if $dF/dy' =$ constant then, provided $F(z)$ is not a constant or a linear function of z, $y'(x)$ is also a constant.

(c) On the stationary path $y'(x)$ is a constant and hence d^2F/dy'^2 is constant and
$$\delta = \frac{1}{2}\epsilon^2 \frac{d^2F}{dy'^2} \int_a^b dx\, g'(x)^2 + O(\epsilon^3).$$

The integral is positive, so δ is positive or negative when d^2F/dy'^2 is positive or negative. That is, $S[y]$ is either a minimum ($d^2F/dy'^2 > 0$) or a maximum ($d^2F/dy'^2 < 0$). If $d^2F/dy'^2 = 0$ the nature of the stationary path can be determined only by expanding to higher order in ϵ.

Solution 1.25

The functional $L[y]$ is the length of the chain, given by the integral (1.9) (page 9). The potential energy, δV, of an element of the rope of length δs centred on a point x is given by mass $\times$ height $\times g$, i.e. $\delta V = (\rho \delta s) y(x) g$. Since $\delta s = \sqrt{1 + y'^2}\delta x$ this gives the length of the chain as
$$L[y] = \int_{-a}^a dx\, \sqrt{1 + y'^2}$$

and the total potential energy as
$$E[y] = \rho g \int_{-a}^a dx\, y\sqrt{1 + y'^2}.$$

Solution 1.26

(a) The cylinder radius, ρ, is constant so, to first order, $\delta x = -\rho\delta\phi\sin\phi$ and $\delta y = \rho\delta\phi\cos\phi$, the distance is
$$\delta s^2 = \delta x^2 + \delta y^2 + \delta z^2 = \rho^2\delta\phi^2 + \delta z^2 = \delta\phi^2\left(\rho^2 + \left(\frac{\delta z}{\delta\phi}\right)^2\right).$$

(b) The length along a curve is just the sum of the small elements which in the limit $\delta\phi \to 0$ becomes the integral
$$L[z] = \int_{\phi_1}^{\phi_2} d\phi\, \sqrt{\rho^2 + z'(\phi)^2}.$$

(c) The functional $L[z]$ is the same as that considered in Section 1.3.1 hence its minimum value is given when $z(\phi)$ is a linear function of ϕ. The boundary conditions give the result quoted.

Solution 1.27

The Cartesian coordinates of a point (ρ, ϕ) on the cone are
$$(x, y, z) = (\rho \cos \phi, \rho \sin \phi, \rho/\tan \alpha)$$
and for the adjacent point at $(\rho + \delta\rho, \phi + \delta\phi)$, or $(x + \delta x, y + \delta y, z + \delta z)$ in Cartesian coordinates, we have, to first order
$$\delta x = \delta\rho \cos\phi - \rho\delta\phi \sin\phi, \quad \delta y = \delta\rho \sin\phi + \rho\delta\phi \cos\phi, \quad \delta z = \delta\rho/\tan\alpha.$$
The distance between the two adjacent points is therefore
$$\delta s^2 = \left(1 + \frac{1}{\tan^2\alpha}\right)\delta\rho^2 + \rho^2 \delta\phi^2 = \frac{\delta\rho^2}{\sin^2\alpha} + \rho^2 \delta\phi^2 = \left(\rho^2 + \frac{1}{\sin^2\alpha}\left(\frac{\delta\rho}{\delta\phi}\right)^2\right)\delta\phi^2.$$
Hence the distance between the points ϕ_1 and ϕ_2 along the curve $\rho(\phi)$ is
$$L[\rho] = \int_{\phi_1}^{\phi_2} d\phi \sqrt{\rho^2 + \frac{\rho'^2}{\sin^2\alpha}}.$$

Solution 1.28

Let the velocity of the boat relative to the water be (u_x, u_y), where $c^2 = u_x^2 + u_y^2$, and we assume that u_x is positive. The velocity of the boat relative to land is therefore $(u_x, v(x) + u_y)$. If the path taken is $y(x)$ it follows that
$$\frac{dy}{dx} = \frac{dy}{dt} \bigg/ \frac{dx}{dt} = \frac{u_y + v}{u_x}.$$
Also, the time of passage is
$$T[y] = \int_0^a \frac{dx}{u_x}.$$
Now we need an expression for u_x in terms of y, v and c. Since $c^2 = u_x^2 + u_y^2$, we have
$$(y'(x)u_x - v)^2 = c^2 - u_x^2$$
which rearranges to the quadratic
$$\left(1 + y'^2\right) u_x^2 - 2vy' u_x - \left(c^2 - v^2\right) = 0,$$
having the solutions
$$u_x = \frac{vy' \pm \sqrt{(vy')^2 + (c^2 - v^2)(1 + y'^2)}}{1 + y'^2}.$$
Because $c > v$ this equation has one positive and one negative root. We need the positive root:
$$u_x = \frac{vy' + \sqrt{(vy')^2 + (c^2 - v^2)(1 + y'^2)}}{1 + y'^2}.$$
This may be put in a more convenient form by noting that, for any X
$$1 + \sqrt{X} = \left(1 + \sqrt{X}\right)\left(\frac{1 - \sqrt{X}}{1 - \sqrt{X}}\right) = \frac{1 - X}{1 - \sqrt{X}}.$$
Hence
$$u_x = \frac{c^2 - v^2}{\sqrt{(vy')^2 + (c^2 - v^2)(1 + y'^2)} - vy'},$$
giving
$$T[y] = \int_0^a dx \, \frac{\sqrt{(vy')^2 + (c^2 - v^2)(1 + y'^2)} - vy'}{c^2 - v^2} = \int_0^a dx \, \frac{\sqrt{c^2(1 + y'^2) - v^2} - vy'}{c^2 - v^2}.$$

CHAPTER 2
The Euler–Lagrange equation

2.1 Introduction

In this chapter we derive the most important result of the calculus of variations and apply it to a few of the examples described in Chapter 1. We show that for the functional

$$S[y] = \int_a^b dx\, F(x,y,y'), \quad y(a) = A, \ y(b) = B, \tag{2.1}$$

where $F(x,u,v)$ is a real function of three real variables, and a *necessary* condition for the function $y(x)$ to be a stationary path is that it satisfies the equation

$$\frac{d}{dx}\left(\frac{\partial F}{\partial y'}\right) - \frac{\partial F}{\partial y} = 0, \quad y(a) = A, \ y(b) = B, \tag{2.2}$$

where the conditions $y(a) = A$, $y(b) = B$ are named as the boundary conditions. This equation is known as either *Euler's equation* or the *Euler–Lagrange equation*.

In order to derive the Euler–Lagrange equation it is helpful to first discuss some preliminary ideas. We start by describing briefly Euler's original analysis, because this provides an intuitive method and indicates why variational problems are different from those of conventional calculus, and also provides a link between the calculus of functions of many variables and the calculus of variations. This leads directly to the idea of the rate of change of a functional, needed in subsequent analysis. Next we prove the *fundamental lemma* of the calculus of variations, which is essential for the derivation of the Euler–Lagrange equation.

In the last two sections of this chapter, before the end of chapter exercises, we apply this theory to two of the problems described in Chapter 1: the surface of minimum area and the brachistochrone problems. There are many exercise in this chapter; you should attempt as many as time permits, but remember that it is unnecessary to attempt them all.

2.2 Preliminary remarks

2.2.1 Relation to differential calculus

Euler was the first to make a systematic study of problems that can be described by functionals, though it was Lagrange who developed the method we now use. Euler studied functionals having the form defined in equation (2.1). He related these functionals to functions of many variables using the simple device of dividing the abscissa into $N+1$ equal intervals,

Leonhard Euler, 1707–1783.

Joseph-Louis Lagrange, 1736–1813.

$$a = x_0, \; x_1, \; x_2, \; \ldots, \; x_N, \quad x_{N+1} = b, \tag{2.3}$$

where $x_{k+1} - x_k = h$, and replacing the curve $y(x)$ with segments of straight lines having the vertices

$$(x_0, A), \; (x_1, y_1), \; (x_2, y_2), \; \ldots, \; (x_N, y_N), \; (x_{N+1}, B), \tag{2.4}$$

where $y_k = y(x_k)$, as shown in Figure 2.1.

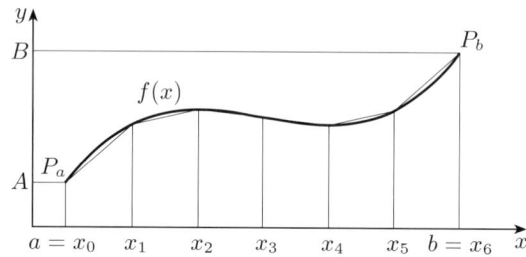

Figure 2.1 Diagram showing the rectification of a curve $y = f(x)$ by a series of six straight-line segments ($N = 5$).

Approximating the derivative at x_k by the difference $(y_k - y_{k-1})/h$, the functional (2.1) is replaced by a function of the N variables $(y_1, y_2, \ldots, y_N)$,

$$S(y_1, y_2, \ldots, y_N) = h \sum_{k=1}^{N+1} F\left(x_k, y_k, \frac{y_k - y_{k-1}}{h}\right), \tag{2.5}$$

where $h = (b-a)/(N+1)$, $y_0 = A$ and $y_{N+1} = B$. This association with ordinary functions of many variables can illuminate the nature of functionals and, if all else fails, it can be used as the basis of a numerical approximation; examples of this procedure are given in Exercises 2.1 and 2.22. The integral (2.1) is obtained from this sum by taking the limit $N \to \infty$; similarly, the Euler–Lagrange equation (2.2) may be derived by taking the same limit of the N algebraic equations $\partial S/\partial y_k = 0$, $k = 1, 2, \ldots, N$. In any mathematical analysis, however, care is usually needed when such limits are taken and this example is no exception; we shall not dwell on these difficulties, although they are important in the further development of this subject.

Euler made extensive use of this method of finite differences. By replacing smooth curves with polygonal lines, he reduced the problem of finding stationary paths of functionals to finding stationary points of a function of N variables; he then obtained exact solutions by taking the limit as $N \to \infty$. In this sense functionals may be regarded as functions of infinitely many variables – that is, the values of the function $y(x)$ at distinct points – and the calculus of variations may be regarded as the corresponding analogue of the differential calculus.

Exercise 2.1

If the functional depends only upon y',

$$S[y] = \int_a^b dx\, F(y'), \quad y(a) = A,\ y(b) = B,$$

show that the corresponding approximation defined by equation (2.5) becomes

$$S(y_1, y_2, \ldots, y_N) = h\left[F\left(\frac{y_1 - A}{h}\right) + F\left(\frac{y_2 - y_1}{h}\right) + \cdots + F\left(\frac{y_k - y_{k-1}}{h}\right)\right.$$
$$\left. + \cdots + F\left(\frac{y_N - y_{N-1}}{h}\right) + F\left(\frac{B - y_N}{h}\right)\right].$$

Show that a stationary point of this function satisfies the equations

$$\frac{\partial S}{\partial y_k} = F'\left(\frac{y_k - y_{k-1}}{h}\right) - F'\left(\frac{y_{k+1} - y_k}{h}\right) = 0, \quad k = 1, 2, \ldots, N.$$

Use these to deduce that

$$F'\left((y_k - y_{k-1})/h\right) = c, \quad k = 1, 2, \ldots, N+1,$$

where c is a constant, independent of k, and provide arguments to show that, provided $F'(z)$ is continuous, $S(y_1, y_2, \ldots, y_N)$ is stationary when the points $(x_k, y(x_k))$ lie on a straight line.

2.2.2 The differentiation of a functional

The calculus of variations is primarily about finding stationary paths of functionals. The subtle mathematical ideas necessary for the rigorous development of the subject are beyond the scope of this introductory course. However, you ought to be aware of some of the potential problems, and in the next few paragraphs we try to explain some of these difficulties, which are not an assessed part of the course.

Consider for a moment the analogous problem of finding the stationary points of a function, $f(x)$, of a real variable x: it is evident that the concept of closeness of two points on the real line and of two values of the function are essential. In this case these measures are the same, namely $|a - b|$ for the distance between two real numbers.

For a function $f(\boldsymbol{x})$ of n real variables $\boldsymbol{x} = (x_1, x_2, \ldots, x_n)$, we need the notion of the distance between two points in the n-dimensional space of variables. Usually, the numbers $(x_1, x_2, \ldots, x_n)$ can be regarded as the Cartesian coordinates of a point in an n-dimensional space, then the distance, $|\boldsymbol{x}|$, of a point $\boldsymbol{x}$ from the origin can be defined by the natural extension of Pythagoras' theorem,

$$|\boldsymbol{x}| = \sqrt{x_1^2 + x_2^2 + \cdots + x_n^2}, \tag{2.6}$$

and the distance between two points is $|\boldsymbol{x} - \boldsymbol{y}|$. However, alternative definitions of distance can be used: two are

$$|\boldsymbol{x}|_1 = |x_1| + |x_2| + \cdots + |x_n| \quad \text{and} \quad |\boldsymbol{x}|_2 = \max_{i=1,2,\ldots,n}\left(|x_i|\right). \tag{2.7}$$

Functions, such as these, that define distances in abstract spaces are called *norms*: any function $N(\boldsymbol{x})$ of the coordinates can be used as a norm provided it satisfies the following three rules:

(a) $N(\boldsymbol{x}) \geq 0$ and $N(\boldsymbol{x}) = 0$ if and only if $\boldsymbol{x} = 0$.

(b) $N(\boldsymbol{x} - \boldsymbol{y}) = N(\boldsymbol{y} - \boldsymbol{x})$.

(c) $N(\boldsymbol{x} - \boldsymbol{y}) + N(\boldsymbol{y} - \boldsymbol{z}) \geq N(\boldsymbol{x} - \boldsymbol{z})$ (triangle inequality).

It can be verified that the three norms defined above satisfy these rules.

Returning to functionals, now the elements corresponding to the points $\boldsymbol{x}$ are functions, $y(x)$, belonging to a particular class. For functionals of the form

$$\int_a^b dx\, F(x, y, y'), \quad y(a) = A,\; y(b) = B, \tag{2.8}$$

we would require functions, $y(x)$, with continuous first derivative that satisfy the boundary conditions. One of the crucial differences between the point $\boldsymbol{x}$ and a function $y(x)$ is that the former belongs to a finite-dimensional space, in the sense that there are at most n independent directions. Functions, however, belong to an infinite-dimensional space: this step, from finite- to infinite-dimensional spaces, is subtle and important in the development of rigorous theory of the calculus of variations, but is not considered in this course.

For the same reason that we need a norm in an n-dimensional Cartesian space, we need norms for the distance between allowed functions. There are several different norms for functions, and the choice of an appropriate norm can be important. An example of a norm for functions continuous on $[a, b]$ is

$$\|y(x)\| = \max_{a \leq x \leq b} |y(x)|. \tag{2.9}$$

But continuous functions may not be differentiable at some points – or indeed at any point – and we need functions that are both continuous and have continuous first derivatives. A suitable norm for such functions is

$$\|y(x)\| = \max_{a \leq x \leq b} |y(x)| + \max_{a \leq x \leq b} |y'(x)|. \tag{2.10}$$

In the rigorous analysis of variational principles, the choice of norms is important, particularly when trying to determine whether or not stationary paths are extrema. In this course we have no need to consider norms any further, although the notion is implicit in all that follows.

It is necessary, however, to restrict the class of functions to a subset of all possible functions that satisfy the boundary conditions. Normally we shall simply refer to this restricted class of functions as the *admissible functions*: in most circumstances admissible functions are defined to satisfy the boundary conditions and have continuous first, or possibly higher, derivatives.

If you are interested in knowing why these details are important, consider the optional Exercise 2.37 (page 74) which gives a simple example showing why care is needed. Further, in Section 2.5, in particular equation (2.59), we shall encounter a real problem for which the path producing the minimum value of the functional does not have a continuous first derivative. These subtleties are not, however, an assessed part of the course, or required in the theory presented here.

We also need the notion of the rate of change of a functional, because this is implicit in the idea of a stationary path. Recall that a real differentiable function of n real variables $G(\boldsymbol{x})$, for $\boldsymbol{x} = (x_1, x_2, \ldots, x_n)$, is stationary at a point if, at this point, all its first partial derivatives are zero, i.e. $\partial G/\partial x_k = 0$,

2.2 Preliminary remarks

$k = 1, 2, \ldots, n$. This result follows by considering the difference between the values of $G(\boldsymbol{x})$ at adjacent points:

$$G(\boldsymbol{x} + \epsilon\boldsymbol{\xi}) - G(\boldsymbol{x}) = \epsilon \sum_{k=1}^{n} \xi_k \frac{\partial G}{\partial x_k} + O(\epsilon^2), \quad |\boldsymbol{\xi}| = 1, \tag{2.11}$$

where $\boldsymbol{\xi} = (\xi_1, \xi_2, \ldots, \xi_n)$. The rate of change of $G(\boldsymbol{x})$ in the direction of $\boldsymbol{\xi}$ is obtained by dividing by ϵ,

$$\frac{G(\boldsymbol{x} + \epsilon\boldsymbol{\xi}) - G(\boldsymbol{x})}{\epsilon} = \sum_{k=1}^{n} \xi_k \frac{\partial G}{\partial x_k} + O(\epsilon), \tag{2.12}$$

and taking the limit as $\epsilon \to 0$. A stationary point is *defined* to be one at which the rate of change of G is zero in *every* direction; it follows that at a stationary point *all* first partial derivatives must be zero.

This idea, embodied in equation (2.12), may be applied to the functional

$$S[y] = \int_a^b dx\, F(x, y, y'), \quad y(a) = A,\ y(b) = B, \tag{2.13}$$

which has a real value for each admissible function $y(x)$. The rate of change of a functional $S[y]$ is obtained by examining the difference between the values on neighbouring admissible paths, i.e. $S[y + \epsilon g] - S[y]$; and, since both $y(x)$ and $y(x) + \epsilon g(x)$ are admissible functions for all ϵ, it follows that $g(a) = g(b) = 0$. This difference is a function of the real variable ϵ, so we define the rate of change of $S[y]$ by the limit

$$\Delta S[y, g] = \lim_{\epsilon \to 0} \frac{S[y + \epsilon g] - S[y]}{\epsilon} = \frac{d}{d\epsilon} S[y + \epsilon g]\bigg|_{\epsilon = 0}, \tag{2.14}$$

which we assume exists. The functional ΔS depends upon both $y(x)$ and $g(x)$, just as the limit of the difference $[G(\boldsymbol{x} + \epsilon\boldsymbol{\xi}) - G(\boldsymbol{x})]/\epsilon$, of equation (2.12), depends upon $\boldsymbol{x}$ and $\boldsymbol{\xi}$.

Definition: The functional $S[y]$ is said to be stationary if $y(x)$ is an admissible function and if $\Delta S[y, g] = 0$ for *all* $g(x)$ which have a continuous first derivative *and* satisfy the boundary conditions $g(a) = g(b) = 0$.

This definition generalises the particular example dealt with in Subsection 1.2.2, in particular see equation (1.17) (page 11).

As with functions of many variables, the nature of a stationary path of a functional is usually determined by the term $O(\epsilon^2)$ in the expansion of $S[y + \epsilon g]$. The theory that determines whether a stationary path is an extremum is difficult and is not contained in this course.

In all our applications, the limit in equation (2.14) can be evaluated by the formula

$$\Delta S[y, g] = \frac{d}{d\epsilon} S[y + \epsilon g]\bigg|_{\epsilon = 0}. \tag{2.15}$$

Also, in our applications, ΔS is linear in g. That is, if c is any constant, then $\Delta S[y, cg] = c\Delta S[y, g]$; in this case it is called the *Gâteaux differential*.

As an example, consider the functional

$$S[y] = \int_a^b dx\, \sqrt{1 + y'^2}, \quad y(a) = A,\ y(b) = B, \tag{2.16}$$

for the distance between (a, A) and (b, B) as discussed in Subsection 1.2.2. We have

$$\frac{d}{d\epsilon} S[y + \epsilon g] = \frac{d}{d\epsilon} \int_a^b dx\, \sqrt{1 + (y' + \epsilon g')^2}$$

$$= \int_a^b dx\, \frac{d}{d\epsilon} \sqrt{1 + (y' + \epsilon g')^2}$$

$$= \int_a^b dx\, \frac{(y'(x) + \epsilon g'(x))}{\sqrt{1 + (y'(x) + \epsilon g'(x))^2}} g'(x). \qquad (2.17)$$

Note that we may change the order of differentiation with respect to ϵ and integration with respect to x because a and b are independent of ϵ and all integrands are assumed to be sufficiently well-behaved functions of x and ϵ. Hence, on taking the limit as $\epsilon \to 0$, we have

$$\Delta S[y, g] = \lim_{\epsilon \to 0} \frac{d}{d\epsilon} S[y + \epsilon g] = \int_a^b dx\, \frac{y'(x)}{\sqrt{1 + y'(x)^2}} g'(x), \qquad (2.18)$$

which is just equation (1.18) (page 11).

Exercise 2.2

Find the Gâteaux differentials of the following functionals.

(a) $S[y] = \int_0^1 dx\, y'^2$ (b) $S[y] = \int_1^2 dx\, x^2 y'^2$ (c) $S[y] = \int_0^{\pi/2} dx\, (y'^2 - y^2)$

(d) $S[y] = \int_a^b dx\, \frac{y'^2}{x^3},\ b > a > 0$ (e) $S[y] = \int_a^b dx\, (y'^2 + y^2 + 2y e^x)$

(f) $S[y] = \int_0^1 dx\, \sqrt{x^2 + y^2} \sqrt{1 + y'^2}$

Exercise 2.3

Find the Gâteaux differentials of the following functionals.

(a) $S_1[y] = \int_a^b dx\, y'$ (b) $S_2[y] = \int_a^b dx\, xy'^2$

(c) $S_3[y] = \int_a^b dx\, (1 + xy')\, y'$

Explain why $\Delta S_2 = \Delta S_3$.

A comment

For our final comment of this section, we consider the approximation defined in equation (2.5) in a little more detail to show the sort of problem that can arise in the limit as $N \to \infty$. By associating the variable $\boldsymbol{\xi}$, in equation (2.12), with the discrete values of the function $g(x)$ of equation (2.14), so $\boldsymbol{\xi} = (g_1, g_2, \ldots, g_N)$, where $g_k = g(x_k)$, we see that the condition $|\boldsymbol{\xi}| = 1$ includes the requirement that $g(x)$ is bounded for $a \leq x \leq b$. However, we also require that $g'(x)$ exists, but the example $g_k = -1/\sqrt{2},\ g_{k+1} = 1/\sqrt{2}$ and $g_j = 0$ for all other components, so that $g'(x_k) \simeq (g_{k+1} - g_k)/h$, shows that although g remains bounded as $h \to 0$, its derivative is unbounded. This example shows why some care is required when we take the limit $N \to \infty$ of equation (2.5), why the cases of finite numbers of variables is different from the infinite case and why the choice of norms, discussed above, is important. Nevertheless, provided caution is exercised, the analogy with functions of several variables can be helpful.

2.3 The fundamental lemma of the calculus of variations

This section contains the essential result upon which the calculus of variations depends. Using the result obtained here, we shall be able to use the stationary condition that $\Delta S[y, g] = 0$, for all suitable $g(x)$, to form a differential equation for the unknown function $y(x)$.

> **The fundamental lemma**
>
> If $z(x)$ is a continuous function of x for $a \leq x \leq b$ and
>
> $$\int_a^b dx\, z(x)\, g(x) = 0 \qquad (2.19)$$
>
> for *all* functions $g(x)$ that are continuous for $a \leq x \leq b$ and zero at $x = a$ and $x = b$, then $z(x) = 0$ for $a \leq x \leq b$.

In order to prove this, we assume, on the contrary, that for some η satisfying $a < \eta < b$, $z(\eta) \neq 0$. Then, since $z(x)$ is continuous, there is an interval $[x_1, x_2]$ around η with

$$a < x_1 \leq \eta \leq x_2 < b, \qquad (2.20)$$

in which $z(x) \neq 0$. We now construct a suitable function $g(x)$ that yields a contradiction. Define $g(x)$ to be

$$g(x) = \begin{cases} (x - x_1)(x_2 - x), & a < x_1 \leq x \leq x_2 < b, \\ 0, & \text{otherwise}, \end{cases} \qquad (2.21)$$

so $g(x)$ is continuous and

$$\int_a^b dx\, z(x)\, g(x) = \int_{x_1}^{x_2} dx\, z(x)(x - x_1)(x_2 - x) \neq 0, \qquad (2.22)$$

since the integrand is continuous and non-zero on $[x_1, x_2]$. However, according to equation (2.19), $\int_a^b dx\, zg = 0$, so we have a contradiction.

The assumptions that $z(x)$ is continuous and $z(x) \neq 0$ for some $x \in (a, b)$ lead to a contradiction and it follows that $z(x) = 0$ for $a < x < b$; because $z(x)$ is continuous, $z(x) = 0$ for $a \leq x \leq b$. This result is called the *fundamental lemma of the calculus of variations*.

This proof assumed only that $g(x)$ is continuous and made no assumptions about its differentiability. In previous applications $g(x)$ had to be differentiable for $x \in (a, b)$. However, for the function $g(x)$ defined above, $g'(x)$ does not exist at x_1 and x_2. The proof is easily modified to deal with this case. If $g(x)$ needs to be n times differentiable, then we use the function

$$g(x) = \begin{cases} (x - x_1)^{n+1}(x_2 - x)^{n+1}, & x_1 \leq x \leq x_2, \\ 0, & \text{otherwise}. \end{cases} \qquad (2.23)$$

2.4 The Euler–Lagrange equation

This section contains the most important result of this block: if $F(x, u, v)$ is a sufficiently differentiable function of three variables, then a *necessary* condition for the functional

$$S[y] = \int_a^b dx\, F(x, y, y'), \quad y(a) = A,\ y(b) = B, \tag{2.24}$$

to be stationary on the path $y(x)$, is that it satisfies the differential equation and boundary conditions,

$$\frac{d}{dx}\left(\frac{\partial F}{\partial y'}\right) - \frac{\partial F}{\partial y} = 0, \quad y(a) = A,\ y(b) = B. \tag{2.25}$$

This is called *Euler's equation* or the *Euler–Lagrange equation*. It is a second-order differential equation, as shown in Exercise 1.16 (page 19), and is the analogue of the necessary condition $\partial G/\partial x_k = 0$, $k = 1, 2, \ldots, n$, for a function of n real variables to be stationary, as discussed in Subsection 2.2.2. We now derive this equation.

The integral (2.24) is defined for functions $y(x)$ that are differentiable for $a \leq x \leq b$. Using equation (2.14), we find that the rate of change of $S[y]$ is

$$\Delta S[y, g] = \frac{d}{d\epsilon} \int_a^b dx\, F(x, y + \epsilon g, y' + \epsilon g') \bigg|_{\epsilon=0}$$

$$= \int_a^b dx\, \frac{d}{d\epsilon} F(x, y + \epsilon g, y' + \epsilon g') \bigg|_{\epsilon=0}. \tag{2.26}$$

The integration limits a and b are independent of ϵ, and we assume that the order of integration and differentiation may be interchanged using Leibniz's rule. Using the chain rule we have

See Block 0, Chapter 2.

$$\frac{d}{d\epsilon} F(x, y + \epsilon g, y' + \epsilon g') \bigg|_{\epsilon=0} = g \frac{\partial F}{\partial y} + g' \frac{\partial F}{\partial y'}, \tag{2.27}$$

where the right-hand side is evaluated at (x, y, y'). Hence

$$\Delta S[y, g] = \int_a^b dx \left(g(x) \frac{\partial F}{\partial y} + g'(x) \frac{\partial F}{\partial y'} \right). \tag{2.28}$$

The second term in this integral can be rewritten using integration by parts, to give

$$\int_a^b dx\, g'(x) \frac{\partial F}{\partial y'} = \left[g(x) \frac{\partial F}{\partial y'} \right]_a^b - \int_a^b dx\, g(x) \frac{d}{dx}\left(\frac{\partial F}{\partial y'}\right). \tag{2.29}$$

But $g(a) = g(b) = 0$, so the boundary term on the right-hand side (that is, the first term) vanishes, and the rate of change of the functional $S[y]$ becomes

$$\Delta S[y, g] = -\int_a^b dx \left[\frac{d}{dx}\left(\frac{\partial F}{\partial y'}\right) - \frac{\partial F}{\partial y} \right] g(x). \tag{2.30}$$

If $S[y]$ is stationary then, by definition, $\Delta S[y, g] = 0$ for all functions $g(x)$. It follows from the fundamental lemma of the calculus of variations (see Section 2.3 and equation (2.19)) that $y(x)$ satisfies the second-order differential equation

$$\frac{d}{dx}\left(\frac{\partial F}{\partial y'}\right) - \frac{\partial F}{\partial y} = 0, \quad y(a) = A,\ y(b) = B. \tag{2.31}$$

If real solutions exist, they provide stationary paths of $S[y]$, but these are *not* necessarily maxima or minima of the functional. It is important to note that

2.4 The Euler–Lagrange equation

the Euler–Lagrange equation is, in most cases, a second-order, nonlinear, boundary-value problem, with possibly no solutions or many solutions. The properties of nonlinear equations are quite different from those of linear equations.

The Euler–Lagrange equation is a second-order differential equation. But if the integrand does not depend explicitly upon x, so has the form

$$S[y] = \int_a^b dx\, G(y, y'), \quad y(a) = A,\ y(b) = B, \qquad (2.32)$$

it can be shown that the Euler–Lagrange equation reduces to the first-order differential equation

$$y' \frac{\partial G}{\partial y'} - G = c, \quad y(a) = A,\ y(b) = B, \qquad (2.33)$$

FIRST INTEGRAL

where c is an arbitrary constant determined by the boundary conditions; see the example below. The expression on the left-hand side of this equation is often called the *first integral* of the Euler–Lagrange equation. This result is important because, when applicable, it saves a great deal of effort, since it is usually far easier to solve the lower-order equation. Two proofs of this result are provided in this course. The first involves deriving an algebraic identity and we suggest doing this yourself in Exercise 2.6. The second proof is given in Subsection 3.5.1 and uses the invariance properties of the integrand $G(y, y')$. In Chapter 4 we show that for simple mechanical systems equation (2.33) is sometimes equivalent to conservation of energy. A warning, however: in some circumstances, a solution of equation (2.33) will *not* be a solution of the original Euler–Lagrange equation – see Exercise 2.5 and also Section 2.5.

Another important consequence of this simplification is that the stationary function, the solution of (2.33), depends only upon the variables $u = x - a$ and $b - a$ (besides A and B), rather than x, a and b independently, as is the case when the integrand depends explicitly upon x. A specific example illustrating this behaviour is given in Exercise 2.4.

As an example, consider the functional

$$S[y] = \int_0^1 dx\, \left(y'^2 - y\right), \quad y(0) = 0,\ y(1) = 1. \qquad (2.34)$$

In this case $G(y, y') = y'^2 - y$ and the Euler–Lagrange equation is the linear equation

$$2\frac{d^2 y}{dx^2} + 1 = 0, \quad y(0) = 0,\ y(1) = 1. \qquad (2.35)$$

The general solution of this equation is $y = -x^2/4 + Ax + B$, where A and B are constants: the boundary condition at $x = 0$ gives $B = 0$ and that at $x = 1$ gives $A = 5/4$. Hence $y = x(5-x)/4$ is the stationary path.

Applying the first integral, equation (2.33), to equation (2.34) gives the nonlinear equation

$$\left(\frac{dy}{dx}\right)^2 + y = c, \quad y(0) = 0,\ y(1) = 1, \qquad (2.36)$$

which is separable and can be rearranged into

$$\int \frac{dy}{\sqrt{c - y}} = \pm \int dx, \qquad (2.37)$$

so integration gives

$$2\sqrt{c - y} = D \mp x, \qquad (2.38)$$

for some constant D. The boundary condition at $x=0$ gives $4c = D^2$ and that at $x=1$ gives $D = \pm 5/2$, both giving the same solution as before.

In this example it is easier to solve the second-order Euler–Lagrange equation than the first-order equation (2.33), which is nonlinear. Normally, both equations are nonlinear and then it is easier to solve the first-order equation. In the examples dealt with in Sections 2.5 and 2.7 it is more convenient to use the first integral.

An observation

You may have noticed that the original functional (2.24) was defined on the class of functions that have just one derivative. More precisely, it requires that $F(x,y,y')$ is integrable, which is a weaker condition, though this difference is not important for this course. However, the Euler–Lagrange equation (2.31) requires the stronger condition that $y''(x)$ is continuous. A theorem due to the German mathematician du Bois-Reymond shows that if:

Emil Heinrich du Bois-Reymond, 1818–1896.

(a) $y(x)$ has a continuous first derivative,
(b) $\Delta S[y,g] = 0$ for all admissible $g(x)$,
(c) $F(x,u,v)$ has continuous first and second derivatives in all variables,
(d) $\partial^2 F/\partial y'^2 \neq 0$ for $a \leq x \leq b$,

then $y(x)$ has a continuous second-order derivative and satisfies the Euler–Lagrange equation (2.31). (You will not be assessed on this result.)

Exercise 2.4

Consider the functional
$$S[y] = \int_a^b dx\,(y'^2 + y), \quad y(a) = A,\ y(b) = B.$$

(a) By forming and solving the Euler–Lagrange equation (2.31), show that the stationary path is
$$y = \tfrac{1}{4}u^2 + \left(\frac{B-A}{b-a} - \tfrac{1}{4}(b-a)\right)u + A, \quad u = x-a.$$

(b) By making the change of variable $u = x - a$ and defining $Y(u) = y(x(u)) = y(u+a)$, use the chain rule to show that $y'(x) = Y'(u)$ and that the functional becomes
$$S[Y] = \int_0^{b-a} du\,(Y'^2 - Y^2), \quad Y(0) = A,\ Y(b-a) = B.$$

Deduce that the stationary path depends only upon $u = x - a$ and $b - a$ (besides A and B).

Exercise 2.5

(a) Show that provided $G_{y'}(y,y')$ exists when $y' = 0$, the differential equation (2.33) has a solution $y(x) = \gamma$, where the constant γ is defined implicitly by the equation $G(\gamma, 0) = -c$.

(b) Under what circumstances is the solution $y(x) = \gamma$ also a solution of the Euler–Lagrange equation (2.31)?

Note that this part of the exercise is harder.

Exercise 2.6

If $G(y, y')$ does not depend explicitly upon x, that is $\partial G/\partial x = 0$, show that

$$y'(x)\left(\frac{d}{dx}\left(\frac{\partial G}{\partial y'}\right) - \frac{\partial G}{\partial y}\right) = \frac{d}{dx}\left(y'\frac{\partial G}{\partial y'} - G\right).$$

Hence show that equation (2.33) is true if either (a) $y(x)$ satisfies the Euler–Lagrange equation, or (b) $y(x)$ is constant.

Exercise 2.7

Show that the Euler–Lagrange equation for the functional

$$S[y] = \int_0^X dx\, [y'^2 - y^2], \quad y(0) = 0,\ y(X) = 1,\ X > 0,$$

is $y'' + y = 0$. Hence show that, provided $X \neq n\pi$, $n = 1, 2, \ldots$, the stationary function is $y = \sin x/\sin X$.

Exercise 2.8

(a) Show that the Euler–Lagrange equation for the functional

$$S[y] = \int_0^1 dx\, [y'^2 + y^2 + 2xy], \quad y(0) = 0,\ y(1) = \alpha,$$

is $y'' - y = x$ and hence that a stationary function is

$$y(x) = (1+\alpha)\frac{\sinh x}{\sinh 1} - x.$$

(b) By considering the term $O(\epsilon^2)$ in the difference $S[y+\epsilon] - S[y]$, show also that this solution makes the functional a minimum.

2.5 Minimal surface of revolution

The problem is to find the function $y(x)$, with given end points $y(a) = A$ and $y(b) = B$, such that the area of the surface formed by rotating the curve $y(x)$ about the x-axis is a minimum. The left-hand diagram in Figure 2.2 shows the construction of this surface.

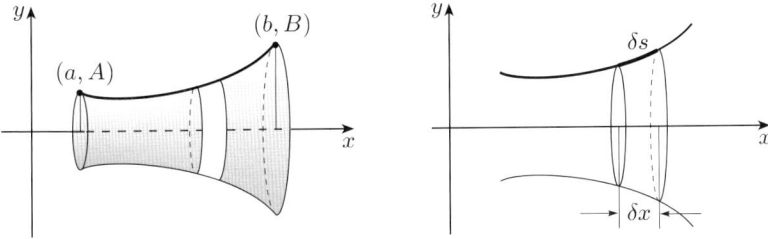

Figure 2.2 Diagram showing the construction of a surface of revolution (left), and the small segment used to construct the integral in equation (2.41) (right).

This section is divided into three parts. First, we derive the functional $S[y]$ giving the required area. Second, we derive the equation that a differentiable function must satisfy to make the functional stationary. Finally, we solve this equation in a simple case in order to show that even this relatively simple problem has pitfalls.

2.5.1 Derivation of the functional

An expression for the area of this surface is obtained by first finding the area of the edge of a thin disk of width δx, as shown in the right-hand diagram in Figure 2.2. The small segment may be approximated by a straight line provided that δx is sufficiently small, so its length δs is given by the analysis in the previous chapter leading to equation (1.7) (page 9),

$$\delta s = \sqrt{1 + y'(x)^2}\, \delta x + O(\delta x^2). \tag{2.39}$$

The area δS traced out by this segment as it rotates about the x-axis is

$$\delta S = 2\pi y(x)\, \delta s = 2\pi y(x) \sqrt{1 + y'(x)^2}\, \delta x + O(\delta x^2). \tag{2.40}$$

Hence the area of the whole surface from $x = a$ to $x = b$ is

$$S[y] = 2\pi \int_a^b dx\, y(x) \sqrt{1 + y'(x)^2}, \quad y(a) = A \geq 0,\ y(b) = B \geq 0. \tag{2.41}$$

Exercise 2.9

Show that the equation of the straight line joining the points (a, A) and (b, B) in the left-hand diagram of Figure 2.2 is

$$y = \frac{B - A}{b - a}(x - a) + A.$$

Use this together with equation (2.41) to show that the surface area of the frustum of the cone shown in Figure 2.3 is given by

$$S = \pi(B + A)\sqrt{(b - a)^2 + (B - A)^2}.$$

Recall that the frustum of a solid is that part of the solid that lies between two parallel planes which cut the solid, and its area does not include the area of the parallel ends.

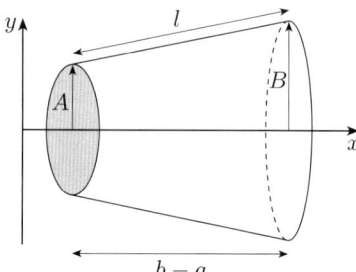

Figure 2.3 Diagram showing the frustum of a cone, the shaded area. The slant height is l, and the radii of the circular ends are A and B.

Show further that this expression may be written in the form $\pi(A + B)l$ where l is the length of the slant height and A and B are the radii of the end circles.

2.5.2 Application of the Euler–Lagrange equation

The integrand of the functional (2.41) does not depend explicitly upon x, hence the integrated form of the Euler–Lagrange equation (2.33) may be used. In this case, apart from an overall factor of 2π,

$$G(y, y') = y\sqrt{1 + y'^2}, \tag{2.42}$$

so

$$\frac{\partial G}{\partial y'} = \frac{yy'}{\sqrt{1 + y'^2}} \quad \text{and} \quad y'\frac{\partial G}{\partial y'} - G = -\frac{y}{\sqrt{1 + y'^2}}. \tag{2.43}$$

Hence the Euler–Lagrange equation integrates to give

$$\frac{y}{\sqrt{1 + y'^2}} = c, \quad y(a) = A \geq 0, \ y(b) = B \geq 0, \tag{2.44}$$

for some constant c; since $y(x) > 0$ we may assume that c is positive. By squaring and rearranging this equation we obtain the simpler first-order equation

$$\frac{dy}{dx} = \pm\frac{\sqrt{y^2 - c^2}}{c}, \quad y(a) = A \geq 0, \ y(b) = B \geq 0. \tag{2.45}$$

The solutions of equation (2.45), if they exist, ensure that the functional (2.41) is stationary. We shall see, however, that suitable solutions do not always exist and that when they do further work is necessary in order to determine the nature of the stationary point.

2.5.3 The solution in a special case

Here we solve the first-order differential equation (2.45) when the ends of the solid have the same radius, that is $A = B > 0$. Even this seemingly simple case has surprises in store and so provides an indication of the sort of difficulties that may be encountered with variational problems: such difficulties are typical of nonlinear boundary-value problems. This analysis is difficult and to understand it you may need to study it in depth; writing your own notes will probably be very useful.

Because the ends have the same radius it is convenient to introduce a symmetry by putting the ends at $x = \pm a$. This change, which is merely a shift along the x-axis, does not affect the differential equation (because its right-hand side is independent of x); the boundary conditions, however, are slightly different. If we denote the required solution by $f(x)$, then it satisfies the differential equation and boundary conditions,

$$\frac{df}{dx} = \pm\frac{\sqrt{f^2 - c^2}}{c}, \quad f(-a) = f(a) = A > 0. \tag{2.46}$$

A solution of this equation is obtained by putting $c = A$ and $f(x) = A$, but this is not a solution of the original Euler–Lagrange equation (see Exercise 2.5).

The structure of equation (2.46) suggests that we can use the identity $\cosh^2 z - 1 = \sinh^2 z$ to define a new dependent variable $\phi(x)$ by the equation $f(x) = c\cosh\phi(x)$. Then the differential equation for $\phi(x)$ is obtained by substituting this into (2.46) and using the chain rule,

$$c\frac{d\phi}{dx}\sinh\phi = \pm\frac{1}{c}\sqrt{c^2\left(\cosh^2\phi - 1\right)} = \pm\sinh\phi, \tag{2.47}$$

and hence

$$\frac{d\phi}{dx} = \pm\frac{1}{c}, \quad \text{which integrates to} \quad \phi = \alpha \pm \frac{x}{c}, \tag{2.48}$$

for some constant α. Therefore the general solution is $f(x) = c\cosh(\alpha \pm x/c)$: the boundary conditions determine the values of the constants c and α, and these give the two equations

$$A = c\cosh\left(\alpha \pm \frac{a}{c}\right) = c\left(\cosh\alpha\cosh\frac{a}{c} \pm \sinh\alpha\sinh\frac{a}{c}\right) \text{ at } x = a,$$

$$A = c\cosh\left(\alpha \mp \frac{a}{c}\right) = c\left(\cosh\alpha\cosh\frac{a}{c} \mp \sinh\alpha\sinh\frac{a}{c}\right) \text{ at } x = -a.$$

Subtracting these gives $0 = 2c\sinh\alpha\sinh(a/c)$. If, however, z is real, $\sinh z = 0$ only if $z = 0$; hence, since $a \neq 0$, we must have $\alpha = 0$, giving the solution

$$f(x) = c\cosh\left(\tfrac{x}{c}\right) \quad \text{with } c \text{ determined by } A = c\cosh\left(\tfrac{a}{c}\right). \tag{2.49}$$

The function $f(x)$ is a solution of the Euler–Lagrange equation, but we need to determine the values of c such that it satisfies the boundary conditions.

The function $\cosh z$ has a single minimum at $z = 0$, where its value is unity, and it increases monotonically for increasing $|z|$. Thus the constant c represents the smallest value of $f(x)$, which is at $x = 0$, as shown in Figure 2.4, where we plot the graph of $f(x)$ for three values of c.

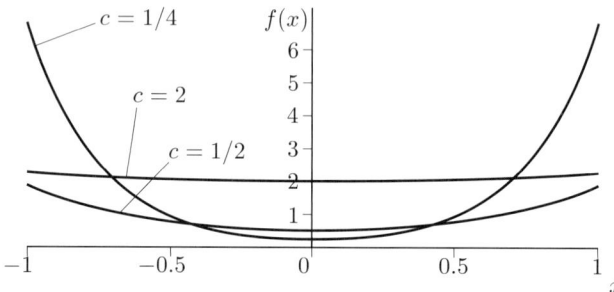

Figure 2.4 Graphs of $f(x) = c\cosh(x/c)$ for $|x| < 1$ and $c = 2$, $1/2$ and $1/4$. Note that the interval $|x| < 1$ is not related to the position of the ends at $x = \pm a$, except if $a = 1$.

From these graphs we see that as c increases, $f(x)$, becomes flatter in the interval shown.

The required solutions are obtained by finding the real values of c satisfying equation (2.49). Unfortunately, it cannot be inverted to express c in terms of known functions of A. Numerical solutions may be found, but first it is necessary to determine those values of a and A for which real solutions exist.

In this problem there are two lengths, a and A, that can be varied independently. Because there is no other length scale we expect the solution to depend only upon the dimensionless ratio of these two lengths. This reduces the number of independent parameters to one and makes the task of understanding how the solutions change with a and A far easier. As an aside, we note that if $A \neq B$, there are two ratios and this makes it far harder to understand the general behaviour of the solutions.

For this reason it is convenient to introduce a new dimensionless variable $\eta = a/c$. Our aim is to rewrite equation (2.49) in terms of η, a and A, so we write the equation for c in the form

$$\frac{A}{a} = g(\eta) \quad \text{where} \quad g(\eta) = \frac{1}{\eta}\cosh\eta. \tag{2.50}$$

This equation shows directly that η depends only upon the dimensionless ratio A/a, that is, the ratio of the end radii to half the distance between the

2.5 Minimal surface of revolution

ends. In terms of η and A, the solution (2.49) becomes

$$f(x) = \frac{a}{\eta}\cosh\left(\eta\frac{x}{a}\right) = A\frac{\cosh\left(\frac{x}{a}\eta\right)}{\cosh\eta}. \tag{2.51}$$

The stationary solutions are found by solving the equation $A/a = g(\eta)$ for η. The graph of $g(\eta)$, depicted in Figure 2.5, shows that $g(\eta)$ has a single minimum and that for $A/a > \min(g)$ there are two real solutions.

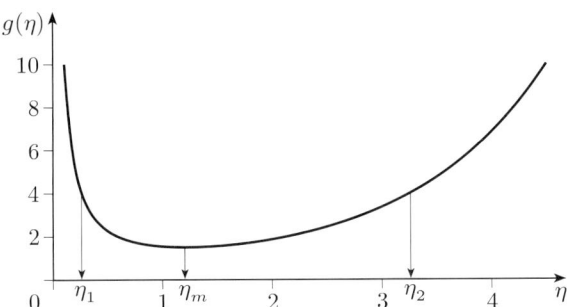

Figure 2.5 Graph of $g(\eta) = (\cosh\eta)/\eta$ showing the solutions of the equation $g(\eta) = A/a$.

This graph also suggests that $g(\eta) \to \infty$ as $\eta \to 0$ and ∞, and this behaviour can be verified with the simple analysis performed in Exercise 2.11, which shows that

$$g(\eta) \sim \frac{1}{\eta} \text{ for } \eta \ll 1 \quad \text{and} \quad g(\eta) \sim \frac{e^\eta}{2\eta} \text{ for } \eta \gg 1. \tag{2.52}$$

The minimum of $g(\eta)$ is at the real root of $\eta\tanh\eta = 1$ (see Exercise 2.12). This may be found numerically, and is at $\eta_m \simeq 1.2$, and here $g(\eta_m) = 1.51$. Hence if $A < 1.51a$ there are no real solutions of equation (2.50), meaning that there are no functions with continuous derivatives making the area stationary. For $A > 1.51a$ there are two real solutions both being stationary values of the functional (2.41); we denote these two solutions by η_1 and η_2 with $\eta_1 < \eta_2$. Because there is no upper bound on the area, neither solution can be a global maximum.

Figure 2.6 shows values of the dimensionless area S/a^2 for these two stationary solutions as functions of A/a when $A/a \geq g(\eta_m) \simeq 1.51$. The area associated with the smaller root, η_1, is denoted by S_1, and S_2 denotes the area associated with η_2. These graphs show that $S_2 > S_1$ for $A > ag(\eta_m) \simeq 1.51a$.

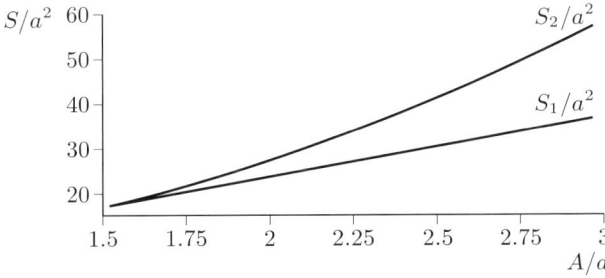

Figure 2.6 Graphs showing how the dimensionless area S/a^2 varies with A/a, for $A > 1.51a$.

The graphs depicted in Figure 2.6 were drawn using the expression for S expressed in terms of η, Exercise 2.10, and by finding η for each A/a by solving equation (2.50) numerically for each root.

It is difficult to find simple approximations for the area $S[f]$ except when $A \gg a$, in which case the results obtained in Exercise 2.11 may be used. We consider the smaller and larger roots separately.

If $A \gg a$ the smaller root, η_1 is seen from Figure 2.5 to be small. The approximation developed in Exercise 2.11(a) gives $\eta_1 \simeq a/A$, so equation (2.51) becomes

$$f_1(x) \simeq A\cosh(x/A) \simeq A, \tag{2.53}$$

since $|x| < a \ll A$ and $\cosh(x/A) \simeq 1$. Because $f_1(x)$ is approximately constant, the original functional, equation (2.41), is easily evaluated to give

$$S_1 = S[f_1] = 4\pi a A \quad \text{or} \quad \frac{S_1}{a^2} = 4\pi \frac{A}{a}. \tag{2.54}$$

The latter expression is the equation of the approximately straight line seen in Figure 2.6. The area S_1 is that of the cylinder formed by joining the end circles at $x = \pm a$ with parallel lines.

For the larger root, η_2, since $\cosh\eta \simeq e^\eta/2$ for large η, equation (2.50) for η becomes

$$\frac{A}{a} = \frac{1}{2\eta}e^\eta \tag{2.55}$$

(see Exercise 2.11(b)). So, from equation (2.51), we have

$$f_2(x) \simeq Ae^{-\eta_2}\left[\exp\left(\frac{x\eta_2}{a}\right) + \exp\left(-\frac{x\eta_2}{a}\right)\right]$$
$$= A\exp\left(-\frac{\eta_2}{a}(a-x)\right) + A\exp\left(-\frac{\eta_2}{a}(a+x)\right). \tag{2.56}$$

For positive x the second term is negligible (because $\eta_2 \gg 1$) provided $x\eta_2 \gg a$. For negative x the first term is negligible, for the same reason. Hence an approximation for $f_2(x)$ is

$$f_2(x) \simeq A\exp\left(-\frac{\eta_2}{a}(a-|x|)\right) \quad \text{provided} \quad |x|\eta_2 \gg a. \tag{2.57}$$

Some discussion of the behaviour of this function is provided after equation (2.59). In Exercise 2.11(b) it is shown that the area is given by

$$S_2 = S[f_2] \simeq 2\pi A^2 \quad \text{or} \quad \frac{S_2}{a^2} = 2\pi\left(\frac{A}{a}\right)^2, \tag{2.58}$$

which is just the area of the ends. The latter expression increases quadratically with A/a, as seen in Figure 2.6.

These approximations show directly that if $A \gg a$ then $S_2 > S_1$, confirming the results shown in Figure 2.6. They also show that when $A \gg a$ the smallest area is given when the surface of revolution approximates that of a cylinder.

In the three parts of Figure 2.7, we show examples of these stationary solutions, computed numerically, for $A = 2a$, $A = 10a$ and $A = 100a$. In the first example, on the left, the ratio $A/a = 2$ is only a little larger than $\min(g(\eta))$, but the two solutions differ substantially, with $f_1(x)$, the solution associated with the smaller value of η, already close to the constant value of A for all x. In the two other examples, the ratio A/a is larger, and now $f_1(x)$ is indistinguishable from the constant A, while $f_2(x)$ is relatively small for most values of x.

2.5 Minimal surface of revolution

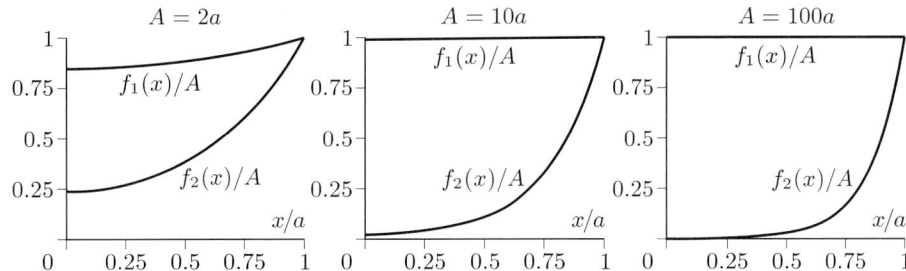

Figure 2.7 Graphs showing the stationary solutions $f(x)/A = \cosh(x\eta/a)/\cosh\eta$ as a function of x/a, for various values of A.

These examples and the preceding analysis show that when the rings are relatively close, i.e. A/a large, $f_1(x) \simeq A$, for all x, and that as $A/a \to \infty$, $f_2(x)$ tends to the function

$$f_2(x) \to f_G(x) = \begin{cases} 0, & |x| < a, \\ A, & |x| = a. \end{cases} \qquad (2.59)$$

This result may be derived from the approximate solution given in equation (2.57). Consider positive values of x, with $x\eta_2 \gg a$. If $x = a(1-\delta)$, where δ is a small positive number, then

$$f_2(x) \simeq A e^{-\delta\eta_2}. \qquad (2.60)$$

However, from equation (2.55), $\ln(A/a) = \eta - \ln(2\eta)$; if $\eta \gg 1$, $\ln(2\eta) \ll \eta$, and $\eta \simeq \ln(A/a)$, the above approximation for $f_2(x)$ thus becomes

$$\frac{f_2(x)}{A} \simeq \left(\frac{a}{A}\right)^\delta, \quad x = a(1-\delta). \qquad (2.61)$$

Hence, provided $\delta > 0$, that is, $x \neq a$, $f_2(x)/A \to 0$ as $A/a \to \infty$.

You will not be expected to reproduce this type of argument in this course.

The surface defined by this limiting function comprises two disks of radius A, a distance $2a$ apart, so has area $S_G = 2\pi A^2$, independent of a. Since this limiting solution has discontinuous derivatives at $x = \pm a$, it is not an admissible function, though there are admissible functions that are arbitrarily close to it. Nevertheless, it is important because if $A < ag(\eta_m) \simeq 1.51a$, it can be shown that this surface gives the *global* minimum of the area and, as will be seen in the next subsection and in Section 2.6, this has physical significance. This solution to the problem was first found by B. C. W. Goldschmidt in 1831, and is hence known as the *Goldschmidt curve* or *Goldschmidt solution*.

2.5.4 Summary of Section 2.5

We have considered the special case where the ends of the solid generated by our surface of revolution are at $x = \pm a$ and each end has the same radius A. In this case the curve $y = f(x)$ is symmetric about $x = 0$ and we have obtained the following results.

1. If the radius of the ends is small by comparison to the distance between them, $A < ag(\eta_m) \simeq 1.51a$, there are no curves described by differentiable functions making the traced out area stationary. In this case it can be shown that the smallest area is given by the Goldschmidt solution, $f_G(x)$, defined in equation (2.59), and that this is the global minimum.
2. If $A > 1.51a$, there are two smooth stationary curves. One of these approaches the Goldschmidt solution as $A/a \to \infty$, and the other approaches the constant function $f(x) \to A$ in this limit. The latter curve gives the smaller area.

The nature of the stationary solutions is not easy to determine. In Figure 2.8 we show the areas S_1/a^2 and S_2/a^2, as in Figure 2.6, and also the area given by the Goldschmidt solution, $S_G/a^2 = 2\pi(A/a)^2$, curve G, and the area of the cylinder, $S_c/a^2 = 4\pi A/a$, curve c.

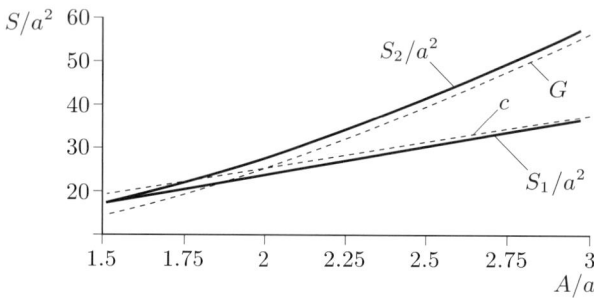

Figure 2.8 Graphs showing how the dimensionless area S/a^2 varies with A/a. Here the curves S_k/a^2 denote the areas as in Figure 2.6, $k = 1, 2$; G is the scaled area of the Goldschmidt curve, $S_G/a^2 = 2\pi(A/a)^2$; and c is the scaled area of the cylinder, $4\pi A/a$.

If $A > ag(\eta_m) \simeq 1.51a$, it can be shown that S_1 is a local minimum of the functional, but the graphs in Figure 2.8 show that for $A < 1.895a$ the Goldschmidt curve yields a smaller area. A physical interpretation of these solutions is given in the next section.

This relatively simple example of a variational problem provides some idea of the possible complications that can arise.

Exercise 2.10

If $f(x) = c\cosh(x/c)$, show that the functional $S[y] = 2\pi \int_a^b dx\, y\sqrt{1+y'^2}$ has the value
$$\frac{S[f]}{a^2} = \frac{2\pi}{\eta^2}\left(\eta + \sinh\eta \cosh\eta\right), \quad \eta = \frac{a}{c}.$$

2.5 Minimal surface of revolution

Exercise 2.11

(a) Use the expansion $\cosh\eta = 1 + \frac{1}{2}\eta^2 + O(\eta^4)$ to show that, for small η, $g(\eta) = 1/\eta + \eta/2 + O(\eta^3)$, where $g(\eta)$ is defined in equation (2.50). Hence show that if $A \gg a$, then $\eta \simeq a/A$, and hence that $c \simeq A$ and $f(x) \simeq A$. Using the result obtained in the previous exercise, or otherwise, show that $S_1 = 4\pi Aa$.

This exercise is more difficult than average.

(b) Show that if η_2 is large, the equation defining it is given approximately by

$$\frac{A}{a} \simeq \frac{1}{2\eta}e^\eta$$

and, using the result obtained in the previous exercise, that

$$\frac{S_2}{a^2} \simeq 2\pi\left(\frac{e^\eta}{2\eta}\right)^2 + \frac{2\pi}{\eta} \simeq 2\pi\left(\frac{e^\eta}{2\eta}\right)^2.$$

Exercise 2.12

(a) Show that the position of the minimum of the function $g(\eta) = (\cosh\eta)/\eta$, $\eta > 0$, is at the real root, η_m, of $\eta\tanh\eta = 1$.

By sketching the graphs of $y = 1/\eta$ and $y = \tanh\eta$, for $\eta > 0$, show that the equation $\eta\tanh\eta = 1$ has only one real root.

(b) If $a/c = \eta_m$ and $A/a = g(\eta_m)$, use the result derived in Exercise 2.10 to show that the area is $S_m = 2\pi A^2 \eta_m$.

Exercise 2.13

Show that a solution of equation (2.46) is obtained by putting $c = A$ and $f(x) = A$. Show also that this function is *not* a solution of the associated second-order Euler–Lagrange equation.

Exercise 2.14

(a) Show that the functional

$$S[y] = \int_{-1}^{1} dx\,\sqrt{y(1+y'^2)}, \quad y(-1) = y(1) = A > 0,$$

is stationary on the two paths

$$y(x) = \frac{1}{4c^2}(4c^4 + x^2) \quad \text{where } c^2 = c_\pm^2 = \frac{1}{2}\left(A \pm \sqrt{A^2-1}\right).$$

This a long, fairly hard exercise that you might consider doing when revising this chapter.

In the following, these solutions are denoted by $y_+(x)$ and $y_-(x)$, respectively.

(b) Show that on these stationary paths

$$S[y] = 2c + \frac{1}{6c^3},$$

and deduce that when $A > 1$, $S[y_-] > S[y_+]$, and that when $A = 1$, $S[y] = 4\sqrt{2}/3$. Show also that if $A \gg 1$

$$S[y_-] \simeq \tfrac{4}{3}A^{3/2} \quad \text{and} \quad S[y_+] \simeq 2\sqrt{A}.$$

(c) Find the value of $S[y]$ for the function

$$y_\delta(x) = \begin{cases} 0, & 0 \le x < 1-\delta,\ 0 < \delta \ll 1, \\ A - \dfrac{A}{\delta}(1-x), & 1-\delta \le x \le 1, \end{cases}$$

with $y_\delta(-x) = y_\delta(x)$. Show that as $\delta \to 0$, and $x > 0$, $y_\delta(x) \to f_G(x)$, the Goldschmidt curve defined in equation (2.59). Show also that

$$\lim_{\delta \to 0} S[y_\delta] = S[f_G] = \tfrac{4}{3}A^{3/2}.$$

2.6 Soap films

An easy way of forming soap films is to dip a loop of wire into soap solution and then to blow on it. Almost everyone will have noticed that the initial flat soap film bounded by the wire forms a segment of a sphere when blown. It transpires that there is a very close connection between these surfaces and problems in the calculus of variations. The exact physics of soap films is complicated, but a fairly simple and accurate approximation shows that the shapes assumed by a soap film are such as to minimise their areas, because the free-energy is approximately proportional to the area and equilibrium positions are given by the minimum of this energy. Thus, in some circumstances the shapes given by the minimum surface of revolution, described in Section 2.5, are those assumed by soap films.

The study of the formation and shapes of soap films has a very distinguished pedigree: Newton, Young, Laplace, Euler, Gauss, Poisson are some of the eminent scientists and mathematicians who have studied the subject. Here we cannot do the subject justice, but the interested reader should try to obtain a copy of Isenberg's *The science of soap films and soap bubbles*. The essential property is that a stable soap film is formed in the shape of a surface of minimal area that is consistent with its wire boundary.

C. Isenberg, *The science of soap films and soap bubbles* (Dover, 1992).

Probably the simplest example is that of a soap film supported by a circular loop of wire. If we distort it by blowing on it gently to form a portion the approximate shape of a sphere, when we stop blowing the surface returns to its previous shape, that is a circular disc. Essentially this is because in each case the free-energy, which is proportional to the area, is smallest in the assumed configuration.

Imagine a framework comprising two equal circles of radius A, held a distance $2a$ apart (like wheels on an axle). What shape soap film can such a frame support? Figure 2.9 illustrates the alternatives suggested by the analysis of the previous section and agree qualitatively with the solutions one would intuitively expect.

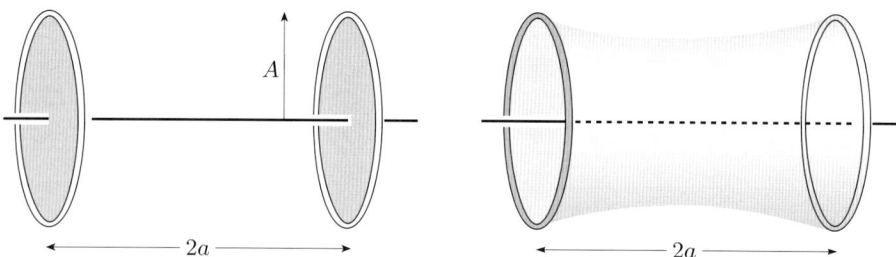

Figure 2.9 Diagrams showing two configurations assumed by soap films on two rings of radius A, a distance $2a$ apart. On the left, $A < 1.89a$, the soap film simply fills the two circular wires because they are too far apart: this is the Goldschmidt solution, equation (2.59). On the right, $A > 1.51a$, the soap film joins the two rings in the shape defined by equation (2.51) with $\eta = \eta_1$.

The left-hand configuration, with two distinct surfaces, is the Goldschmidt solution, equation (2.59), and it gives an absolute minimum area if $A < 1.89a$. The shape on the right is a catenoid of revolution and represents the absolute minimum if $A > 1.89a$; it is a *local* minimum if $1.51a < A < 1.89a$ and does not exist if $A < 1.51a$. When $1.51a < A < 1.89a$ the catenoid is unstable and we have only to disturb it slightly, by blowing on it for instance, and it may suddenly jump to the Goldschmidt solution which has a smaller area, as seen in Figure 2.8.

2.6 Soap films

The methods discussed previously provide the shape of the right-hand film, but the matter of determining whether these stationary positions are extrema, local or global, is of a different order of difficulty and beyond the scope of this course. The complexity of this physical problem is further compounded when one realises that there can be minimum energy solutions of a quite unexpected form. Figure 2.10 illustrates a possible configuration of this kind. We do not expect the theory described in the previous section to find such a solution because the mathematical formulation of the physical problem makes no allowance for this type of behaviour.

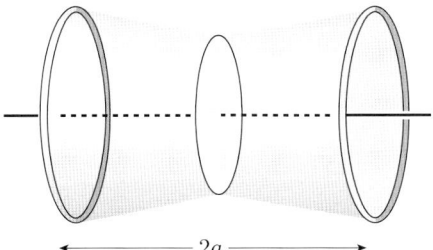

Figure 2.10 Diagram showing a possible soap film. In this example a circular film, perpendicular to the axis, is formed in the centre and this is joined to both outer rings by a catenoid.

The relationship between soap films and some problems in the calculus of variations can certainly add to our intuitive understanding, but this example should provide a salutary warning against dependence on intuition.

As an example of the complex shapes that soap bubbles can form, consider a cubical frame of wire. When dipped into a soap solution then taken out, films of local minimum free-energy, that is minimum area, will form on this frame: it transpires that the possible shapes formed are many, varied and often counter-intuitive. Some of the possible shapes are shown in Figures 2.11–2.13.

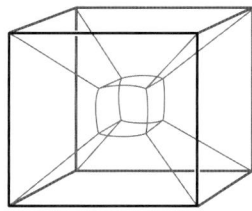 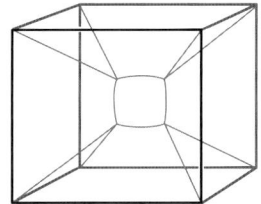 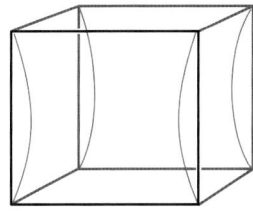

Figure 2.11   *Figure 2.12*   *Figure 2.13*

The above examples illustrate the fact that there may be more than one area of minimal surface with the same boundary conditions and shows that some physical problems that are simple to state can have bizarre solutions which are difficult to describe mathematically.

2.7 The brachistochrone

The problem, described in Subsection 1.5.1 (page 21), is to find the smooth curve joining two given points P_a and P_b, lying in a vertical plane, such that a bead sliding on the curve, without friction but under the influence of gravity, travels from P_a to P_b in the shortest possible time, the initial speed at P_a being given. It was pointed out in Subsection 1.5.1 that Johann Bernoulli made this problem famous in 1696 and that several solutions were published in 1697: Newton's comprised the simple statement that the solution was a cycloid, giving no proof. There are several variants of this problem that may be treated using the methods described here; these include the addition of speed dependent resisting forces and constraints placed on the position of one or both end points. Here we consider the simplest problem: we shall prove algebraically that the stationary path is a cycloid, so first we need to define the cycloid. Even if you are familiar with this curve, you should find the following subsection of interest besides being a reminder of the mathematical description of the cycloid.

2.7.1 The cycloid

The cycloid is one of a class of curves formed by a point fixed on a circle that rolls, without slipping, on another curve. A cycloid is formed when the fixed point is on the circumference of the circle and the circle rolls on a straight line, as shown in Figure 2.14.

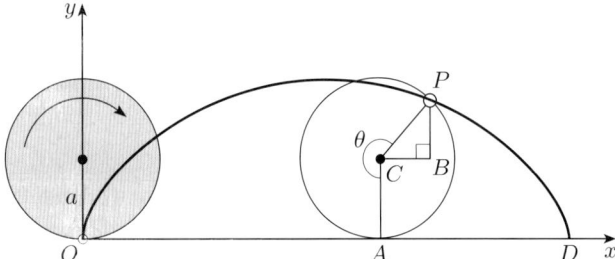

Figure 2.14 Diagram showing how the cycloid OPD is traced out by a point on the circumference of a circle rolling along a straight line.

In this figure a circle of radius a rolls along the x-axis, starting with its centre on the y-axis. Fix attention on the point P attached to the circle, initially at the origin O. As the circle rolls, P traces out the curve OPD called the *cycloid*.

The cycloid has been studied by many mathematicians from the time of Galileo (1564–1642), and was the cause of so many controversies and quarrels in the 17th century that it became known as 'the Helen of geometers'. Galileo named the cycloid but knew insufficient mathematics to make progress. He tried to find the area between it and the x-axis, but the best he could do was to trace the curve on paper, cut out the arc and weigh it, to conclude that its area was a little less than three times that of the generating circle – in fact it is exactly three times the area of this circle, as you can show in Exercise 2.15. Galileo abandoned his study of the cycloid, suggesting only that the cycloid would make an attractive arch for a bridge. This suggestion was implemented in 1764 with the building of a bridge with three cycloidal arches over the river Cam in the grounds of Trinity College, Cambridge.

2.7 The brachistochrone

Figure 2.15 Essex's bridge over the Cam, in the grounds of Trinity College, with its three cycloidal arches.

The reason why cycloidal arches were used is no longer known, all records and original drawings having been lost. However, it seems likely that the architect, Essex, chose this shape to impress Smith, the Master of Trinity College, who was keen to promote the study of applied mathematics.

James Essex, 1722–1784.
Robert Smith, 1689–1768.

The area under a cycloid was first found by Roberval in 1634. In 1638 he also found the tangent to the curve at any point, a problem solved at about the same time by Fermat and Descartes. Indeed, it was at this time that Fermat gave the modern definition of a tangent to a curve. Later, in 1658, Wren, the architect of St. Paul's Cathedral, determined the length of a cycloid.

Gilles Personne de Roberval, 1602–1675.
Pierre de Fermat, 1601–1665.
René Descartes, 1596–1650.
Christopher Wren, 1632–1723.
Blaise Pascal, 1623–1662.

Pascal's last mathematical work, in 1658, was on the cycloid and, having found certain areas, volumes and centres of gravity associated with the cycloid, he proposed a number of such questions to the mathematicians of his day with a first and second prize for their solution. However, publicity and timing were so poor that only two solutions were submitted and because these contained errors no prizes were awarded, which caused a degree of aggravation among the two contenders Antoine de Lalouvère and John Wallis.

Antoine de Lalouvère,1600–1664.
John Wallis, 1616–1703.
Christiaan Huygens, 1629–1695.

At about the time of this contest Huygens designed the first pendulum clock, which was made by Salomon Closter in 1658, but was aware that the period of the pendulum depended upon the amplitude of the swing; a short discussion of this problem is given in Subsection 4.5.1. It occurred to Huygens to consider the motion of an object sliding on an inverted cycloidal arch and he found that the object reaches the lowest point in a time independent of the starting point. The question that remained was how to persuade a pendulum to oscillate in a cycloidal, rather than a circular arc. Huygens then made the remarkable discovery illustrated in Figure 2.16. If one suspends from a point P at the cusp, between two inverted cycloidal arcs PQ and PR, then a pendulum of the same length as one of the semi-arcs will swing in an cycloidal arc QSR which has the same size and shape as the cycloidal arcs of which PQ and PR are parts. Such a pendulum will have a period independent of the amplitude of the swing.

Huygens made a pendulum clock with cycloidal jaws, but found that in practice it was no more accurate than an ordinary pendulum clock: his results on the cycloid were published in 1673 when his *Horologium oscillatorium* appeared. However, the discovery illustrated in Figure 2.16 was significant in the development of the mathematical understanding of curves in space.

A more detailed account of Huygens' work is given in *Unrolling time* by J. G. Yoder (Cambridge University Press, 1988).

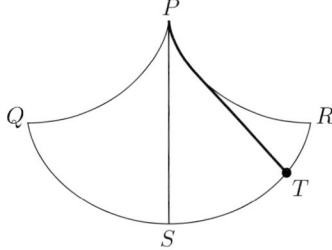

Figure 2.16 Diagram showing how Huygens' cycloidal pendulum, PT, swings between two fixed, similar cycloidal arcs PR and PQ.

The equation of the cycloid is obtained by finding the coordinates of P, in Figure 2.14, after the circle has rolled through an angle θ, so the length of the longer circular arc PA is $a\theta$. Because there is no slipping, $OA = PA = a\theta$, and the coordinates of the circle centre are $C = (a\theta, a)$. The lengths PB and BC are $PB = -a\cos\theta$ and $BC = -a\sin\theta$ and hence the coordinates of P are

$$x = a(\theta - \sin\theta), \quad y = a(1 - \cos\theta), \tag{2.62}$$

which are the parametric equations of the cycloid. For $|\theta| \ll 1$, x and y are related, approximately, by $y = (a/2)(6x/a)^{2/3}$ (see Exercise 2.17).

If we assume that the y-axis in Figure 2.16 is in the direction PS, that is pointing downwards, the upper arc QPR, with the cusp at P is given by these equations with $-\pi \leq \theta \leq \pi$ and it can be shown that the lower arc is described by $x = a(\theta + \sin\theta)$, $y = a(3 + \cos\theta)$, and the same range of θ.

Exercise 2.15

Show that the area under the arc OPD in Figure 2.14 is $3\pi a^2$ and that the length of the cycloidal arc OP is $s(\theta) = 8a\sin^2(\theta/4)$.

Exercise 2.16

Show that the gradient of the cycloid is given by

$$\frac{dy}{dx} = \frac{1}{\tan(\theta/2)}.$$

Deduce that the cycloid intersects the x-axis perpendicularly when $\theta = 0$ and 2π.

Exercise 2.17

By using the Taylor series of $\sin\theta$ and $\cos\theta$, show that for small $|\theta|$, $x \simeq a\theta^3/6$ and $y \simeq a\theta^2/2$. By eliminating θ from these equations, show that near the origin $y \simeq (a/2)(6x/a)^{2/3}$.

2.7.2 Formulation of the problem

In this section we formulate the variational principle for the brachistochrone problem by obtaining an expression for the time of passage from given points $P_a = (a, A)$ to $P_b = (b, B)$ along a curve $y(x)$, using energy conservation.

Define a coordinate system Oxy with the y-axis, oriented vertically upwards, as the starting point and the final point lying on the horizontal x-axis, so $a = B = 0$, as in Figure 2.17.

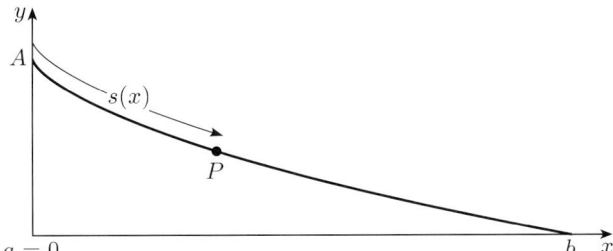

Figure 2.17 Diagram showing the curve $y(x)$ through $P_a = (0, A)$ and $P_b = (b, 0)$ on which the bead slides. Here $s(x)$ is the distance along the curve from P_a to a point $P = (x, y(x))$ on it.

At a point $P = (x, y(x))$ on this curve, let $s(x)$ be the distance along the curve from P_a to P, so the speed of the bead is defined to be $v = ds/dt$. The kinetic energy of a bead having mass m at P is $\frac{1}{2}mv^2$ and its potential energy is mgy; because the bead is sliding without friction, energy conservation gives

$$\tfrac{1}{2}mv^2 + mgy = E \tag{2.63}$$

where the value of the energy E is given by the initial conditions,

$$E = \tfrac{1}{2}mv_0^2 + mgA, \tag{2.64}$$

where v_0 is the speed at $P_a = (0, A)$. From Figure 1.5 (page 9), small changes in s are given by $\delta s^2 = \delta x^2 + \delta y^2$, and so

$$\left(\frac{ds}{dt}\right)^2 = \left(\frac{dx}{dt}\right)^2 + \left(\frac{dy}{dt}\right)^2 = \left(\frac{dx}{dt}\right)^2 \left(1 + y'(x)^2\right). \tag{2.65}$$

Thus on rearranging equation (2.63) we obtain

$$\left(\frac{ds}{dt}\right)^2 = \frac{2E}{m} - 2gy \quad \text{or} \quad \frac{dx}{dt}\sqrt{1 + y'(x)^2} = \sqrt{\frac{2E}{m} - 2gy(x)}. \tag{2.66}$$

Now the time of passage from $x = 0$ to $x = b$ is given by the integral

$$T = \int_0^T dt = \int_0^b dx \, \frac{1}{dx/dt}. \tag{2.67}$$

Thus on rearranging equation (2.66) to express dx/dt in terms of the function $y(x)$, we obtain the required functional

$$T[y] = \int_0^b dx \sqrt{\frac{1 + y'^2}{\frac{2E}{m} - 2gy}}. \tag{2.68}$$

This functional may be put in a slightly more convenient form by noting that the energy and the initial conditions are related by equation (2.64), so by defining the new dependent variable

$$z(x) = A + \frac{v_0^2}{2g} - y(x), \tag{2.69}$$

we obtain
$$T[z] = \int_0^b dx \sqrt{\frac{1+z'(x)^2}{2gz(x)}}. \tag{2.70}$$

Exercise 2.18

Use equation (2.68) to show that the time taken for a particle of mass m to slide down the curve $y = Ax$ from the point (X, AX), where it is stationary, to the origin is
$$T = 2\sqrt{X\frac{1+A^2}{2gA}}.$$

Show also that if the point (X, AX) lies on the circle of radius R and with centre at $(0, R)$, so the equation of the circle is $x^2 + (y-R)^2 = R^2$, as shown in Figure 2.18, then the time taken to slide along the straight line from (X, AX) to the origin, O, is *independent* of X and is given by
$$T = 2\sqrt{\frac{R}{g}}.$$

This result was known by Galileo and was one reason why he thought that the solution to the brachistochrone problem was a circle.

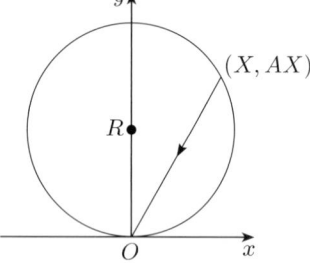

Figure 2.18

2.7.3 A solution

The integrand of the functional (2.70) is independent of x, so we may use equation (2.33) (page 49) to write Euler's equation in the form

Note that we ignore the factor $(2g)^{-1/2}$.

$$z'\frac{\partial F}{\partial z'} - F = \text{constant}, \quad \text{where } F(z, z') = \sqrt{\frac{1+z'^2}{z}}, \tag{2.71}$$

and $z(0) = v_0^2/2g$, $z(b) = A + z(0)$. Since
$$\frac{\partial F}{\partial z'} = \frac{z'}{\sqrt{z(1+z'^2)}}, \tag{2.72}$$
this gives
$$\frac{z'^2}{\sqrt{z(1+z'^2)}} - \sqrt{\frac{1+z'^2}{z}} = -\frac{1}{c} \tag{2.73}$$
for some positive constant c. Note that c must be positive because the left-hand side of equation (2.73) is negative. Rearranging the last expression gives
$$z(1+z'^2) = c^2 \quad \text{or} \quad \frac{dz}{dx} = \pm\sqrt{\frac{c^2}{z} - 1}. \tag{2.74}$$
This first-order differential equation is separable and can be solved. First, however, note that because the y-axis is vertically upwards, we expect the solution $y(x)$ to decrease away from $x = 0$, that is, $z(x)$ will increase so we take the positive sign in equation (2.74) and integration gives
$$x = \int dz \sqrt{\frac{z}{c^2 - z}}. \tag{2.75}$$
Now substitute $z = c^2 \sin^2\phi$ to give
$$x = 2c^2\int d\phi \sin^2\phi = c^2\int d\phi (1 - \cos 2\phi). \tag{2.76}$$

2.7 The brachistochrone

Integration then gives the solution in the parametric form

$$x = \tfrac{1}{2}c^2(2\phi - \sin 2\phi) + d \quad \text{and} \quad z = c^2 \sin^2 \phi = \tfrac{1}{2}c^2(1 - \cos 2\phi), \quad (2.77)$$

where d is a constant. Both c and d are determined by the values of A, b and the initial speed, v_0. Comparing these equations with the parametric equations (2.62), we see that the required stationary curve is a cycloid. It may also be shown that this solution is the global minimum; a simple method of doing this is explored in Exercise 2.36.

In the case that the particle starts from rest, $v_0 = 0$, these solutions give

$$x = d + \tfrac{1}{2}c^2\left(2\phi - \sin 2\phi\right), \quad y = A - \tfrac{1}{2}c^2\left(1 - \cos 2\phi\right), \quad (2.78)$$

where c and d are constants determined by the known end points of the curve.

At the starting point $y = A$ so here $\phi = 0$; and since $x = 0$ it follows that $d = 0$. This means that initially the particle falls vertically downwards. At the final point of the curve, $x = b$, $y = 0$, let $\phi = \phi_b$. Then

$$\frac{2b}{c^2} = 2\phi_b - \sin 2\phi_b, \quad \frac{2A}{c^2} = 1 - \cos 2\phi_b, \quad (2.79)$$

giving two equations for c and ϕ_b.

We now show that these equations have a solution and that it is unique. Consider the cycloid

$$u = 2\theta - \sin 2\theta, \quad v = 1 - \cos 2\theta, \quad 0 \leq \theta \leq \pi. \quad (2.80)$$

The value of ϕ_b is given by the value of θ where this cycloid intersects the straight line $Au = bv$. The graphs of these two curves are shown in Figure 2.19.

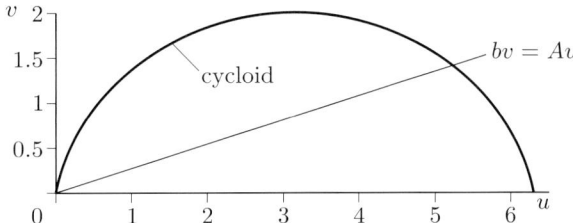

Figure 2.19 Graphs of the cycloid defined in equation (2.80) and the straight line $Au = bv$.

Because the gradient of the cycloid at $\theta = 0$ ($u = v = 0$) is infinite this graph shows that there is a single value of ϕ_b for all positive values of the ratio A/b. By dividing the second of equations (2.80) by the first we see that this root is given by solving the equation

$$\frac{2\theta - \sin 2\theta}{2\sin^2 \theta} = \frac{b}{A}, \quad 0 < \theta < \pi. \quad (2.81)$$

This equation is most conveniently solved numerically; however, if b/A is small a satisfactory approximation can be found. Once ϕ_b is known, the value of c is given from the equation $2A/c^2 = 1 - \cos 2\phi_b$, which may be put in the more convenient form $c^2 = A/\sin^2 \phi_b$.

Exercise 2.19

Use the solution defined in equation (2.77) to show that on the stationary path the time of passage is

$$T[z] = \sqrt{\frac{2A}{g}} \frac{\phi_b}{\sin \phi_b}.$$

This problem is slightly harder than average.

We end this section by showing a few graphs of these solutions and quoting some formulae that may help you to understand them. The analysis that follows is not an assessed part of the course, and you are not expected to be able to derive the quoted formulae.

Figure 2.20 shows graphs of the stationary paths for $A = 1$ and various values of b, ranging from small to large, so all curves start and end at the points $(0, 1)$ and $(b, 0)$, respectively, with $0.1 \leq b \leq 4$.

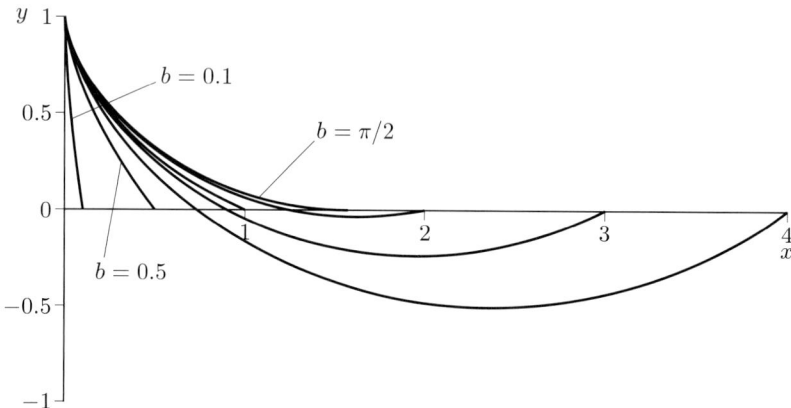

Figure 2.20 Graphs showing the stationary paths for joining the points $(0, 1)$ and $(b, 0)$ for $b = 0.1, 1/2, 1, \pi/2, 2, 3$ and 4.

From this figure we see that for small b the stationary path is close to that of a straight line, as would be expected. In this case ϕ_b is small and it can be shown that

$$\phi_b = \frac{3b}{2A} - \frac{9b^3}{20A^3} + O(b^5) \quad \text{and} \quad \frac{y}{A} \simeq 1 - \left(\frac{x}{b}\right)^{2/3}. \qquad (2.82)$$

Also, the time of passage is

$$T = \sqrt{\frac{2A}{g}} \left(1 + \frac{3b^2}{8A^2} - \frac{81b^4}{640A^4} + O(b^6)\right). \qquad (2.83)$$

By comparison, if a particle slides down the straight line joining $(0, A)$ to $(b, 0)$, that is $y/A + x/b = 1$, so $z = Ax/b$, then the time of passage is

$$T_{SL} = \sqrt{\frac{2(A^2 + b^2)}{Ag}} = \begin{cases} \sqrt{\frac{2A}{g}} \left(1 + \frac{b^2}{2A^2} + O(b^4)\right), & b \ll A, \\ b\sqrt{\frac{2}{Ag}} \left(1 + \frac{A^2}{2b^2} + O(b^{-4})\right), & b \gg A. \end{cases} \qquad (2.84)$$

Thus, for small b, the relative difference is

$$T_{SL} - T = T\frac{b^2}{8A^2} + O(b^4). \qquad (2.85)$$

Returning to Figure 2.20 we see for small b the stationary paths cross the x-axis at the terminal point. At some critical value of b the stationary path is tangential to the x-axis at the terminal point. We can see from the equation for $x(\phi)$ that this critical path occurs when $y'(\phi) = 0$, that is when $\phi_b = \pi/2$ and, from equation (2.81), we see that this gives $b = A\pi/2$. On this path the time of passage is

$$T = \frac{\pi}{2}\sqrt{\frac{2A}{g}} \quad \text{and also} \quad T_{SL} = T\sqrt{1 + \frac{4}{\pi^2}} = 1.185T. \tag{2.86}$$

For $b > A\pi/2$ the stationary path dips below the x-axis and approaches the terminal point from below. For $b \gg A\pi/2$ it can be shown that $\phi_b = \pi - \sqrt{A\pi/b} + O(b^{-3/2})$, and that the path is given approximately by

$$x \simeq \frac{b}{2\pi}(2\phi - \sin 2\phi), \quad y \simeq A - \frac{b}{\pi}\sin^2 \phi, \tag{2.87}$$

and that

$$T = \sqrt{\frac{2\pi b}{g}}\left(1 - \sqrt{\frac{A}{b\pi}} + \frac{\sqrt{\pi}}{6}\left(\frac{A}{b}\right)^{3/2} + \cdots\right). \tag{2.88}$$

Thus the time of passage increases with $\sqrt{b}$, compared to the time to slide down the straight line, which is proportional to b, for large b. Further, the stationary path reaches its lowest point when $\phi = \pi/2$, where $y = A - b/\pi$, in other words it falls about $1/3$ as far below the x-axis as it travel along it, provided $b \gg A\pi$. That is, the particle first accelerates to a high speed, reaching a speed $v \simeq \sqrt{2gb/\pi}$, before slowing to reach the terminal point at speed $v = \sqrt{2gA}$: on the straight line path the particle accelerates uniformly to this speed.

2.8 Further Exercises

Exercise 2.20
Show that the Euler–Lagrange equation for the functional

$$S[y] = \int_0^1 dx \left(y'^2 - y^2 - 2xy\right), \quad y(0) = y(1) = 0,$$

is $y'' + y = -x$. Hence show that the stationary function is

$$y(x) = \frac{\sin x}{\sin 1} - x.$$

Exercise 2.21
Consider the functional

$$S[y] = \int_a^b dx \left(y'^2 + y^2\right), \quad y(a) = A, \, y(b) = B.$$

(a) By forming and solving the Euler–Lagrange equation, show that the stationary path is

$$y = A\cosh u + \frac{B - A\cosh(b-a)}{\sinh(b-a)}\sinh u, \quad u = x - a.$$

(b) By making the change of variable $u = x - a$ and defining $Y(u) = y(x(u)) = y(u+a)$, show that $y'(x) = Y'(u)$ and that the functional becomes

$$S[Y] = \int_0^{b-a} du \left(Y'^2 + Y^2\right), \quad Y(0) = A, \, Y(b-a) = B.$$

Deduce that the stationary path depends only upon $u = x - a$ and $b - a$ (besides A and B).

Exercise 2.22

Euler's original method for finding solutions of variational problems is described in equation (2.5) (page 42). Consider approximating the functional defined in Exercise 2.20 using the polygon passing through the points $(0,0)$, $(\frac{1}{2}, y_1)$ and $(1,0)$, so there is one variable y_1 and two segments.

This polygon can be defined by the straight-line segments

$$y(x) = \begin{cases} 2y_1 x, & 0 \le x \le \frac{1}{2}, \\ 2y_1(1-x), & \frac{1}{2} \le x \le 1. \end{cases}$$

By substituting this function into the functional, show that the corresponding polygon approximation to the functional becomes

$$S(y_1) = \tfrac{11}{3} y_1^2 - \tfrac{1}{2} y_1$$

and hence that the stationary polygon is given by $y(1/2) = y_1 = 3/44$. Note that this gives $y(1/2) \simeq 0.0682$ by comparison to the exact value 0.0697.

Exercise 2.23

Find the functions that make the following functionals stationary.

(a) $S[y] = \int_0^1 dx \, (y'^2 + 12xy), \quad y(0) = 0, \ y(1) = 2.$

(b) $S[y] = \int_1^2 dx \, \dfrac{y'^2}{x^2}, \quad y(1) = A, \ y'(2) = B.$

(c) $S[y] = \int_0^1 dx \, (2y^2 y'^2 - (1+x)y^2), \quad y(0) = 1, \ y(1) = 2.$

(d) $S[y] = \int_0^a dx \, \dfrac{y}{y'^2}, \quad y'(0) = 1, \ y(a) = A^2, \ A > \sqrt{2a}.$

Exercise 2.24

Find the general solution of the Euler–Lagrange equation corresponding to the functional

$$S[y] = \int_a^b dx \, w(x) \sqrt{1 + y'^2},$$

and find explicit solutions in the special cases $w(x) = \sqrt{x}$ and $w(x) = x$.

Exercise 2.25

What is the equivalent of the fundamental lemma of the calculus of variations in the theory of stationary points of functions $f(x_1, x_2, \ldots, x_n)$ of several real variables?

Exercise 2.26

Consider the functional

$$S[y] = \int_0^1 dx \, (y'^2 - 1)^2, \quad y(0) = 0, \ y(1) = A > 0.$$

(a) Show that the Euler–Lagrange equation reduces to $y'^2 = m^2$, where m is a constant.

(b) Show that the equation $y'^2 = m^2$, with $m > 0$, has the following solutions that fit the boundary conditions, $y_1(x) = Ax$:

$$y_2(x) = \begin{cases} mx, & 0 \le x \le \dfrac{A+m}{2m}, \\ A + m(1-x), & \dfrac{A+m}{2m} \le x \le 1, \end{cases} \quad m \ge A$$

2.8 Further Exercises

and

$$y_3(x) = \begin{cases} -mx, & 0 \leq x \leq \dfrac{m-A}{2m}, \\ A - m(1-x), & \dfrac{m-A}{2m} \leq x \leq 1. \end{cases} \quad m \geq A$$

Show also that on these solutions the functional has the values

$$S[y_1] = (A^2 - 1)^2 \quad \text{and} \quad S[y_2] = S[y_3] = (m^2 - 1)^2.$$

(c) Deduce that if $A \geq 1$, the minimum value of $S[y]$ is $(A^2 - 1)^2$, and that this occurs on the curve $y_1(x)$, but if $A < 1$, the minimum value of $S[y]$ is zero, and this occurs on the curves $y_2(x)$ and $y_3(x)$ with $m = 1$.

Exercise 2.27

Show that the following functionals do not have stationary values, where, in all cases, $y(0) = 0$ and $y(1) = 1$.

(a) $\displaystyle\int_0^1 dx\, y'$ (b) $\displaystyle\int_0^1 dx\, yy'$ (c) $\displaystyle\int_0^1 dx\, xyy'$

Exercise 2.28

Show that the Euler–Lagrange equations for the functionals

$$S_1[y] = \int_a^b dx\, F(x,y,y') \quad \text{and} \quad S_2[y] = \int_a^b dx\left(F(x,y,y') + \frac{d}{dx}G(x,y)\right)$$

are identical.

Exercise 2.29

Show that the functional

$$S[y] = \int_{-1}^1 dx\, y^2 (2x - y')^2, \quad y(-1) = 0,\ y(1) = 1,$$

achieves its minimum value when

$$y(x) = \begin{cases} 0, & -1 \leq x \leq 0, \\ x^2, & 0 \leq x \leq 1, \end{cases}$$

a function for which the second derivative does not exist at $x = 0$. Show that, despite the fact that $y''(x)$ does not exist everywhere, the Euler–Lagrange equation is satisfied.

Note: It can be shown that if $y(x)$ is a solution of the Euler–Lagrange equation for the functional with the integrand $F(x, y, y')$, and if F has continuous first and second derivatives with respect to all its arguments, then $y(x)$ has a continuous second derivative at all points where $\partial^2 F/\partial y'^2 \neq 0$. In the present case, $\partial^2 F/\partial y'^2 = -2y^2$ (see page 50).

2.9 Harder Exercises

Exercise 2.30

Use the approximation (2.5) (page 42) to show that the equations for the values of $\boldsymbol{y} = (y_1, y_2, \ldots, y_N)$, where $y_{k+1} = y_k + h$, that make $S[\boldsymbol{y}]$ stationary are

$$\frac{\partial S}{\partial y_n} = h\frac{\partial}{\partial u}F(z_n) + \frac{\partial}{\partial v}F(z_n) - \frac{\partial}{\partial v}F(z_{n+1}) = 0, \quad n = 1, 2, \ldots, N,$$

where

$$z_n = (x_n, u, v), \quad u = y_n, \quad v = \frac{y_n - y_{n-1}}{h},$$

and $y_0 = A$ and $y_{N+1} = B$.

Show also that

$$z_{n+1} = z_n + h\left(1, y_n', y_n''\right) + O(h^2),$$

and hence that

$$\frac{\partial S}{\partial y_n} = h\left(\frac{\partial F}{\partial u} - \frac{\partial^2 F}{\partial x \partial v} - y_n'\frac{\partial^2 F}{\partial u \partial v} - y_n''\frac{\partial^2 F}{\partial v^2}\right) + O(h^2)$$

$$= -h\left(\frac{d}{dx}\left(\frac{\partial F}{\partial v}\right) - \frac{\partial F}{\partial u}\right) + O(h^2),$$

where F and its derivatives are evaluated at $z = z_n$. Hence derive the Euler–Lagrange equations.

Exercise 2.31

This exercise is a continuation of Exercise 2.22 and uses a set of N variables to define the polygon. Take a set of $N+2$ equally-spaced points on the x-axis, $x_k = k/(N+1)$, $k = 0, 1, \ldots, N+1$ with $x_0 = 0$ and $x_{N+1} = 1$, and a polygon passing through the points (x_k, y_k). Since $y(0) = y(1) = 0$ we have $y_0 = y_{N+1} = 0$, leaving N unknown variables.

Show that the functional defined in Exercise 2.20 approximates to

$$S = \sum_{k=0}^{N}\left[\frac{(y_{k+1} - y_k)^2}{h} - h\left(y_k^2 + \frac{2k}{N+1}y_k\right)\right], \quad h = \frac{1}{N+1}.$$

(a) For $N = 1$, the case treated in Exercise 2.22, show that this reduces to

$$S(y_1) = \tfrac{7}{2}y_1^2 - \tfrac{1}{2}y_1.$$

Explain the difference between this and the previous expression for $S(y_1)$, given in Exercise 2.22.

(b) For $N = 2$ show that this becomes

$$S = \tfrac{17}{3}y_1^2 + \tfrac{17}{3}y_2^2 - 6y_1y_2 - \tfrac{2}{9}y_1 - \tfrac{4}{9}y_2,$$

and hence that the equations for y_1 and y_2 are

$$34y_1 - 18y_2 = \tfrac{2}{3}, \quad 34y_2 - 18y_1 = \tfrac{4}{3}.$$

Solve these equations to show that $y(1/3) \simeq 35/624 \simeq 0.0561$ and $y(2/3) \simeq 43/624 \simeq 0.0689$. Note that these compare favourably with the exact values, $y(1/3) = 0.0555$ and $y(2/3) = 0.0682$ obtained from the solution of Exercise 2.22.

2.9 Harder Exercises

Exercise 2.32

Consider the functional
$$S[y] = \int_a^b dx\, F(y''(x)),$$
where $F(z)$ is a differentiable function and the admissible functions are at least twice differentiable and satisfy the boundary conditions
$$y(a) = A_1, \quad y(b) = B_1, \quad y'(a) = A_2, \quad y'(b) = B_2.$$

(a) Show that the function making $S[y]$ stationary satisfies the equation
$$\frac{\partial F}{\partial y''} = c(x-a) + d,$$
where c and d are constants. [Hint: Integrate the Gâteaux differential by parts twice.]

(b) In the case that $F(z) = \tfrac{1}{2}z^2$, show that the solution is
$$y(x) = \tfrac{1}{6}c(x-a)^3 + \tfrac{1}{2}d(x-a)^2 + A_2(x-a) + A_1,$$
where c and d satisfy the equations
$$\tfrac{1}{6}cD^3 + \tfrac{1}{2}dD^2 = B_1 - A_1 - A_2 D, \quad \text{where } D = b - a,$$
$$\tfrac{1}{2}cD^2 + dD = B_2 - A_2.$$

(c) Show that this stationary function is also a minimum of the functional.

Exercise 2.33

The theory described in the text considered functionals with integrands depending only upon x, $y(x)$ and $y'(x)$. However, functionals depending upon higher derivatives also exist and are important, for example, in the theory of stiff beams; the equivalent Euler–Lagrange equation may be derived using a direct extension of the methods described in this chapter.

Consider the functional
$$S[y] = \int_a^b dx\, F(x, y, y', y''), \quad y(a) = A_1,\; y'(a) = A_2,\; y(b) = B_1,\; y'(b) = B_2.$$

Show that the Gâteaux derivative of this functional is
$$\Delta S[y, g] = \int_a^b dx \left(g \frac{\partial F}{\partial y} + g' \frac{\partial F}{\partial y'} + g'' \frac{\partial F}{\partial y''} \right).$$

Using integration by parts show that
$$\int_a^b dx\, g'' \frac{\partial F}{\partial y''} = \int_a^b dx\, g \frac{d^2}{dx^2}\left(\frac{\partial F}{\partial y''}\right),$$
being careful to describe the necessary properties of $g(x)$. Hence show that $S[y]$ is stationary for the functions that satisfy the fourth-order differential equation
$$\frac{d^2}{dx^2}\left(\frac{\partial F}{\partial y''}\right) - \frac{d}{dx}\left(\frac{\partial F}{\partial y'}\right) + \frac{\partial F}{\partial y} = 0,$$
with the boundary conditions $y(a) = A_1,\; y'(a) = A_2,\; y(b) = B_1,\; y'(b) = B_2$.

Exercise 2.34

Using the result derived in the previous exercise, find the stationary functions of the following functionals.

(a) $S[y] = \displaystyle\int_0^1 dx\, (1 + y''^2), \quad y(0) = 0,\; y'(0) = 1,\; y(1) = 1,\; y'(1) = 1$

(b) $S[y] = \displaystyle\int_0^{\pi/2} dx\, (y''^2 - y^2 + x^2), \quad y(0) = 1,\; y'(0) = 0,\; y(\tfrac{\pi}{2}) = 0,\; y'(\tfrac{\pi}{2}) = -1$

Exercise 2.35

(a) Show that the Euler–Lagrange equation for the functional
$$S[y] = \int_a^b dx\, \left[y'(x)^2 - y(x)^2\right], \quad y(a) = A,\ y(b) = B,$$
is
$$\frac{d^2 y}{dx^2} + y = 0, \quad y(a) = A,\ y(b) = B.$$

(b) Second-order equations of the above form occur frequently, but the boundary conditions are sometimes different, involving linear combinations of y and y'. Thus a typical equation is
$$\frac{d^2 y}{dx^2} + y = 0, \quad h_a y(a) + y'(a) = 0, \quad h_b y(b) + y'(b) = 0, \tag{2.89}$$
where h_a and h_b are constants.

Show, from first principles, that the functional
$$S[y] = h_b y(b)^2 - h_a y(a)^2 + \int_a^b dx\, \left[y'(x)^2 - y(x)^2\right]$$
is stationary on the path that satisfies equation (2.89).

Exercise 2.36

In this exercise you will show that the cycloid is a minimum for the brachistochrone problem. The proof is in two stages.

(a) Show that the functional $T[z]$, defined in equation (2.70) (page 66) with the boundary conditions $z(0) = A > 0$, $z(b) = 0$, can be re-expressed in the form
$$T[x] = \int_0^A dz\, \sqrt{\frac{x'^2 + 1}{z}}$$
if the roles of x and z are interchanged, that is, if z is taken to be the independent variable and x the dependent variable.

Here we ignore the external factor $1/\sqrt{2g}$.

(b) Consider the varied path $x(z) + \epsilon g(z)$, and show that
$$T[x + \epsilon g] - T[x] = \frac{\epsilon^2}{2} \int_0^A dz\, \frac{g'(z)^2}{\sqrt{z}(1 + x'^2)^{3/2}} + O(\epsilon^3)$$
$$= \epsilon^2 c \int_0^{\phi_b} d\phi\, \cos^4 \phi\, g'(z)^2,$$
where $z = c^2 \sin^2 \phi$ and $x = \frac{1}{2} c^2 (2\phi - \sin 2\phi)$. Deduce that $T[x + \epsilon g] > T[x]$, for $\epsilon \neq 0$ and all $g(x)$, and hence that the stationary path is actually a minimum.

Exercise 2.37

In this exercise you will examine a simple example that emphasises the need to be very careful with the choice of admissible functions.

Consider the functional
$$S[y] = \int_0^1 dx\, \sqrt{1 + y'^2}, \quad y(0) = y(1) = 0,$$
for the distance between the two points $(0, 0)$ and $(1, 0)$ on the x-axis. If the admissible functions are everywhere differentiable then we saw in Chapter 1 that the minimum distance is given by the straight line joining the end points. Here you will show that we can construct a continuous function arbitrarily close to this path with a length larger than any given number.

Consider an isosceles triangle ABC, where A and C are on the x-axis at $x = 0$ and 1, respectively, as shown in the left-hand diagram in Figure 2.21, with height h, and AB of length l.

2.9 Harder Exercises

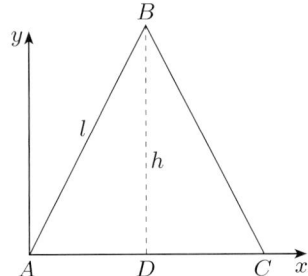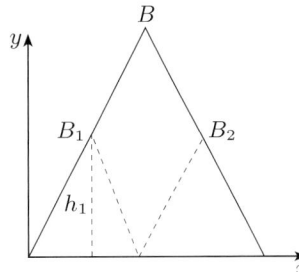

Figure 2.21

(a) Construct the two smaller triangles AB_1D and DB_2C by halving the height and width of the triangle ABC, as shown in the right-hand diagram of Figure 2.21. Show that $AB_1 = l/2$ and $h_1 = h/2$. Hence show that the lengths of the lines AB_1DB_2C and ABC are the same and equal to $2l$.

(b) Show that after n such divisions there are 2^n similar triangles of height $2^{-n}h$, and that the total length of the curve is $2l$. Deduce that arbitrarily close to AC, the shortest distance between A and C, we may find a continuous curve every point of which is arbitrarily close to AC, but which has any given length.

Solutions to Exercises in Chapter 2

Solution 2.1

The first result follows directly from equation (2.5) because F is independent of x and y, $y(a) = y_0 = A$ and $y(b) = y_{N+1} = B$. The variable y_k for each $k = 1, 2, \ldots, N$ appears in only two terms of the sum, so

$$\frac{\partial S}{\partial y_k} = h \frac{\partial}{\partial y_k}\left[F\left(\frac{y_k - y_{k-1}}{h}\right) + F\left(\frac{y_{k+1} - y_k}{h}\right)\right]$$

and hence, since F depends only upon y' and not y, the stationary points are given by the equations

$$\frac{\partial S}{\partial y_k} = F'\left(\frac{y_k - y_{k-1}}{h}\right) - F'\left(\frac{y_{k+1} - y_k}{h}\right) = 0, \quad k = 1, 2, \ldots, N.$$

Thus $F'((y_k - y_{k-1})/h)$ is independent of k, for all $k = 1, 2, \ldots, N+1$, so must equal a constant c, independent of k. Because this is true for all k, if $F'(z)$ is continuous, $y_k - y_{k-1} = $ constant and hence the points (x_k, y_k) lie on a straight line.

Solution 2.2

(a) Since $S[y + \epsilon g] = \int_0^1 dx\, (y' + \epsilon g')^2$, we have

$$\frac{d}{d\epsilon}S[y + \epsilon g] = 2\int_0^1 dx\, (y' + \epsilon g')g' \quad \text{and} \quad \Delta S[y, g] = 2\int_0^1 dx\, y'g'.$$

(b) Since $S[y + \epsilon g] = \int_1^2 dx\, x^2(y' + \epsilon g')^2$, we have

$$\frac{d}{d\epsilon}S[y + \epsilon g] = 2\int_1^2 dx\, x^2(y' + \epsilon g')g' \quad \text{and} \quad \Delta S[y, g] = 2\int_1^2 dx\, x^2 y'g'.$$

(c) Since $S[y + \epsilon g] = \int_0^{\pi/2} dx\, \left[(y' + \epsilon g')^2 - (y + \epsilon g)^2\right]$, we have

$$\frac{d}{d\epsilon}S[y + \epsilon g] = 2\int_0^{\pi/2} dx\, [(y' + \epsilon g')g' - (y + \epsilon g)g]$$

and $\quad \Delta S[y, g] = 2\int_0^{\pi/2} dx\, (y'g' - yg).$

(d) Since $S[y + \epsilon g] = \int_a^b dx\, \frac{(y' + \epsilon g')^2}{x^3}$, we have

$$\frac{d}{d\epsilon}S[y + \epsilon g] = 2\int_a^b dx\, \frac{(y' + \epsilon g')}{x^3}g' \quad \text{and} \quad \Delta S[y, g] = 2\int_a^b dx\, \frac{y'g'}{x^3}.$$

(e) Since $S[y + \epsilon g] = \int_a^b dx\, \left[(y' + \epsilon g')^2 + (y + \epsilon g)^2 + 2e^x(y + \epsilon g)\right]$, we have

$$\frac{d}{d\epsilon}S[y + \epsilon g] = 2\int_a^b dx\, [(y' + \epsilon g')g' + (y + \epsilon g)g + e^x g]$$

and $\quad \Delta S[y, g] = 2\int_a^b dx\, [y'g' + (y + e^x)g].$

(f) Since $S[y+\epsilon g] = \int_0^1 dx\, \sqrt{x^2+(y+\epsilon g)^2}\sqrt{1+(y'+\epsilon g')^2}$, we have

$$\frac{d}{d\epsilon}S[y+\epsilon g] = \int_0^1 dx\left[\frac{(y+\epsilon g)g}{\sqrt{x^2+(y+\epsilon g)^2}}\sqrt{1+(y'+\epsilon g')^2} \right.$$
$$\left. + \frac{\sqrt{x^2+(y+\epsilon g)^2}(y'+\epsilon g')g'}{\sqrt{1+(y'+\epsilon g')^2}}\right]$$

and $\quad \Delta S[y,g] = \int_0^1 dx\left[\frac{y\sqrt{1+y'^2}}{\sqrt{x^2+y^2}}g + \frac{\sqrt{x^2+y^2}\,y'}{\sqrt{1+y'^2}}g'\right].$

Solution 2.3

(a) We have $S_1[y+\epsilon g] = \int_a^b dx\,(y'+\epsilon g')$, giving

$$\Delta S_1 = \int_a^b dx\, g' = g(b) - g(a) = 0.$$

(b) We have $S_2[y+\epsilon g] = \int_a^b dx\, x(y'+\epsilon g')^2$, giving

$$\Delta S_2 = 2\int_a^b dx\, xy'g'.$$

(c) Since $S_3 = S_1 + S_2$ and Gâteaux differentiation is a linear operation,

$$\Delta S_3 = \Delta S_1 + \Delta S_2 = \Delta S_2.$$

Solution 2.4

(a) Since $\partial F/\partial y' = 2y'$ and $\partial F/\partial y = 1$, the Euler–Lagrange equation is

$$2y'' - 1 = 0.$$

The general solution of this equation can be written in the form

$$y = \tfrac{1}{4}(x-a)^2 + \alpha(x-a) + \beta.$$

Putting $x=a$ shows that $\beta = A$, and then putting $x=b$ gives

$$B - A = \tfrac{1}{4}(b-a)^2 + \alpha(b-a), \quad \text{so} \quad \alpha = \frac{B-A}{b-a} - \tfrac{1}{4}(b-a),$$

giving the solution

$$y = \tfrac{1}{4}u^2 + \left(\frac{B-A}{b-a} - \tfrac{1}{4}(b-a)\right)u + A, \quad u = x-a.$$

(b) If $u = x-a$, then the chain rule gives

$$\frac{dY}{du} = \frac{dy}{dx}\frac{dx}{du} = \frac{dy}{dx}$$

and the functional becomes

$$S[Y] = \int_0^{b-a} du\left[Y'^2 + Y\right], \quad Y(0) = A,\ Y(b-a) = B.$$

The independent variable of the associated Euler–Lagrange equation is u, and a and b occur only in the boundary conditions and as the difference $b-a$. Hence the solution depends upon the three variables x, a and b only in the two combinations $u = x-a$ and $b-a$.

Solution 2.5

(a) If γ is a constant and $y(x) = \gamma$, equation (2.33) becomes $G(\gamma, 0) = -c$.

(b) The second-order Euler–Lagrange equation is
$$\frac{\partial F^2}{\partial y'^2} y'' + \frac{\partial F^2}{\partial y \partial y'} y' + \frac{\partial F^2}{\partial x \partial y'} - \frac{\partial F}{\partial y} = 0.$$

If $F(x, y, y') = G(y, y')$, the third term is zero, and if $y = \gamma$, this equation becomes $G_y(\gamma, 0) = 0$, assuming that $G_{y'y'}(\gamma, 0)$ and $G_{yy'}(\gamma, 0)$ exist.

Let $g(y) = G(y, 0)$ be a function of y. The equation $G_y(\gamma, 0) = g(\gamma) = 0$ shows that γ must be at a stationary point of $g(y)$, whereas the equation $G(\gamma, 0) = -c$, found in part (a), imposes the weaker restriction that c lies in the domain of $G(\gamma, 0)$.

Thus, in general the constant solution $y = \gamma$ of the first integral is not a solution of the Euler–Lagrange equation.

Solution 2.6

Using the result of Exercise 1.16, we see that if G does not depend explicitly upon x, $\partial G/\partial x = 0$, and
$$y'\left(\frac{d}{dx}\left(\frac{\partial G}{\partial y'}\right) - \frac{\partial G}{\partial y}\right) = \frac{\partial^2 G}{\partial y'^2} y'y'' + \frac{\partial^2 G}{\partial y \partial y'} y'^2 - \frac{\partial G}{\partial y} y'.$$

But, using the chain rule,
$$\frac{d}{dx}\left(y'\frac{\partial G}{\partial y'}\right) = \frac{\partial}{\partial y}\left(y'\frac{\partial G}{\partial y'}\right) y' + \frac{\partial}{\partial y'}\left(y'\frac{\partial G}{\partial y'}\right) y'' = \frac{\partial^2 G}{\partial y \partial y'} y'^2 + \frac{\partial G}{\partial y'} y'' + \frac{\partial^2 G}{\partial y'^2} y'y'',$$

so the right-hand side of the previous equation becomes
$$\frac{d}{dx}\left(y'\frac{\partial G}{\partial y'}\right) - \frac{\partial G}{\partial y'} y'' - \frac{\partial G}{\partial y} y' = \frac{d}{dx}\left(y'\frac{\partial G}{\partial y'}\right) - \frac{dG}{dx} = \frac{d}{dx}\left(y'\frac{\partial G}{\partial y'} - G\right).$$

The left-hand side of this equation is zero if either $y'(x) = 0$ or $y(x)$ satisfies the Euler–Lagrange equations. In the latter case, setting the right-hand side to zero and integrating gives
$$y'\frac{\partial G}{\partial y'} - G = c = \text{constant}.$$

This is a first-order differential equation: its general solution will depend upon one other arbitrary constant, d, and to find the solution of the original problem we need to express the constants c, d in terms of the constants A, B defined in equation (2.33) (page 49). Often this is difficult, because it involves the solutions of nonlinear algebraic equations, and frequently there are real solutions only for some values of A and B.

Solution 2.7

In this case $F = y'^2 - y^2$, giving $\partial F/\partial y' = 2y'$ and $\partial F/\partial y = -2y$ which leads to the Euler–Lagrange equation $y'' + y = 0$. The general solution of this equation is $y = A\cos x + B\sin x$, where A and B are arbitrary constants determined by the boundary conditions. The boundary condition at $x = 0$ gives $A = 0$, and that at $x = X$ gives the solution
$$y(x) = \frac{\sin x}{\sin X},$$
provided that $\sin X \neq 0$. If $\sin X = 0$, that is, $X = n\pi$, $n = 1, 2, \ldots$, there is no solution.

Solution 2.8

(a) In this case $F = y'^2 + y^2 + 2xy$, $\partial F/\partial y' = 2y'$ and $\partial F/\partial y = 2y + 2x$, giving the Euler–Lagrange equation $y'' - y = x$ (for this type of equation see Block 0, Subsection 1.3.2). The general solution of this second-order inhomogeneous equation is $y = A\cosh x + B\sinh x - x$, where A and B are arbitrary constants determined by the boundary conditions. The boundary condition at $x = 0$ gives $A = 0$, and that at $x = 1$ gives the solution

$$y(x) = (1+\alpha)\frac{\sinh x}{\sinh 1} - x.$$

(b) Consider the difference $\delta S = S[y + \epsilon g] - S[y]$, where y is the above solution:

$$\delta S = \epsilon^2 \int_0^1 dx \, (g'^2 + g^2) > 0$$

for all non-zero $g(x)$. Hence the functional has a minimum along this solution.

Solution 2.9

The general equation of a straight line can be written as $y = m(x - a) + c$. The line passes through (a, A), so $c = A$, and through (b, B), so $B = m(b - a) + A$. Hence

$$y = \frac{(B-A)}{b-a}(x-a) + A$$

is the required equation.

Substituting this into equation (2.41), with $u = x - a$, gives

$$S[y] = 2\pi \int_0^{b-a} du \left(\frac{B-A}{b-a}u + A\right) \sqrt{1 + \left(\frac{B-A}{b-a}\right)^2}$$

$$= 2\pi \frac{\sqrt{(b-a)^2 + (B-A)^2}}{(b-a)^2} \int_0^{b-a} du \, [(B-A)u + A(b-a)]$$

$$= \pi(B+A)\sqrt{(b-a)^2 + (B-A)^2}.$$

Pythagoras' theorem gives $l^2 = (b-a)^2 + (B-A)^2$, hence $S = \pi(B+A)l$.

Solution 2.10

If $f(x) = c\cosh(x/c)$, $f'(x) = \sinh(x/c)$ and the functional (2.41) (page 52), with the appropriate change to the limits, becomes

$$S[f] = 2\pi c \int_{-a}^{a} dx \, \cosh(x/c)\sqrt{1 + \sinh^2(x/c)}$$

$$= 4\pi c^2 \int_0^{\eta} du \, \cosh^2 u, \quad u = \frac{x}{c}, \quad \eta = \frac{a}{c}$$

$$= 2\pi c^2 \left(\eta + \sinh\eta \cosh\eta\right),$$

where we have used the relations $\cosh^2 u = \cosh 2u + 1$ to evaluate the integral, and $\sinh 2u = 2\sinh u \cosh u$ to cast the result in this form. Dividing this by a^2, we see that the dimensionless area $S[f]/a^2$ depends only upon η:

$$\frac{S[f]}{a^2} = \frac{2\pi}{\eta^2}\left(\eta + \sinh\eta\cosh\eta\right).$$

Solution 2.11

(a) Using the expansion $\cosh\eta = 1 + \frac{1}{2}\eta^2 + O(\eta^4)$, we obtain the small η expansion of $g(\eta)$

$$g(\eta) = \frac{1}{\eta}\cosh\eta = \frac{1}{\eta} + \frac{1}{2}\eta + O(\eta^3),$$

so for small η the solution of the equation $g(\eta) = A/a$ is $\eta \simeq a/A$. But since $\eta = a/c$ this gives $c \simeq A$ and

$$f(x) = A\cosh(x/A) \simeq A \quad \text{since } |x| \le a \ll A.$$

With $f(x) = A$ the area is $S = 4\pi Aa$. Alternatively, since $\eta = a/A \ll 1$, so $\cosh\eta \simeq 1$ and $\sinh\eta \simeq \eta$ the result derived in the previous exercise gives $S[f_1]/a^2 = 4\pi/\eta$, and hence $S_1 = 4\pi Aa$.

(b) The equation for η can be written as

$$\frac{A}{a} = \frac{1}{2\eta}\left(e^\eta + e^{-\eta}\right) = \frac{e^\eta}{2\eta}\left(1 + e^{-2\eta}\right).$$

If $\eta \gg 1$ the $e^{-2\eta}$ term is negligible by comparison to 1; for instance, if $\eta = 3$, $e^{-2\eta} = 0.0025$, and if $\eta = 5$, $e^{-2\eta} = 0.00005$. Hence the equation becomes

$$\frac{A}{a} = \frac{1}{2\eta}e^\eta \quad (\eta \gg 1).$$

For large η,

$$\cosh\eta = \tfrac{1}{2}e^\eta\left(1 + e^{-2\eta}\right) \simeq \tfrac{1}{2}e^\eta \quad \text{and} \quad \sinh\eta = \tfrac{1}{2}e^\eta\left(1 - e^{-2\eta}\right) \simeq \tfrac{1}{2}e^\eta,$$

so

$$\frac{S[f]}{a^2} \simeq \frac{2\pi}{\eta^2}\left(\eta + \tfrac{1}{4}e^{2\eta}\right) = 2\pi\left(\frac{e^\eta}{2\eta}\right)^2 + \frac{2\pi}{\eta}.$$

Since $\eta \gg 1$, $e^{2\eta} \gg \eta$, that is, $e^{2\eta}/\eta^2 \gg 1/\eta$, so the first term dominates.

Solution 2.12

(a) The derivative of $g(\eta)$ is $g'(\eta) = \eta^{-1}\sinh\eta - \eta^{-2}\cosh\eta$, which is zero when $\eta\tanh\eta = 1$. The graphs of $y = \tanh\eta$ and $y = 1/\eta$, for $\eta > 0$, are shown in Figure 2.22: $\tanh\eta$ increases monotonically from zero to unity as η increases from 0 to infinity, and $1/\eta$ decreases monotonically from infinity to zero over the same range of η, hence there is one and only one positive real root of $\eta\tanh\eta = 1$.

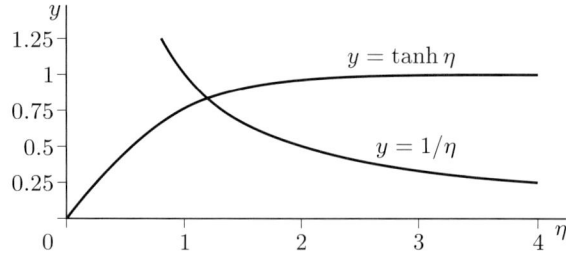

Figure 2.22 Graphs of $y = \tanh\eta$ and $y = 1/\eta$.

A numerical calculation shows that $g'(\eta) = 0$ at $\eta = \eta_m = 1.1997$ and here $g(\eta_m) = 1.5089$.

(b) At the stationary point the area is, using the result obtained in Exercise 2.10

$$S = 2\pi a^2 \left(\frac{1}{\eta_m} + \frac{1}{\eta_m^2} \sinh \eta_m \cosh \eta_m\right)$$
$$= \frac{2\pi a^2}{\eta_m} \left(1 + \sinh^2 \eta_m\right)$$
$$= \frac{2\pi a^2}{\eta_m} \cosh^2 \eta_m$$

since $\eta_m \sinh \eta_m = \cosh \eta_m$. But, by definition,

$$\frac{A}{a} = g(\eta_m) = \frac{1}{\eta_m} \cosh \eta_m$$

hence

$$S = 2\pi a^2 \eta_m g(\eta_m)^2 = 2\pi A^2 \eta_m.$$

Solution 2.13

With $c = A$ and $f(x) = A$, equation (2.46) and the boundary conditions are satisfied. The Euler–Lagrange equation is, since $G = y\sqrt{1 + y'^2}$,

$$\frac{d}{dx}\left(\frac{yy'}{\sqrt{1 + y'^2}}\right) - \sqrt{1 + y'^2} = 0.$$

With $y = $ constant this reduces to $1 = 0$; hence the function $y(x) = A$ is not a solution.

Solution 2.14

(a) The functional does not depend explicitly upon x, so we may use the first integral of the Euler–Lagrange equation $y' \partial F/\partial y' - F = $ constant, where $F(y, y') = \sqrt{y(1 + y'^2)}$. This gives $\sqrt{y} = c\sqrt{1 + y'^2}$, where c is a positive constant. Rearranging this equation then gives the first-order differential equation

$$c^2 \left(\frac{dy}{dx}\right)^2 = y - c^2, \quad y(-1) = y(1) = A.$$

This equation is separable, so can be written in terms of two integrals as

$$c \int \frac{dy}{\sqrt{y - c^2}} = \int dx,$$

and integration gives

$$2c\sqrt{y - c^2} = x + \alpha \quad \text{or} \quad y = c^2 + \frac{(x + \alpha)^2}{4c^2}$$

for some constant α. The boundary conditions at $x = \pm 1$ give

$$A = c^2 + \frac{(\alpha + 1)^2}{4c^2} = c^2 + \frac{(\alpha - 1)^2}{4c^2}.$$

Hence $\alpha = 0$ and $A = c^2 + 1/4c^2$. This last equation is a quadratic in c^2 so gives

$$c^2 = c_\pm^2 = \tfrac{1}{2}\left(A \pm \sqrt{A^2 - 1}\right).$$

Hence, if $A > 1$ there are two solutions of the Euler–Lagrange equation, but none if $A < 1$. The two solutions are

$$y_\pm(x) = \frac{1}{4c_\pm^2}\left(4c_\pm^4 + x^2\right).$$

Typical graphs of $y_\pm(x)$ are shown in Figure 2.23: note that for large values of A, $y_+(x) \simeq A$.

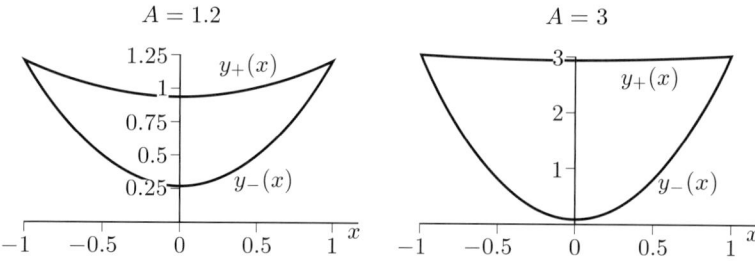

Figure 2.23 Graphs of $y_\pm(x)$ for $A = 1.2$ (left) and $A = 3$ (right).

(b) Substituting the general solution (for any c) into the functional gives

$$S[y] = \frac{1}{2c}\int_{-1}^{1} dx\,\sqrt{4c^4+x^2}\sqrt{1+\frac{x^2}{4c^4}}$$
$$= \frac{1}{4c^3}\int_{-1}^{1} dx\,(4c^4+x^2)$$
$$= 2c + \frac{1}{6c^3}. \tag{2.90}$$

In order to determine which path gives the largest value of $S[y]$, we consider the difference

$$S[y_-] - S[y_+] = 2(c_- - c_+) + \tfrac{1}{6}\left(\frac{1}{c_-^3} - \frac{1}{c_+^3}\right)$$
$$= (c_+ - c_-)\left[\frac{c_+^2 + c_+c_- + c_-^2}{6(c_+c_-)^3} - 2\right]$$
$$= \tfrac{4}{3}(c_+ - c_-)(A-1) > 0 \quad \text{if } A > 1,$$

where we have used the relations $c_+c_- = \tfrac{1}{2}$ and $c_+^2 + c_-^2 = A$, which follow directly from the original quadratic equation for $c_\pm^2$. This relation shows that $S[y_-] > S[y_+]$ for $A > 1$.

If $A = 1$, $c_+ = c_- = 1/\sqrt{2}$ and $S = 4\sqrt{2}/3$. Further, if $A \gg 1$ we have

$$c_\pm^2 = \frac{A}{2}\left(1 \pm \sqrt{1-\frac{1}{A^2}}\right) = \frac{A}{2}\left[1 \pm \left(1 - \frac{1}{2A^2} - \frac{1}{8A^4} + \cdots\right)\right],$$

where we have used the binomial expansion $\sqrt{1-x} = 1 - \tfrac{1}{2}x - \tfrac{1}{8}x^2 + \cdots$. Hence

$$c_+^2 = A\left(1 - \frac{1}{4A^2} - \frac{1}{16A^4} + \cdots\right),$$

and on taking the square root

$$c_+ = \sqrt{A}\left(1 - \frac{1}{4A^2}\left(1 + \frac{1}{4A^2} + \cdots\right)\right)^{1/2} = \sqrt{A}\left(1 - \frac{1}{8A^2} + \cdots\right).$$

Similarly,

$$c_-^2 = \frac{1}{4A}\left(1 + \frac{1}{4A^2} + \cdots\right) \quad \text{giving} \quad c_- = \frac{1}{2\sqrt{A}}\left(1 + \frac{1}{8A^2} + \cdots\right).$$

Putting $c_+ = \sqrt{A}$ and $c_- = 1/2\sqrt{A}$, we obtain the approximations

$$y_+ \simeq A + \frac{x^2}{2A} \simeq A \quad \text{and} \quad y_- \simeq \frac{1}{4A} + Ax^2 \simeq Ax^2, \quad A \gg 1.$$

Substituting these approximations for c into the integral (2.90) for S we obtain

$$S[y_+] \simeq 2\sqrt{A} \quad \text{and} \quad S[y_-] \simeq \tfrac{4}{3}A^{3/2}, \quad A \gg 1.$$

For A close to 1, we find the value of $S[y_\pm]$ by setting $A^2 = 1 + B^2$, where B is a small positive constant. This gives

$$c_\pm^2 = \tfrac{1}{2}\left(\sqrt{1+B^2} \pm B\right), \quad \text{giving} \quad c_\pm = \sqrt{2}\left(\frac{1}{2} \pm \frac{B}{4} + \frac{B^2}{16} \mp \frac{B^3}{32} + \cdots\right),$$

which gives
$$S[y_\pm] = \frac{4}{3}\sqrt{2} + \frac{B^2}{\sqrt{2}} \mp \frac{B^3}{3\sqrt{2}} + \cdots.$$

This shows that, as expected, at $A = 1$ ($B = 0$), $S[y_-] = S[y_+]$, but also that the two curves join tangentially at $A = 1$, as shown in Figure 2.24.

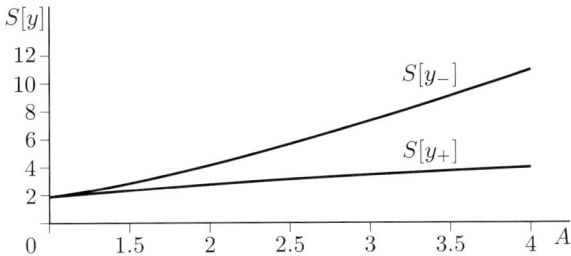

Figure 2.24   Graphs of $S[y_\pm]$.

(c) The Goldschmidt curve is defined by
$$f_G(x) = \begin{cases} 0, & |x| < 1, \\ A, & |x| = 1, \end{cases}$$

so $y'(x)$ is not defined at $|x| = 1$. Hence we define a function that approaches $f_G(x)$ as $\delta \to 0$ for some parameter δ. We need consider only positive values of x:
$$y_\delta(x) = \begin{cases} 0, & 0 \leq x < 1 - \delta, \ 0 < \delta \ll 1, \\ A - \dfrac{A}{\delta}(1 - x), & 1 - \delta \leq x \leq 1. \end{cases}$$

The graph of this function is shown in Figure 2.25.

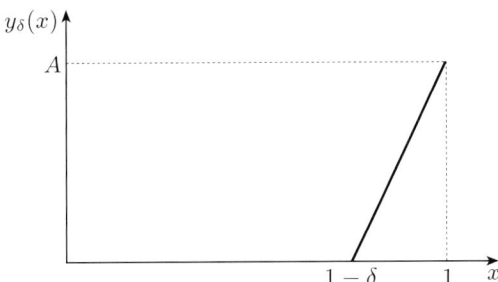

Figure 2.25   Graph of the function $y_\delta(x)$.

The value of the functional on $y_\delta(x)$ is
$$\begin{aligned}
S[y_\delta] &= 2\int_{1-\delta}^{1} dx\, \sqrt{A - \frac{A}{\delta}(1-x)}\sqrt{1 + \frac{A^2}{\delta^2}} \\
&= \frac{2}{\delta}\sqrt{A(A^2 + \delta^2)} \int_0^\delta dv\, \sqrt{1 - \frac{v}{\delta}}, \quad \text{where } v = 1 - x, \\
&= \tfrac{4}{3}\sqrt{A(A^2 + \delta^2)} \to \tfrac{4}{3}A^{3/2} \quad \text{as } \delta \to 0.
\end{aligned}$$

Solution 2.15

For a curve defined parametrically by the functions $x(\theta)$, $y(\theta)$, the area under it and between $\theta = \theta_1$ and θ_2 is

$$A = \int_{x(\theta_1)}^{x(\theta_2)} dx\, y(x) = \int_{\theta_1}^{\theta_2} d\theta\, \frac{dx}{d\theta} y(\theta).$$

For the cycloid, $x(\theta) = a(\theta - \sin\theta)$, $y(\theta) = a(1 - \cos\theta)$, therefore

$$A = a^2 \int_0^{2\pi} d\theta\, (1 - \cos\theta)^2 = a^2 \int_0^{2\pi} d\theta\, (1 - 2\cos\theta + \cos^2\theta) = a^2(2\pi + \pi) = 3\pi a^2.$$

For the length of a curve, we use a variant of equation (1.7) (page 9). Suppose that θ increases from θ to $\theta + \delta\theta$, then to $O(\delta\theta)$, x and y increase by $x'(\theta)\,\delta\theta$ and $y'(\theta)\,\delta\theta$, respectively. Hence the length of the small element of the curve is, using Pythagoras' theorem (see Figure 1.5, page 9),

$$\delta s = \sqrt{x'(\theta)^2 + y'(\theta)^2}\,\delta\theta + O(\delta\theta^2),$$

and the length of the curve between θ_1 and θ_2 is

$$s = \int_{\theta_1}^{\theta_2} d\phi\, \sqrt{x'(\phi)^2 + y'(\phi)^2}.$$

For the cycloid, $x'(\theta) = a(1 - \cos\theta)$, $y'(\theta) = a\sin\theta$ and the length of the arc OP is

$$s = a \int_0^{\theta} d\phi\, \sqrt{(1 - \cos\phi)^2 + \sin^2\phi}$$

$$= a \int_0^{\theta} d\phi\, \sqrt{2 - 2\cos\phi},$$

$$= 2a \int_0^{\theta} d\phi\, \sin(\phi/2)$$

$$= 4a\,(1 - \cos(\theta/2)) = 8a\sin^2(\theta/4),$$

where we have used the identity $\cos z = 1 - 2\sin^2(z/2)$ twice.

Solution 2.16

The gradient is

$$\frac{dy}{dx} = \frac{dy}{d\theta} \bigg/ \frac{dx}{d\theta} = \frac{a\sin\theta}{a(1-\cos\theta)} = \frac{1}{\tan(\theta/2)},$$

where we have used the identities $\sin 2w = 2\sin w \cos w$ and $\cos 2w = 1 - 2\sin^2 w$. The cycloid is perpendicular to the x-axis when the gradient is infinite, that is, when $\tan(\theta/2) = 0$, or $\theta/2 = n\pi$, $n = 0, 1, \ldots$.

Solution 2.17

The Taylor series for $\sin\theta$ and $\cos\theta$ are given in the Handbook; the first few terms are

$$\sin\theta = \theta - \tfrac{1}{6}\theta^3 + O(\theta^5), \quad \cos\theta = 1 - \tfrac{1}{2}\theta^2 + O(\theta^4).$$

Hence

$$x = a(\theta - \sin\theta) = \tfrac{1}{6}a\theta^3 + O(\theta^5) \quad \text{and} \quad y = a(1 - \cos\theta) = \tfrac{1}{2}a\theta^2 + O(\theta^4).$$

The first equation gives $\theta = (6x/a)^{1/3}$, and substituted into the equation for y this gives

$$y = \frac{a}{2}(6x/a)^{2/3}.$$

Solution 2.18

(a) The initial energy is $E = mgAX$, and since x decreases during the fall, equation (2.68) becomes

$$T = -\int_X^0 dx \, \frac{\sqrt{1+A^2}}{\sqrt{2gAX - 2gAx}}$$

$$= \sqrt{\frac{1+A^2}{2gA}} \int_0^X dx \, \frac{1}{\sqrt{X-x}}$$

$$= 2\sqrt{\frac{1+A^2}{2gA}} \sqrt{X}.$$

(b) The initial point (X, Y), where $Y = AX$, satisfies the equation

$$X^2 + (Y-R)^2 = R^2,$$

which becomes $(1+A^2)X = 2AR$. Substituting this into the above equation for T gives the required, rather surprising, result.

Solution 2.19

It is convenient to write $z(\phi)$ in the form $z = c^2 \sin^2 \phi$, use the fact that $z'(x) = z'(\phi)/x'(\phi)$, and express the integrand of the functional in terms of ϕ:

$$T = \int_0^{\phi_b} d\phi \, \frac{dx}{d\phi} \sqrt{\frac{1 + z'(\phi)^2/x'(\phi)^2}{2gz(\phi)}}.$$

But

$$\frac{z'}{x'} = \frac{2c^2 \sin\phi\cos\phi}{c^2(1 - \cos 2\phi)} = \frac{1}{\tan\phi},$$

so

$$T = \frac{2c}{\sqrt{2g}} \int_0^{\phi_b} d\phi = \sqrt{\frac{2A}{g}} \frac{\phi_b}{\sin\phi_b}.$$

Solution 2.20

In this case $F = y'^2 - y^2 - 2xy$, so $\partial F/\partial y' = 2y'$, $\partial F/\partial y = -2y - 2x$, and the Euler–Lagrange equation is the linear second-order inhomogeneous equation (see Block 0, Subsection 1.3.2) $y'' + y + x = 0$. The general solution of this equation is

$$y = A\cos x + B\sin x - x.$$

The boundary condition at $x = 0$ gives $A = 0$, and at $x = 1$ we have $0 = B\sin 1 - 1$, giving the required solution.

Solution 2.21

(a) $\partial F/\partial y' = 2y'$ and $\partial F/\partial y = 2y$, so the Euler–Lagrange equation is $y'' - y = 0$. The general solution of this equation can be written in the form

$$y = \alpha \cosh(x-a) + \beta \sinh(x-a).$$

Putting $x = a$ shows that $\alpha = A$, then putting $x = b$ gives

$$B = A\cosh(b-a) + \beta\sinh(b-a) \quad \text{so } \beta = \frac{B - A\cosh(b-a)}{\sinh(b-a)},$$

giving the solution

$$y = A\cosh u + \frac{B - A\cosh(b-a)}{\sinh(b-a)} \sinh u, \quad u = x - a.$$

(b) If $u = x - a$, then the chain rule gives

$$\frac{dY}{du} = \frac{dy}{dx}\frac{dx}{du} = \frac{dy}{dx},$$

and the functional becomes
$$S[Y] = \int_0^{b-a} du\,(Y'^2 + Y^2), \quad Y(0) = A,\ Y(b-a) = B.$$

The independent variable of the associated Euler–Lagrange equation is u, and a and b occur only in the boundary conditions and as the difference $b - a$. Hence the solution depends upon the three variables x, a and b only in the two combinations $u = x - a$ and $b - a$.

Solution 2.22

Using the given trial function, the functional becomes
$$S(y_1) = \int_0^{1/2} dx\,\left[4y_1^2 - 4y_1^2 x^2 - 4y_1 x^2\right]$$
$$+ \int_{1/2}^1 dx\,\left[4y_1^2 - 4y_1^2(1-x)^2 - 4y_1 x(1-x)\right]$$
$$= \left[\tfrac{11}{6}y_1^2 - \tfrac{1}{6}y_1\right] + \left[\tfrac{11}{6}y_1^2 - \tfrac{1}{3}y_1\right] = \tfrac{11}{3}y_1^2 - \tfrac{1}{2}y_1.$$

This function is stationary at the root of $S'(y_1) = 22y_1/3 - 1/2$, that is, $y_1 = 3/44 \simeq 0.0682$.

Solution 2.23

(a) In this example $F = y'^2 + 12xy$ and $\partial F/\partial y' = 2y'$, $\partial F/\partial y = 12x$. Hence the Euler–Lagrange equation is $y'' = 6x$, $y(0) = 0$, $y(1) = 2$, having the general solution $y = x^3 + Ax + B$, which satisfies the condition at $x = 0$ if $B = 0$ and the condition at $x = 1$ if $A = 1$. Hence the stationary path is $y = x^3 + x$.

(b) In this case $F = y'^2/x^2$ and $\partial F/\partial y' = 2y'/x^2$ giving the Euler–Lagrange equation
$$\frac{d}{dx}\left(\frac{y'}{x^2}\right) = 0 \quad \text{or} \quad \frac{dy}{dx} = \alpha x^2,$$
for some constant α. The general solution is therefore $y(x) = \alpha x^3/3 + \beta$. The boundary conditions give
$$A = \tfrac{1}{3}\alpha + \beta, \quad B = 4\alpha \quad \text{so} \quad \alpha = \tfrac{1}{4}B \quad \text{and} \quad \beta = A - \tfrac{1}{12}B.$$
Hence $y(x) = B\left(x^3 - 1\right)/12 + A$.

(c) In this example $F = 2y^2 y'^2 - (1+x)y^2$ and
$$\frac{\partial F}{\partial y'} = 4y^2 y', \quad \frac{\partial F}{\partial y} = 4yy'^2 - 2(1+x)y.$$
The Euler–Lagrange equation is
$$2\frac{d}{dx}\left(y^2 \frac{dy}{dx}\right) - 2y\left(\frac{dy}{dx}\right)^2 + (1+x)y = 0,$$
which simplifies to $(yy')' + \tfrac{1}{2}(1+x) = 0$. Integrating this gives
$$y\frac{dy}{dx} = \frac{1}{2}\frac{d}{dx}\left(y^2\right) = -\frac{1}{4}(1+x)^2 + \frac{A}{2},$$
and integrating again, $y(x)^2 = B + Ax - \tfrac{1}{6}(1+x)^3$. The boundary conditions then give $y(0)^2 = B - \tfrac{1}{6} = 1$, so $B = \tfrac{7}{6}$, and $y(1)^2 = \tfrac{7}{6} + A - \tfrac{8}{6} = 4$, so $A = \tfrac{25}{6}$. Hence the solution is
$$y(x)^2 = \frac{1}{6}(1+x)\left(25 - (1+x)^2\right) - 3 = -3 + \frac{1}{6}(1+x)(6+x)(4-x).$$

The solution is written in this way because it is easier to understand. The cubic $f = (1+x)(6+x)(4-x)$ is zero at $x = -6, -1$ and 4; f is positive for $x < -6$ and negative for $x > 4$. It follows that y is real only for $x < x_1$, for some $x_1 < -6$, and possibly for some x in the interval $-1 < x < 4$, depending upon the magnitude of f in this interval. Numerical calculations, which you are not expected to do, show that $x_1 \simeq -6.33$ and that y is real in the interval $(-0.264, 3.59)$.

(d) In this case $F = y/y'^2$ is independent of x, $\partial F/\partial y' = -2y/y'^3$ and we may use the first integral, equation (2.33), to give $y/y'^2 = \alpha$, $y'(0) = 1$, $y(a) = A^2$, where α is a constant. The boundary condition at $x = 0$ gives $y(0) = \alpha$. At $x = a$, $y(a) = A^2 > 0$ and hence we assume, for the present, that $\alpha > 0$ and put $\alpha = c^2$, with $c > 0$. The differential equation becomes, on taking the positive square root, because $y'(0) = 1 > 0$ and $c > 0$,

$$\frac{dy}{dx} = \frac{\sqrt{y}}{c} \quad \text{or} \quad \int_{c^2}^{y} \frac{dy}{\sqrt{y}} = \frac{x}{c}.$$

Hence the solution that fits the boundary condition at $x = 0$ is $\sqrt{y} = c + x/2c$. The boundary condition at $x = a$ then gives $2c^2 - 2Ac + a = 0$ giving

$$c = \frac{1}{2}\left(A \pm \sqrt{A^2 - 2a}\right).$$

Both roots are positive and real provided $A > \sqrt{2a}$. In this case we have the two solutions

$$y_\pm(x) = \left(c_\pm + \frac{x}{2c_\pm}\right)^2 \quad \text{and} \quad c_\pm = \frac{1}{2}\left(A \pm \sqrt{A^2 - 2a}\right).$$

Since $y/y'^2 = \alpha = c^2$, the value of the functional on these solutions is $S[y_\pm] = ac_\pm^2$, and so $S[y_+] > S[y_-]$.

Finally, consider the possibility that $y/y'^2 = -c^2$, $c > 0$, so $y(0) = -c^2 < 0$. The general solution that satisfies the boundary condition at $x = 0$ is $y = -(c + x/(2c))^2$, but this cannot be made to satisfy the boundary condition at $x = a$.

Solution 2.24

The Euler–Lagrange equation is

$$\frac{d}{dx}\left(\frac{w(x)y'(x)}{\sqrt{1+y'(x)^2}}\right) = 0,$$

which integrates to $w(x)y'(x) = A\sqrt{1+y'(x)^2}$, where A is a constant. Rearranging this and integrating again gives the general solution

$$y(x) = B \pm A\int_a^x du \, \frac{1}{\sqrt{w(u)^2 - A^2}}.$$

If $w(x) = \sqrt{x}$, this becomes $y(x) = B \pm A\int_a^x du \, \frac{1}{\sqrt{u - A^2}}$ and hence $y(x) = C \pm 2A\sqrt{x - A^2}$, where C is a constant.

If $w(x) = x$ the general solution becomes $y(x) = B \pm A\int_a^x du \, \frac{1}{\sqrt{u^2 - A^2}}$, giving $y(x) = C \pm A\cosh^{-1}(x/A)$.

Solution 2.25

For a function of N variables a stationary point is where

$$\sum_{k=1}^{N} \xi_k \frac{\partial G}{\partial x_k} = 0 \quad \text{for all } \xi_k. \tag{2.91}$$

The fact that the sum is zero for *all* ξ_k, means that $\partial G/\partial x_k = 0$ for all k; this is the equivalent of the fundamental lemma of the calculus of variations.

Solution 2.26

(a) Since $F = (y'^2 - 1)^2$ and $\partial F/\partial y' = 4y'(y'^2 - 1)$, the first integral of the Euler–Lagrange equation, equation (2.33), is $(y'^2 - 1)(3y'^2 + 1) = \text{constant}$. Hence $y'^2 = m^2$ for some constant m, which we assume positive.

(b) The solutions of the equation $y'(x)^2 = m^2$ that satisfy the boundary condition $y(0) = 0$ are $y(x) = \pm mx$, $m > 0$. Hence one solution that fits the boundary condition at $x = 1$ is $y = y_1 = Ax$ and on this path $S[y_1] = (A^2 - 1)^2$.

Another solution has the form
$$y(x) = \begin{cases} mx, & 0 \le x \le \xi \le 1 \\ c - mx, & \xi \le x \le 1, \end{cases}$$
where m, c and ξ are constants. The boundary condition at $x = 1$ gives $c = A + m$. Since the solution needs to be continuous at $x = \xi$ we also have $m\xi = c - m\xi$ and hence $\xi = (A + m)/2m$.

Because $m > 0$ and $\xi \le 1$ it follows that $m \ge A$; for $m = A$ we regain the solution $y = Ax$, but for $m > A$ we obtain $y_2(x)$.

Another solution is
$$y(x) = \begin{cases} -mx, & 0 \le x \le \xi \\ c + mx, & \xi \le x \le 1. \end{cases}$$

The boundary condition at $x = 1$ gives $c = A - m$ and the continuity condition gives $-m\xi = c + m\xi$, and hence $\xi = -c/2m = (m - A)/2m$. Since $\xi \ge 0$ this gives $m \ge A$, as before. This gives the solution $y_3(x)$.

(c) If $A \ge 1$, the minimum value of the functional is $(A^2 - 1)^2$ and this is given by the solution $y = Ax$.

If $A < 1$, we may choose $m = 1$, for y_2, and $m = -1$ for y_3 to give the minimum value of zero.

Solution 2.27

(a) This integral can be evaluated directly, $S[y] = y(1) - y(0)$, and its value is independent of the path, regardless of the boundary values.

(b) Similarly $S[y] = \frac{1}{2}(y(1)^2 - y(0)^2)$.

(c) Since $F = xyy'$, $\partial F/\partial y' = xy$, $\partial F/\partial y = xy'$ and the Euler–Lagrange equation is $y = 0$, which does not satisfy the boundary conditions.

Alternatively we have
$$S[y] = \frac{1}{2}\int_0^1 dx\, x \frac{d}{dx}y^2 = \left[\frac{1}{2}xy(x)^2\right]_0^1 - \frac{1}{2}\int_0^1 dx\, y^2 = \frac{1}{2} - \frac{1}{2}\int_0^1 dx\, y^2.$$

The Euler–Lagrange equation for the functional on the right-hand side of this equation is again $y = 0$.

Solution 2.28

We expect the Euler–Lagrange equations for these two functionals to be identical because
$$S_2[y] = S_1[y] + [G(x, y(x))]_{x=a}^{x=b}$$
and the boundary term is independent of the path. Now we derive the result directly.

Consider the Euler–Lagrange equation for $S_2[y]$. First define
$$\mathcal{F}(x, y, y') = F(x, y, y') + \frac{dG}{dx} = F(x, y, y') + \frac{\partial G}{\partial x} + \frac{\partial G}{\partial y}y'$$

so that
$$\frac{\partial \mathcal{F}}{\partial y} = \frac{\partial F}{\partial y} + \frac{\partial^2 G}{\partial x \partial y} + \frac{\partial^2 G}{\partial y^2}y' \quad \text{and} \quad \frac{\partial \mathcal{F}}{\partial y'} = \frac{\partial F}{\partial y'} + \frac{\partial G}{\partial y}.$$

Hence the Euler–Lagrange equation for $\mathcal{F}$ is
$$\frac{d}{dx}\left(\frac{\partial \mathcal{F}}{\partial y'}\right) - \frac{\partial \mathcal{F}}{\partial y} = \frac{d}{dx}\left(\frac{\partial F}{\partial y'}\right) - \frac{\partial F}{\partial y} + \frac{d}{dx}\left(\frac{\partial G}{\partial y}\right) - \frac{\partial^2 G}{\partial x \partial y} - \frac{\partial^2 G}{\partial y^2}y'.$$

But,
$$\frac{d}{dx}\left(\frac{\partial G}{\partial y}\right) = \frac{\partial^2 G}{\partial x \partial y} + \frac{\partial^2 G}{\partial y^2}y',$$

so the Euler–Lagrange equations for $\mathcal{F}$ and F are identical, as expected.

Solution 2.29

Clearly $S[y] \geq 0$ and for the given solution the integrand is identically zero, so for this solution $S = 0$, its minimum value. The Euler–Lagrange equation is
$$(y'' - 2)y^2 + 2(y' - 2x)yy' - (y' - 2x)^2 y = 0,$$

which is satisfied by the functions $y(x) = 0$ and $y(x) = x^2$. Thus the given function satisfies the Euler–Lagrange equation except at $x = 0$ where $y''(x)$ is not defined.

Solution 2.30

If $\boldsymbol{y} = (y_1, y_2, \cdots, y_n)$ with $y_0 = A$ and $y_{N+1} = B$ and $y_{k+1} = y_k + h$, then y_n occurs only in the $k = n$ and $k = n + 1$ terms and
$$\frac{\partial S}{\partial y_n} = h\frac{\partial}{\partial y_n}F\left(x_n, y_n, \frac{y_n - y_{n-1}}{h}\right) + h\frac{\partial}{\partial y_n}F\left(x_{n+1}, y_{n+1}, \frac{y_{n+1} - y_n}{h}\right)$$
$$= h\frac{\partial}{\partial u}F(z_n) + \frac{\partial}{\partial v}F(z_n) - \frac{\partial}{\partial v}F(z_{n+1}), \quad z = \left(x, y_n, \frac{y_n - y_{n-1}}{h}\right).$$

Now we need to express $(y_{n+1} - y_n)/h$ in terms of $(y_n - y_{n-1})/h$. First, write
$$\frac{y_{n+1} - y_n}{h} = \frac{y_n - y_{n-1}}{h} + \frac{y_{n+1} - 2y_n + y_{n-1}}{h},$$

and use the Taylor expansion
$$y_{n+1} - 2y_n + y_{n-1} = y(x_n + h) - 2y(x_n) + y(x_n - h)$$
$$= y(x_n) + hy'(x_n) + \tfrac{1}{2}h^2 y''(x_n) + \tfrac{1}{6}h^3 y'''(x_n) + O(h^4) - 2y(x_n)$$
$$+ y(x_n) - hy'(x_n) + \tfrac{1}{2}h^2 y''(x_n) - \tfrac{1}{6}h^3 y'''(x_n) + O(h^4)$$
$$= h^2 y''(x_n) + O(h^4).$$

Hence
$$z_{n+1} = \left(x_n + h, y_n + hy'_n + O(h^2), \frac{y_n - y_{n-1}}{h} + hy''_n + O(h^3)\right)$$
$$= z_n + (1, y'_n, y''_n)h + O(h^2),$$

which gives
$$F(z_{n+1}) = F(z_n) + h\left(\frac{\partial F}{\partial x} + y'_n\frac{\partial F}{\partial u} + y''_n\frac{\partial F}{\partial v}\right) + O(h^2).$$

It follows that the equation for $\partial S/\partial y_n$ becomes
$$\frac{\partial S}{\partial y_n} = h\left[\frac{\partial F}{\partial u} - \frac{\partial}{\partial v}\left(\frac{\partial F}{\partial x} + y'_n\frac{\partial F}{\partial u} + y''_n\frac{\partial F}{\partial v}\right)\right] + O(h^2),$$
$$= h\left[\frac{\partial F}{\partial u} - \left(\frac{\partial G}{\partial x} + y'_n\frac{\partial G}{\partial u} + y''_n\frac{\partial G}{\partial v}\right)\right] + O(h^2), \quad G = \frac{\partial F}{\partial v},$$
$$\frac{\partial S}{\partial y_n} = h\left[\frac{\partial F}{\partial u} - \frac{d}{dx}\left(\frac{\partial F}{\partial v}\right)\right] + O(h^2), \quad n = 1, 2, \ldots, N.$$

Since $\partial S/\partial y_n = 0$ it follows that
$$\frac{d}{dx}\left(\frac{\partial F}{\partial v}\right) - \frac{\partial F}{\partial u} = O(h), \quad n = 1, 2, \ldots, N,$$
and that as $h \to 0$ we obtain the Euler–Lagrange equation.

Solution 2.31

In this more general case we use the approximations
$$\int_0^1 dx\, z(x) \simeq h\sum_{k=0}^N z(x_k) \quad \text{and} \quad \int_0^1 dx\, z'(x)^2 \simeq h\sum_{k=0}^N \left(\frac{z(x_{k+1}) - z(x_k)}{h}\right)^2,$$
where $z(x)$ is any function and the set of equally spaced points $x_k = k/(N+1)$ defined in the question. Hence the functional becomes
$$S = \sum_{k=0}^N \frac{(y_{k+1} - y_k)^2}{h} - h\sum_{k=0}^N y_k^2 - 2h\sum_{k=0}^N x_k y_k, \quad h = \frac{1}{N+1},$$
$$= \sum_{k=0}^N \left[\frac{(y_{k+1} - y_k)^2}{h} - h\left(y_k^2 + \frac{2k}{N+1}y_k\right)\right].$$

(a) If $N = 1$ there are two terms in the sum; the first is y_1^2/h, since $y_0 = 0$, and the second is $(1/h - h)y_1^2 - hy_1$, and since $h = 1/2$ this gives
$$S(y_1) = \tfrac{7}{2}y_1^2 - \tfrac{1}{2}y_1.$$

This function is stationary where $\partial S/\partial y_1 = 7y_1 - 1/2 = 0$, that is $y_1 = 1/14 = 0.0714$, compared to the exact value of $y(1/2) = 0.0697$.

The difference between this approximation to S and that obtained in Exercise 2.22 is because the approximations to the functional are different. In both cases we approximate the solution by the same type of polygon. However, in the first case we evaluated the integrals exactly, whereas in the second case we made an additional approximation to evaluate the integrals. For the approximation used in Exercise 2.22 we have
$$\int_0^1 dx\, y'(x)^2 = 4y_1^2\left[\int_0^{1/2} dx + \int_{1/2}^1 dx\right] = 4y_1^2,$$
$$\int_0^1 dx\, y(x)^2 = 4y_1^2\left[\int_0^{1/2} dx\, x^2 + \int_{1/2}^1 dx\, (1-x)^2\right] = \frac{1}{3}y_1^2,$$
$$\int_0^1 dx\, 2xy(x) = 4y_1\left[\int_0^{1/2} dx\, x^2 + \int_{1/2}^1 dx\, x(1-x)\right] = \frac{1}{2}y_1.$$

For the approximation used here, these integrals are approximated by
$$\int_0^1 dx\, y'(x)^2 = 2\sum_{k=0}^1 (y_{k+1} - y_k)^2 = 4y_1^2$$
$$\int_0^1 dx\, y(x)^2 = \frac{1}{2}\sum_{k=0}^1 y_k^2 = \frac{1}{2}y_1^2 \quad \text{and} \quad \int_0^1 dx\, 2xy(x) = \sum_{k=0}^1 \frac{k}{2}y_k = \frac{1}{2}y_1.$$

(b) If $N = 2$, then $h = 1/3$, $y_3 = 0$ and
$$S = 3y_1^2 + \left[3(y_2 - y_1)^2 - \frac{1}{3}\left(y_1^2 + \frac{2}{3}y_1\right)\right] + \left[3y_2^2 - \frac{1}{3}\left(y_2^2 + \frac{4}{3}y_2\right)\right]$$
$$= \frac{17}{3}y_1^2 + \frac{17}{3}y_2^2 - 6y_1y_2 - \frac{2}{9}y_1 - \frac{4}{9}y_2.$$

The stationary points are at the solutions of
$$\frac{\partial S}{\partial y_1} = \frac{34}{3}y_1 - 6y_2 - \frac{2}{9} = 0 \quad \text{and} \quad \frac{\partial S}{\partial y_2} = \frac{34}{3}y_2 - 6y_1 - \frac{4}{9} = 0$$

which simplify to the given equations. These have the solutions $y_1 = 35/624 \simeq 0.0561$ and $y_2 = 43/624 \simeq 0.0689$, which are the approximate values of the solution at $x = 1/3$ and $2/3$ respectively.

Solution 2.32

(a) The Gâteaux derivative, equation (2.14) (page 45), of this functional is

$$\Delta S[y, g] = \frac{d}{d\epsilon} S[y + \epsilon g]\bigg|_{\epsilon=0} = \int_a^b dx\, g''(x) \frac{\partial F}{\partial y''}.$$

Integrating by parts twice gives

$$\Delta S = \left[g'(x) \frac{\partial F}{\partial y''} \right]_a^b - \int_a^b dx\, g'(x) \frac{d}{dx}\left(\frac{\partial F}{\partial y''} \right),$$

$$= \left[g'(x) \frac{\partial F}{\partial y''} - g(x) \frac{d}{dx}\left(\frac{\partial F}{\partial y''} \right) \right]_a^b + \int_a^b dx\, g(x) \frac{d^2}{dx^2}\left(\frac{\partial F}{\partial y''} \right).$$

But $g(x)$ and $g'(x)$ are both zero at $x = a$ and b, so for the functional to be stationary we need

$$\frac{d^2}{dx^2}\left(\frac{\partial F}{\partial y''} \right) = 0. \quad \text{Integrating this twice gives} \quad \frac{\partial F}{\partial y''} = c(x - a) + d,$$

for some constants c and d.

(b) If $F(z) = \frac{1}{2} z^2$ the differential equation for $y(x)$ is $y''(x) = c(x - a) + d$. Integrating this twice gives

$$y'(x) = \tfrac{1}{2} c(x - a)^2 + d(x - a) + \alpha \quad \text{and}$$
$$y(x) = \tfrac{1}{6} c(x - a)^3 + \tfrac{1}{2} d(x - a)^2 + \alpha(x - a) + \beta.$$

The boundary conditions at $x = a$ give $y'(a) = A_2 = \alpha$ and $y(a) = A_1 = \beta$, so

$$y(x) = \tfrac{1}{6} c(x - a)^3 + \tfrac{1}{2} d(x - a)^2 + A_2(x - a) + A_1,$$

and the constants c and d are determined from the boundary conditions at $x = b$. Setting $D = b - a$ the two equations $y(b) = B_1$ and $y'(b) = B_2$ become, respectively,

$$\tfrac{1}{6} cD^3 + \tfrac{1}{2} dD^2 + A_2 D + A_1 = B_1 \quad \text{and} \quad \tfrac{1}{2} cD^2 + dD + A_2 = B_2,$$

which simplify to the quoted equations.

(c) Consider the general functional $S[y] = \int_a^b dx\, F(y'')$, so

$$S[y + \epsilon g] = S[y] + \epsilon \int_a^b dx\, g''(x) \frac{\partial F}{\partial y''} + \frac{1}{2} \epsilon^2 \int_a^b dx\, g''(x)^2 \frac{\partial^2 F}{\partial y''^2} + \cdots$$

and on the stationary path

$$S[y + \epsilon g] - S[y] = \frac{1}{2} \epsilon^2 \int_a^b dx\, g''(x)^2 \frac{\partial^2 F}{\partial y''^2} + \cdots.$$

Since $g''(x)^2 \geq 0$ the sign of this integral depends upon $\partial^2 F/\partial y''^2$. In the present case, however, $F(z) = z^2/2$, $F''(z) = 1$ and hence the integral is positive and the stationary path is a minimum.

Solution 2.33

First, note that if $y(x)$ and $y(x) + \epsilon g(x)$ are both admissible functions then $g(x)$ and its derivative, $g'(x)$, are zero at $x = a$ and b. The Gâteaux derivative, $\Delta S[y, g]$ (see Equation (2.14)), is

$$\lim_{\epsilon \to 0} \frac{d}{d\epsilon} S[y + \epsilon g] = \int_a^b dx \, \frac{d}{d\epsilon} F(x, y + \epsilon g, y' + \epsilon g', y'' + \epsilon g'') \bigg|_{\epsilon=0}.$$

Thus

$$\Delta S[y, g] = \int_a^b dx \left(g \frac{\partial F}{\partial y} + g' \frac{\partial F}{\partial y'} + g'' \frac{\partial F}{\partial y''} \right).$$

Integration by parts gives

$$\int_a^b dx \, g' \frac{\partial F}{\partial y'} = \left[g \frac{\partial F}{\partial y'} \right]_a^b - \int_a^b dx \, g \frac{d}{dx} \left(\frac{\partial F}{\partial y'} \right).$$

Since $g(a) = g(b) = 0$ the boundary term vanishes. Similarly,

$$\int_a^b dx \, g'' \frac{\partial F}{\partial y''} = \left[g' \frac{\partial F}{\partial y''} \right]_a^b - \int_a^b dx \, g' \frac{d}{dx} \left(\frac{\partial F}{\partial y''} \right),$$

$$= \left[g' \frac{\partial F}{\partial y''} - g \frac{d}{dx} \left(\frac{\partial F}{\partial y''} \right) \right]_a^b + \int_a^b dx \, g \frac{d^2}{dx^2} \left(\frac{\partial F}{\partial y''} \right).$$

Again the boundary terms vanish because $g'(a) = g'(b) = 0$ and $g(a) = g(b) = 0$. Hence

$$\Delta S[y, g] = \int_a^b dx \, g(x) \left[\frac{\partial F}{\partial y} - \frac{d}{dx} \left(\frac{\partial F}{\partial y'} \right) + \frac{d^2}{dx^2} \left(\frac{\partial F}{\partial y''} \right) \right].$$

Using the fundamental theorem of the calculus of variations we see that a necessary condition for the functional to be stationary on a function $y(x)$ is that it satisfies the equation

$$\frac{d^2}{dx^2} \left(\frac{\partial F}{\partial y''} \right) - \frac{d}{dx} \left(\frac{\partial F}{\partial y'} \right) + \frac{\partial F}{\partial y} = 0,$$

with the given boundary conditions.

Solution 2.34

(a) If $F = 1 + y''(x)^2$ the required derivatives are $\partial F/\partial y'' = 2y''(x)$ and $\partial F/\partial y' = \partial F/\partial y = 0$, so the equation for the stationary function is $d^4y/dx^4 = 0$. The general solution of this equation is the cubic $y(x) = ax^3 + bx^2 + cx + d$, where the constants a, b, c and d are determined by the boundary condition. Those at $x = 0$ give $y(0) = d = 0$ and $y'(0) = c = 1$; those at $x = 1$ then give $y(1) = a + b + 1 = 1$ and $y'(1) = 3a + 2b + 1 = 1$, so that $a = b = d = 0$, $c = 1$ and the solution is $y(x) = x$.

(b) In this case $\partial F/\partial y'' = 2y''$, $\partial F/\partial y' = 0$, $\partial F/\partial y = -2y$, so the equation for the stationary function is

$$\frac{d^4 y}{dx^4} - y = 0, \quad y(0) = 1, \quad y'(0) = 0, \quad y\left(\tfrac{\pi}{2}\right) = 0, \quad y'\left(\tfrac{\pi}{2}\right) = -1.$$

The general solution of this is $y(x) = A \cos x + B \sin x + D \cosh x + E \sinh x$. The boundary conditions at $x = 0$ give

$$y(0) = A + D = 1 \quad \text{and} \quad y'(0) = B + E = 0$$

and those at $x = \pi/2$ give

$$y\left(\tfrac{\pi}{2}\right) = B + Dc + Es = 0, \quad y'\left(\tfrac{\pi}{2}\right) = -A + Ds + Ec = -1,$$

where $c = \cosh(\pi/2)$ and $s = \sinh(\pi/2)$. Using the first two equations to substitute for D and E in the second two gives

$$(s - 1)B + Ac = c \quad \text{and} \quad Bc + (s + 1)A = s + 1.$$

These equations have the solution $A = 1$ and $B = 0$, hence $E = D = 0$, and the required solution is $y(x) = \cos x$.

Solution 2.35

(a) In this case $F = y'^2 - y^2$ so $\partial F/\partial y' = 2y'$, $\partial F/\partial y = -2y$ giving the Euler–Lagrange equation $y'' + y = 0$.

(b) Suppose that $y(x)$ and $y(x) + \epsilon g(x)$ are two admissible functions with $y(x)$ satisfying the two given boundary conditions. Then if $\Delta = S[y + \epsilon g] - S[y]$,

$$\Delta = 2\epsilon\left[h_b g(b) y(b) - h_a g(a) y(a)\right] + 2\epsilon \int_a^b dx\, (g'y' - gy) + O(\epsilon^2).$$

Integrate by parts to put this in the form

$$\Delta S = 2\epsilon \left[g(b)\left(y'(b) + h_b y(b)\right) - g(a)\left(y'(a) + h_a y(a)\right)\right] \tag{2.92}$$
$$- 2\epsilon \int_a^b dx\, (y'' + y)\, g + O(\epsilon^2).$$

But $h_a y(a) + y'(a) = 0$ and $h_b y(b) + y'(b) = 0$, so

$$\Delta = -2\epsilon \int_a^b dx(y'' + y)g + O(\epsilon^2).$$

The fundamental lemma of the calculus of variations shows that a necessary condition for $S[y]$ to be stationary is that $y(x)$ satisfies the differential equation $y'' + y = 0$.

Note that use of the fundamental lemma here needs some care. As stated earlier (see page 47), the function $g(x)$ needs to be zero at $x = a$ and b. Here, however, since both $y(x)$ and $y(x) + \epsilon g(x)$ are admissible functions we have

$$h_a g(a) + g'(a) = 0 \quad \text{and} \quad h_b g(b) + g'(b) = 0$$

so the lemma does not apply directly. But the same method can be used on the open interval $a < x < b$, so we may deduce that on this interval $y'' + y = 0$, and use continuity to extend the interval to $a \leq x \leq b$.

Solution 2.36

(a) Since $dz/dx = 1/(dx/dz)$ we have $\sqrt{1 + z'^2} = \sqrt{1 + x'^2}/x'$ and hence, to within an irrelevant multiplicative constant,

$$T[x] = \int_A^0 dz\, \frac{dx}{dz} \frac{\sqrt{1 + x'^2}}{|x'|\sqrt{z}} = \int_0^A dz\, \sqrt{\frac{x'^2 + 1}{z}}$$

since $x'(z) < 0$.

(b) Use the result given in Exercise 1.7 (page 14) and the fact that the term $O(\epsilon)$ is, by definition, zero on the stationary path to cast the difference in the first required form. Now change the integration variable from z to ϕ, using the result $x'(z) = x'(\phi)/z'(\phi) = \tan \phi$. The integral exists and is positive and hence the stationary path is a minimum.

Solution 2.37

(a) The triangles ABC and $AB_1 D$ are similar and AC is, by construction, twice AD. Hence AB is twice AB_1 and $h = 2h_1$.

(b) At each iterations of this procedure the number of similar triangles doubles. After n iterations we therefore have 2^n similar triangles: the heights are $2^{-n}h$ and the length of the side is $l2^{-n}$. But since there are 2×2^n sides the total length is $2l$, the original distance along ABC.

Since l is arbitrary the distance along the curve may be made as long as required and by choosing n sufficiently large, we will never stray further than any prescribed small distance from the straight line AC.

CHAPTER 3
Further developments of the theory

3.1 Introduction

This chapter contains three main topics which continue the development of the general theory of the calculus of variations in the direction needed for the reformulation of Newton's equations as a variational principle.

First we consider the effects of changing variables, both dependent and independent variables, normally denoted by y and x respectively in previous chapters. This is important because one of the principal means of solving complicated differential equations is to find new variables that simplify the problem. Here we show that when a problem can be formulated as a variational principle, the algebra involved in changing variables is made easier.

Secondly, we describe how the theory is generalised to include many dependent variables. This generalisation is essential when we reformulate Newton's equations because separate variables are required for each degree of freedom; for instance, if a particle moves in a plane, two dependent variables are necessary to describe its position, whereas two particles moving in a plane require four variables. This generalisation is relatively simple and involves no new ideas.

Section 3.5, the final, most difficult (and optional) part of this chapter deals with invariance properties. These are important because, when present, drastic simplification of the Euler–Lagrange equations becomes possible. A precursor to part of this theory was introduced in the previous chapter where it was shown, in Exercise 2.6 (page 51), that under certain circumstances integrals of the motion exist, so the second-order Euler–Lagrange equation can be integrated directly to give a simpler first-order equation. This simplification was used to help solve the minimum surface area problems and the brachistochrone, in Sections 2.5 and 2.7 respectively. In Section 3.5 we describe a more general principle from which the first integral may be derived, and in Subsection 3.5.2 this principle is used to derive a natural generalisation of the elementary result obtained in Exercise 2.6. Students knowing some dynamics will be aware of how important conservation of energy, linear and angular momentum can be: the theory described in Subsection 3.5.2 unifies all these conservation laws. This theory is, however, difficult and is not assessed; we therefore suggest only a cursory reading if it is found to be too hard.

3.2 Invariance of the Euler–Lagrange equation

In this section we consider the effect of changing both the dependent and independent variables, and show that in both cases the form of the Euler–Lagrange equation remains unchanged. Such a theory is important because one of the principal methods of solving a differential equation is to change variables with the aim of converting it to a standard recognisable form. For instance, the unfamiliar equation

$$z\frac{d^2y}{dz^2} + (1-a)\frac{dy}{dz} + a^2 z^{2a-1} y = 0 \qquad (3.1)$$

becomes, on setting $z = x^{1/a}$ ($x \geq 0$), the familiar equation

$$\frac{d^2y}{dx^2} + y = 0. \qquad (3.2)$$

It is rarely easy to find suitable new variables, but if the equation can be derived from a variational principle the task is made easier because the algebra is usually simpler: you will see why in Exercise 3.1, where the above example is treated for $a = 2$; see also Exercise 3.24.

Here we deal with the case of only one dependent variable, but important examples involving many dependent variables are given in the next chapter. Indeed the full power of this technique becomes apparent mainly in the advanced study of dynamical systems.

When deriving the Euler–Lagrange equation we represented the path between end points in a Cartesian coordinate system. This coordinate system is neither always the most convenient nor the most useful: for instance, rather than using the Cartesian coordinates (x, y), it may be better to represent a path in terms of polar coordinates (r, θ), where $x = r\cos\theta$ and $y = r\sin\theta$. Indeed, it is precisely this change of coordinates that prompted Euler's concern with the problem of the invariance of his necessary conditions. In modern parlance, this translates to the form of the Euler–Lagrange equations being invariant; we shall see exactly what this means in the following two sections.

In the first section we consider the simpler types of transformation in which only the independent variable is changed. In the second section we consider Euler's original problem in which paths are described in both Cartesian and polar coordinates, the aim being to understand the changes induced in the Euler–Lagrange equations. In both cases we shall derive the important result that the form of the Euler–Lagrange equation remains unchanged.

3.2.1 Changing the independent variable

The easiest way of understanding why the form of the Euler–Lagrange equation is invariant under a coordinate change is to consider the effect of changing only the independent variable x. Thus for the functional

$$S[y] = \int_a^b dx\, F(x, y(x), y'(x)), \tag{3.3}$$

we change to a new independent variable u, where $x = g(u)$ for a known function $g(u)$, assumed to be invertible for $a \leq x \leq b$.

We note that a continuous function $f(x)$ is invertible for $a \leq x \leq b$ if there is an inverse function $g(x)$ such that $f(g(x)) = x$. Often the notation $f^{-1}(x)$ is used to denote the inverse $g(x)$. For instance, if $f(x) = \sqrt{x}$, for $x \geq 0$, then $f^{-1}(x) = x^2$.

With this change of variable $y(x)$ becomes a function of u and it is convenient to define

$$Y(u) = y(g(u)) = y(x). \tag{3.4}$$

The chain rule gives

$$\frac{dy}{dx} = \frac{dy}{du}\frac{du}{dx} = \frac{Y'(u)}{dx/du} = \frac{Y'(u)}{g'(u)}, \tag{3.5}$$

so the functional becomes

$$S[Y] = \int_c^d du\, g'(u) F\left(g(u), Y(u), \frac{Y'(u)}{g'(u)}\right), \tag{3.6}$$

where the integration limits, c and d, are defined implicitly by the equations $a = g(c)$ and $b = g(d)$. Note, in order to avoid a proliferation of symbols we use the same symbol, S, for the original and the transformed functionals.

The integrand of the original functional depends upon x, $y(x)$ and $y'(x)$. The integrand of the transformed functional depends upon u, $Y(u)$ and $Y'(u)$, so if we define

$$\mathcal{F}(u, Y(u), Y'(u)) = g'(u)\, F\left(g(u), Y(u), \frac{Y'(u)}{g'(u)}\right), \tag{3.7}$$

the functional becomes

$$S[Y] = \int_c^d du\, \mathcal{F}(u, Y(u), Y'(u)). \tag{3.8}$$

Because this functional depends upon u, the new dependent variable, $Y(u)$, and its first derivative the derivation of the associated Euler–Lagrange equation is exactly the same as for the original functional. The Euler–Lagrange equation in the new variable, u, is therefore

$$\frac{d}{du}\left(\frac{\partial \mathcal{F}}{\partial Y'}\right) - \frac{\partial \mathcal{F}}{\partial Y} = 0, \tag{3.9}$$

in contrast to the original Euler–Lagrange equation

$$\frac{d}{dx}\left(\frac{\partial F}{\partial y'}\right) - \frac{\partial F}{\partial y} = 0. \tag{3.10}$$

These two equations have the same form, in the sense that the formula (3.9) is obtained from (3.10) by replacing the explicit occurrence of x, y, y' and F by u, Y, Y' and $\mathcal{F}$ respectively. The new second-order differential equation for Y, obtained from (3.9) is, however, normally quite different from the equation for y derived from (3.10), because $\mathcal{F}$ and F have different functional forms.

Consider, for instance, the functional

$$S[y] = \int_1^2 dx\, \frac{y'^2}{x^2}, \quad y(1) = 1,\ y(2) = 2, \tag{3.11}$$

which is similar to the example dealt with in Exercise 2.23(c) (page 70). The general solution for the associated Euler–Lagrange equation is derived in the solution for this exercise and is $y(x) = \beta + \alpha x^3/3$; the boundary conditions give $\beta = 6/7$ and $\alpha = 3/7$.

Now make the transformation $x = u^a$, for some constant a. The chain rule gives

$$\frac{dy}{dx} = \frac{dy}{du}\frac{du}{dx} = \frac{Y'(u)}{au^{a-1}}, \quad \text{where } Y(u) = y(u^a) = y(x), \tag{3.12}$$

and the functional becomes

$$S[Y] = a\int_1^{2^{1/a}} du\, u^{a-1}\frac{1}{u^{2a}}\left(\frac{Y'(u)}{au^{a-1}}\right)^2 = \frac{1}{a}\int_1^{2^{1/a}} du\, \frac{Y'(u)^2}{u^{3a-1}}. \tag{3.13}$$

Now choose $3a = 1$, so the functional simplifies to

$$S[Y] = 3\int_1^8 du\, Y'^2, \quad Y(1) = 1,\ Y(8) = 2. \tag{3.14}$$

The Euler–Lagrange equation for this functional is $Y''(u) = 0$, having the general solution $Y = C + Du$. The boundary conditions give $C + D = 1$ and $C + 8D = 2$ and hence $S[Y]$ is stationary when

$$Y(u) = \tfrac{1}{7}(6 + u) \quad \text{giving} \quad y(x) = Y(u(x)) = \tfrac{1}{7}\left(6 + x^3\right). \tag{3.15}$$

In this example little was gained, because the Euler–Lagrange equation is equally easily solved in each representation. This is not always the case, as the next exercise shows.

Exercise 3.1

The functional

$$S[y] = \int_0^X dx\, \left(y'^2 - \omega^2 y^2\right), \quad X > 0,$$

where ω is a constant, gives rise to the Euler–Lagrange equation $y'' + \omega^2 y = 0$.

(a) Show that changing the independent variable to u where $x = u^2$ gives the functional

$$S[Y] = \tfrac{1}{2}\int_0^U du\, \left(\frac{Y'(u)^2}{u} - 4\omega^2 u Y^2\right), \quad \text{where } Y(u) = y(u^2)$$

and $U = \sqrt{X}$, with the associated Euler–Lagrange equation

$$u\frac{d^2Y}{du^2} - \frac{dY}{du} + 4\omega^2 u^3 Y = 0.$$

Show that this is the same as equation (3.1) when $a = 2$ and $\omega = 1$.

(b) Show that if $x = u^2$, then

$$\frac{d^2 y}{dx^2} = \frac{1}{4u^3}\left(u\frac{d^2Y}{du^2} - \frac{dY}{du}\right)$$

and hence derive the above Euler–Lagrange equation directly.

Note that the method described in part (a) for transforming the differential equation requires only that we compute dy/dx and avoids the need to calculate the more difficult second derivative, d^2y/dx^2, required by the method used in part (b).

Exercise 3.2

A simpler type of transformation involves a change of the dependent variable. Consider the functional

$$S[y] = \int_a^b dx\, y'^{\,2}.$$

(a) Show that the Euler–Lagrange equation for this functional is $y''(x) = 0$.

(b) Define a new dependent variable z related to y by the differentiable function $y = G(z)$ and show that the functional becomes

$$S[z] = \int_a^b dx\, G'(z)^2 z'^{\,2}.$$

Show also that the Euler–Lagrange equation for this functional is

$$\left[G'(z)\, z'' + G''(z)\, z'^{\,2}\right] G'(z) = 0,$$

and that this is equivalent to the original Euler–Lagrange equation provided $G'(z) \neq 0$. Note that the condition $G'(z) \neq 0$ means that the equation $y = G(z)$ may be inverted to give z as a function of y.

3.2.2 Changing both the dependent and independent variables

In the previous section, particularly in Exercise 3.1, it was seen that when changing the independent variable the algebra is simpler if the transformation is made to the functional rather than to the associated Euler–Lagrange equation. The reason for this is simply that changing the functional involves only first derivatives, whereas transforming the equation directly involves second derivatives.

For the same reason it is far easier to apply more general transformations to the functional than to the Euler–Lagrange equation. The most general transformation we need to consider will be between the Cartesian coordinates (x, y) and two new variables (u, v): such transformations are defined by two equations

$$x = f(u, v), \quad y = g(u, v), \tag{3.16}$$

taking each point (u, v) to a unique point (x, y), and vice versa. Rather than deal with this general case, however, it is easier to consider a particular example that highlights all relevant aspects of the analysis. Thus we consider the transformation between the Cartesian coordinates (x, y) and the plane polar coordinates (r, θ) where

$$x = r\cos\theta, \quad y = r\sin\theta, \quad r \geq 0,\ -\pi < \theta \leq \pi. \tag{3.17}$$

The inverse transformation is given by

$$r^2 = x^2 + y^2, \quad \cos\theta = \frac{x}{r}, \quad \sin\theta = \frac{y}{r}. \tag{3.18}$$

In Cartesian coordinates we normally choose x to be the independent variable, so points on the curve joining (a, A) to (b, B), depicted in Figure 3.1, are given by the Cartesian coordinates $(x, y(x))$.

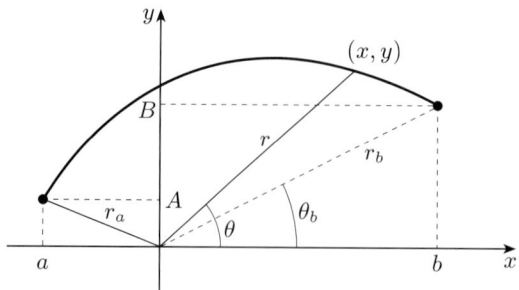

Figure 3.1 Diagram showing the relation between the Cartesian and polar representations of a curve joining (a, A) and (b, B).

Alternatively, we can define each point on the curve by expressing r as a function of θ, and then the curve is defined by the polar coordinates $(r(\theta), \theta)$. If the end points of the curve have the Cartesian coordinates (a, A) and (b, B), then equation (3.18) gives the polar coordinates of these points as

$$r_a = \sqrt{a^2 + A^2}, \quad \cos\theta_a = \frac{a}{r_a}, \quad \sin\theta_a = \frac{A}{r_a},$$
$$r_b = \sqrt{b^2 + B^2}, \quad \cos\theta_b = \frac{b}{r_b}, \quad \sin\theta_b = \frac{B}{r_b}. \tag{3.19}$$

The aim is to transform a functional

$$S[y] = \int_a^b dx\, F(x, y(x), y'(x)), \quad y(a) = A,\ y(b) = B, \tag{3.20}$$

to an integral over θ in which x, $y(x)$ and $y'(x)$ are replaced by expressions involving θ, $r(\theta)$ and $r'(\theta)$. First we change to the new independent variable θ: then since $x = r\cos\theta$ and $y = r\sin\theta$, we have

$$S = \int_{\theta_a}^{\theta_b} d\theta\, \frac{dx}{d\theta} F\left(r\cos\theta, r\sin\theta, y'(x)\right). \tag{3.21}$$

The derivative $dx/d\theta$ is obtained from the relation $x = r\cos\theta$ using the product rule and remembering that r depends upon θ:

$$\frac{dx}{d\theta} = \frac{dr}{d\theta}\cos\theta - r\sin\theta. \tag{3.22}$$

It remains only to express $y'(x)$ in terms of r, θ and $r'(\theta)$. First, note that

$$\frac{dy}{dx} = \frac{dy}{d\theta}\frac{d\theta}{dx} = \frac{dy}{d\theta}\bigg/\frac{dx}{d\theta}. \tag{3.23}$$

Now use the product rule again to give

$$\frac{dy}{d\theta} = \frac{dr}{d\theta}\sin\theta + r\cos\theta. \tag{3.24}$$

Hence

$$\frac{dy}{dx} = \frac{r'(\theta)\sin\theta + r\cos\theta}{r'(\theta)\cos\theta - r\sin\theta}, \tag{3.25}$$

and the functional becomes

$$S[r] = \int_{\theta_a}^{\theta_b} d\theta\, \mathcal{F}(\theta, r(\theta), r'(\theta)), \quad r(\theta_a) = r_a,\ r(\theta_b) = r_b, \tag{3.26}$$

where

$$\mathcal{F} = \left(r'(\theta)\cos\theta - r\sin\theta\right) F\left(r\cos\theta, r\sin\theta, \frac{r'(\theta)\sin\theta + r\cos\theta}{r'(\theta)\cos\theta - r\sin\theta}\right), \tag{3.27}$$

and where (r_a, θ_a) and (r_b, θ_b) are the polar coordinates of the end points of the curve.

3.2 Invariance of the Euler–Lagrange equation

The new functional depends only upon θ, $r(\theta)$ and the first derivative $r'(\theta)$, so the Euler–Lagrange equation is

$$\frac{d}{d\theta}\left(\frac{\partial \mathcal{F}}{\partial r'}\right) - \frac{\partial \mathcal{F}}{\partial r} = 0, \qquad (3.28)$$

which is the transformed version of

$$\frac{d}{dx}\left(\frac{\partial F}{\partial y'}\right) - \frac{\partial F}{\partial y} = 0. \qquad (3.29)$$

This analysis shows that the transformation to polar coordinates keeps the form of the Euler–Lagrange equation invariant because the transformation of the functional introduces only first-order derivatives, via equations (3.22) and (3.25), so does not alter the derivation of the Euler–Lagrange equation. The same transformation applied to equation (3.29) involves finding a suitable expression for the second derivative, d^2y/dx^2, which is harder.

As an example of the reverse transformation, consider the functional

$$S[r] = \int_{\theta_a}^{\theta_b} d\theta \sqrt{r^2 + r'(\theta)^2}, \qquad (3.30)$$

already expressed in polar coordinates. Here,

$$F(r, r') = \sqrt{r^2 + r'^2}, \qquad (3.31)$$

so

$$\partial F/\partial r = r/\sqrt{r^2 + r'^2} \quad \text{and} \quad \partial F/\partial r' = r'/\sqrt{r^2 + r'^2}. \qquad (3.32)$$

The independent variable is θ, so the Euler–Lagrange equation is

$$\frac{d}{d\theta}\left(\frac{r'}{\sqrt{r^2 + r'^2}}\right) - \frac{r}{\sqrt{r^2 + r'^2}} = 0. \qquad (3.33)$$

Expanding this gives

$$\frac{r''}{\sqrt{r^2 + r'^2}} - \frac{r'(rr' + r'r'')}{(r^2 + r'^2)^{3/2}} - \frac{r}{\sqrt{r^2 + r'^2}} = 0, \qquad (3.34)$$

which, on multiplying through by $\left(r^2 + r'^2\right)^{3/2}$, simplifies to

$$r\frac{d^2r}{d\theta^2} - 2\left(\frac{dr}{d\theta}\right)^2 - r^2 = 0. \qquad (3.35)$$

In order to transform the functional to Cartesian coordinates we need the derivative $d\theta/dx$, which can be obtained by differentiating $\tan\theta = y/x$ with respect to x:

$$\frac{1}{\cos^2\theta}\frac{d\theta}{dx} = \frac{y'(x)}{x} - \frac{y}{x^2}, \qquad (3.36)$$

giving

$$\frac{d\theta}{dx} = \frac{xy' - y}{x^2 + y^2}, \qquad (3.37)$$

since $1/\cos^2\theta = 1 + \tan^2\theta = (x^2 + y^2)/x^2$. Also, $r^2 = x^2 + y^2$, and differentiation with respect to x gives

$$r\frac{dr}{d\theta}\frac{d\theta}{dx} = x + yy', \qquad (3.38)$$

and hence

$$\frac{dr}{d\theta} = \frac{(yy' + x)r}{xy' - y}. \qquad (3.39)$$

Thus the functional becomes

$$S[y] = \int_a^b dx\, \frac{d\theta}{dx} \sqrt{r^2 + r'(\theta)^2}$$

$$= \int_a^b dx\, \frac{xy' - y}{r^2} \sqrt{r^2 + \left(\frac{yy' + x}{xy' - y}\right)^2 r^2}$$

$$= \int_a^b dx\, \frac{1}{r} \frac{xy' - y}{|xy' - y|} \sqrt{(xy' - y)^2 + (yy' + x)^2}. \tag{3.40}$$

Now assume that $xy' - y \neq 0$, a condition we shall discuss a little later, so

$$\frac{xy' - y}{|xy' - y|} = \pm 1, \tag{3.41}$$

and remember that the Euler–Lagrange equation is unaffected if the functional is multiplied by a constant factor. Also, we have

$$(xy' - y)^2 + (yy' + x)^2 = (x^2 + y^2) y'^2 + (x^2 + y^2)$$
$$= r^2 (1 + y'^2). \tag{3.42}$$

Hence, the functional is equivalent to

$$S[y] = \int_a^b dx\, \sqrt{1 + y'(x)^2}. \tag{3.43}$$

This is the functional treated in Subsection 1.2.2, where it was shown (page 10) that the Euler–Lagrange equation reduces to $y'(x) = \text{constant}$, with general solution $y = mx + c$. On this solution the condition $xy' - y \neq 0$ becomes $c \neq 0$, which means that the path must not pass through the origin. The reason why this condition has occurred here is simply because the transformation to polar coordinates, $x = r\cos\theta$, $y = r\sin\theta$, is not well behaved at the origin which is represented by $r = 0$, for *all* θ, so here the transformation is not invertible. It is worth noting that the equation $y''(x) = 0$, when expressed in polar coordinates, is just equation (3.35).

This example illustrates the simplification that can occur when suitable transformations are made: the art is to find such transformations and frequently one is guided by geometric insight, but we do not expect you to master this art here.

Exercise 3.3

(a) Show that, in polar coordinates, the functional

$$S[y] = \int_a^b dx\, \sqrt{x^2 + y^2} \sqrt{1 + y'^2} \quad \text{becomes} \quad S[r] = \int_{\theta_a}^{\theta_b} d\theta\, r \sqrt{r^2 + r'^2},$$

and that the resulting Euler–Lagrange equation is

$$\frac{d^2 r}{d\theta^2} - \frac{3}{r} \left(\frac{dr}{d\theta}\right)^2 - 2r = 0.$$

Note: you will need to assume that $r'(\theta)\cos\theta - r(\theta)\sin\theta \neq 0$, which is equivalent to assuming that $|y'(x)|$ is bounded.

(b) Show that this equation may be written in the form

$$\frac{d^2}{d\theta^2}\left(\frac{1}{r^2}\right) + \frac{4}{r^2} = 0,$$

and hence that the equations for the stationary paths are

$$\frac{1}{r^2} = A\cos 2\theta + B\sin 2\theta, \quad \text{that is} \quad A(x^2 - y^2) + 2Bxy = 1,$$

where A and B are constants. The following identities are useful:
$$\cos 2\theta = \cos^2\theta - \sin^2\theta \quad \text{and} \quad \sin 2\theta = 2\sin\theta\cos\theta.$$

Exercise 3.4

A simple transformation is obtained by interchanging the roles of the dependent and independent variables. For instance, consider the elementary functional
$$S[y] = \int_a^b dx\, F(y'), \quad y(a) = A,\ y(b) = B,$$
which depends only upon y', and was shown in Chapter 1 to be stationary along the straight line joining the points (a, A) and (b, B).

Using the fact that $dy/dx = \dfrac{1}{dx/dy}$, show that if y becomes the independent variable, the functional becomes
$$S[x] = \int_A^B dy\, G(x'), \quad x(A) = a,\ y(B) = b,$$
where $x' = dx/dy$ and $G(u) = u\, F(1/u)$.

3.3 Functionals containing many dependent variables

In Chapters 1 and 2 we considered functionals of the type
$$S[y] = \int_a^b dx\, F(x, y, y'), \quad y(a) = A,\ y(b) = B, \tag{3.44}$$
which involve one independent variable, x, a single dependent variable, $y(x)$, and its first derivative. There are many useful and important extensions to this type of functional and here we discuss the generalisation needed for the re-formulation of Newtonian dynamics described in the next chapter: others will not be considered but, in order to provide some idea of other possible developments, we list a few of the more important examples.

(a) The integrand of the functional (3.44) depends upon the independent variable x and a single dependent variable $y(x)$, which is determined by the requirement that $S[y]$ be stationary. A simple generalisation is to integrands that depend upon several dependent variables $y_k(x)$, $k = 1, 2, \ldots, n$, and their first derivatives. This type of functional is required in the next chapter and the necessary theory is described here.

(b) The integrand of (3.44) depends upon $y(x)$ and its first derivative. Another generalisation involves functionals depending upon second or higher derivatives. Some examples of this type are treated in Exercises 2.32, 2.33 and 3.33, though we do not consider this type of problem any further.

(c) The integral defining the functional may be over a surface rather than along a line,
$$J[y] = \iint_{\mathcal{D}} dx_1\, dx_2\, F\left(x_1, x_2, y, \frac{\partial y}{\partial x_1}, \frac{\partial y}{\partial x_2}\right), \tag{3.45}$$
where $\mathcal{D}$ is a region in the (x_1, x_2)-plane, so the functional depends upon two independent variables, x_1 and x_2, rather than just one. Boundary

conditions are, of course, also needed to specify a unique stationary function. In this case the Euler–Lagrange equation becomes a partial differential equation. Many of the standard equations of mathematical physics can be derived from such functionals – the wave equation being one important example and this is considered briefly in the last, optional, section of Chapter 4 (page 190). There is also a natural extension to integrals over higher-dimensional spaces, but we do not consider these types of problems.

(d) Broken extremals: a broken extremal is a continuous solution of the Euler–Lagrange equations with a first derivative which is discontinuous at a finite number of points. That such solutions are important is clear from the observation of soap bubbles: when two or more bubbles form a composite shape this usually comprises spherical segments such that across any common boundary the normal to the surface changes direction discontinuously. A simple example of such a solution is the Goldschmidt curve defined by equation (2.59) (page 57); see also Exercise 2.29.

3.3.1 Two dependent variables

In this Subsection we find the necessary conditions for a functional depending on two functions to be stationary; however, because we are ultimately interested in functionals depending upon any finite number of variables, we shall often use the notation for which this further generalisation becomes almost automatic. Most examples of functionals with two or more dependent variables arise in the study of dynamics and some will be introduced in the next chapter – here we consider only the general theory.

We shall find necessary conditions for a functional depending upon two functions, $y_1(x)$ and $y_2(x)$, and a single independent variable x to be stationary. The functional is

$$S[y_1, y_2] = \int_a^b dx\, F\left(x, y_1, y_2, y_1', y_2'\right), \qquad (3.46)$$

where each function satisfies the boundary conditions

$$y_k(a) = A_k, \quad y_k(b) = B_k, \quad k = 1, 2. \qquad (3.47)$$

We require functions $y_1(x)$ and $y_2(x)$ that make this functional stationary and proceed in the same manner as before. Let $y_1(x)$ and $y_2(x)$ be two admissible functions – that is, functions having continuous first derivatives and satisfying the boundary conditions – and consider the difference

$$\delta = S[y_1 + \epsilon g_1, y_2 + \epsilon g_2] - S[y_1, y_2], \quad |\epsilon| \ll 1, \qquad (3.48)$$

where $y_k(x) + \epsilon g_k(x)$, $k = 1, 2$ are also admissible functions; as in Chapter 2 (page 45) this means that $g_k(a) = g_k(b) = 0$, $k = 1, 2$. The analysis starts by finding the first term in the expansion of δ as a power series in ϵ, and for this we need first to expand the integrand,

$$F\left(x, y_1 + \epsilon g_1, y_2 + \epsilon g_2, y_1' + \epsilon g_1', y_2' + \epsilon g_2'\right) - F\left(x, y_1, y_2, y_1', y_2'\right)$$
$$= \epsilon \left(g_1 \frac{\partial F}{\partial y_1} + g_1' \frac{\partial F}{\partial y_1'} + g_2 \frac{\partial F}{\partial y_2} + g_2' \frac{\partial F}{\partial y_2'} \right) + O(\epsilon^2), \qquad (3.49)$$

where all functions on the right-hand side are evaluated at $\epsilon = 0$. Hence, on writing δ in the form $\delta = \epsilon \Delta S + O(\epsilon^2)$, we find that

$$\Delta S = \int_a^b dx \left(\frac{\partial F}{\partial y_1} g_1 + \frac{\partial F}{\partial y_1'} g_1' + \frac{\partial F}{\partial y_2} g_2 + \frac{\partial F}{\partial y_2'} g_2' \right), \qquad (3.50)$$

3.3 Functionals containing many dependent variables

with the integrand evaluated at $\epsilon = 0$. Notice that ΔS is the appropriate generalisation of the Gâteaux differential defined in equation (2.14) (page 45), that is,

$$\Delta S = \frac{d}{d\epsilon} S[y_1 + \epsilon g_1, y_2 + \epsilon g_2]\bigg|_{\epsilon=0}. \qquad (3.51)$$

The expression (3.50) for ΔS is simplified when integrating by parts those terms involving $g_1'(x)$ and $g_2'(x)$: thus, for the first such term,

$$\int_a^b dx \, \frac{\partial F}{\partial y_1'} g_1' = \left[g_1 \frac{\partial F}{\partial y_1'} \right]_a^b - \int_a^b dx \, g_1 \frac{d}{dx}\left(\frac{\partial F}{\partial y_1'}\right), \qquad (3.52)$$

with a similar expression for the term involving $g_2'(x)$. Since $g_k(a) = g_k(b) = 0$, $k = 1, 2$, the expression for ΔS becomes

$$\Delta S = -\int_a^b dx \, g_1 \left[\frac{d}{dx}\left(\frac{\partial F}{\partial y_1'}\right) - \frac{\partial F}{\partial y_1} \right]$$
$$\quad -\int_a^b dx \, g_2 \left[\frac{d}{dx}\left(\frac{\partial F}{\partial y_2'}\right) - \frac{\partial F}{\partial y_2} \right]. \qquad (3.53)$$

For a stationary path we need, by definition (Chapter 2, page 45), $\Delta S = 0$ for *all* $g_1(x)$ and $g_2(x)$. On setting $g_2(x) = 0$ the above equation becomes the same as equation (2.30) (page 48) with y and g replaced by y_1 and g_1 respectively. Hence we may use the fundamental lemma of the calculus of variations, Section 2.3, to obtain the second-order differential equation

$$\frac{d}{dx}\left(\frac{\partial F}{\partial y_1'}\right) - \frac{\partial F}{\partial y_1} = 0, \quad y_1(a) = A_1, \; y_1(b) = B_1. \qquad (3.54)$$

This equation looks the same as equation (2.31) (page 48), but remember that here F also depends upon the unknown function $y_2(x)$.

Similarly, by setting $g_1(x) = 0$, we obtain another second-order equation

$$\frac{d}{dx}\left(\frac{\partial F}{\partial y_2'}\right) - \frac{\partial F}{\partial y_2} = 0, \quad y_2(a) = A_2, \; y_2(b) = B_2. \qquad (3.55)$$

Equations (3.54) and (3.55) are the Euler–Lagrange equations for the functional. These two equations will normally involve both $y_1(x)$ and $y_2(x)$, so are named *coupled* differential equations; this usually makes them far harder to solve than the Euler–Lagrange equations of Chapter 2, which contained only one dependent variable.

An example should help to make this clear. Consider the functional

$$S[y_1, y_2] = \int_0^{\pi/2} dx \, \left(y_1'^2 + y_2'^2 + 2 y_1 y_2 \right) \qquad (3.56)$$

so that

$$\frac{\partial F}{\partial y_1'} = 2y_1' \quad \text{and} \quad \frac{\partial F}{\partial y_1} = 2y_2. \qquad (3.57)$$

Equation (3.54) becomes

$$\frac{d^2 y_1}{dx^2} - y_2 = 0, \qquad (3.58)$$

which involves both $y_1(x)$ and $y_2(x)$. Also,

$$\frac{\partial F}{\partial y_2'} = 2y_2' \quad \text{and} \quad \frac{\partial F}{\partial y_2} = 2y_1, \qquad (3.59)$$

and equation (3.55) becomes

$$\frac{d^2 y_2}{dx^2} - y_1 = 0, \qquad (3.60)$$

which also involves both $y_1(x)$ and $y_2(x)$.

Equations (3.58) and (3.60) now have to be solved. Coupled differential equations are normally very difficult to solve and their solutions can behave in bizarre ways, including chaotically; but these equations are linear which makes the task of solving them much easier. Also, solutions of linear equations are generally better behaved. One method is to use the first equation to write $y_2 = y_1''$, so the second equation becomes the fourth-order linear equation

$$\frac{d^4 y_1}{dx^4} - y_1 = 0. \tag{3.61}$$

This equation was treated in Block 0, example 1.28, so here we provide a mere outline of the method of solution. Suppose that the solution has the form $y_1 = \alpha \exp(\lambda x)$, where α and λ are constants; substituting this into the equation gives $\lambda^4 = 1$, showing that there are four solutions obtained by setting $\lambda = \pm 1, \pm i$: the general solution is a linear combination of these functions,

$$y_1(x) = A\cos x + B\sin x + C\cosh x + D\sinh x, \tag{3.62}$$

where A, B, C and D are constants. Since $y_2 = y_1''$, we also have

$$y_2(x) = -A\cos x - B\sin x + C\cosh x + D\sinh x. \tag{3.63}$$

This general solution needs to satisfy the boundary conditions of the problem: if a solution exists then the values of these constants are determined from the boundary conditions, as demonstrated in Exercise 3.5, below. It should be noted, however, that a solution may not exist, as shown in Exercise 3.6.

Exercise 3.5

Find the values of the constants A, B, C and D if the functional (3.56) has the following boundary conditions

$$y_1(0) = 0, \quad y_1\left(\tfrac{\pi}{2}\right) = 1, \quad y_2(0) = 0, \quad y_2\left(\tfrac{\pi}{2}\right) = -1.$$

Exercise 3.6

Consider the functional defined in equation (3.56), but with x in the range $0 \le x \le \pi$. Show that if the boundary conditions are $y_1(0) = 0$, $y_1(\pi) = \alpha$, $y_2(0) = 0$, $y_2(\pi) = \beta$, then no solution exists unless $\alpha = \beta$, and then there are infinitely many solutions.

Exercise 3.7

Show that the Euler–Lagrange equations for the functional

$$S[y_1, y_2] = \int_0^1 dx \, \left(y_1'^2 + y_2'^2 + y_1' y_2'\right),$$

with the boundary conditions

$$y_1(0) = 0, \quad y_1(1) = 1, \quad y_2(0) = 1, \quad y_2(1) = 2,$$

integrates to

$$2y_1' + y_2' = a_1 \quad \text{and} \quad 2y_2' + y_1' = a_2,$$

where a_1 and a_2 are constants. Deduce that the stationary path is given by the equations

$$y_1(x) = x \quad \text{and} \quad y_2(x) = x + 1.$$

3.3 Functionals containing many dependent variables

Exercise 3.8

By defining a new variable $z_1 = y_1 + y_2/2$, show that the functional defined in the previous exercise becomes

$$S[z_1, y_2] = \int_0^1 dx \left(z_1'^2 + \tfrac{3}{4} y_2'^2 \right),$$

with boundary conditions

$$z_1(0) = \tfrac{1}{2}, \quad z_1(1) = 2, \quad y_2(0) = 1, \quad y_2(1) = 2,$$

and show that the corresponding Euler–Lagrange equations are

$$\frac{d^2 z_1}{dx^2} = 0 \quad \text{and} \quad \frac{d^2 y_2}{dx^2} = 0.$$

Solve these equations to derive the solution obtained in the previous exercise.

Note that by using the variables (z_1, y_2) in the above exercise each of the new Euler–Lagrange equations depends upon only one of the dependent variables and is therefore far easier to solve. Such systems of equations are said to be *uncoupled*, and one of the main methods of solving coupled Euler–Lagrange equations is to find a transformation that converts them to uncoupled equations. In real problems, finding such transformations is difficult and often relies upon understanding the symmetries of the problem; then the methods described in Section 3.2 and Subsection 3.5.2 can be useful.

Exercise 3.9

Evaluate the expression defined in equation (3.51) to derive equation (3.50).

3.3.2 Many dependent variables

The extension of the above analysis to functionals involving any number, n, of dependent variables, their first derivatives and a single independent variable is straightforward. It is helpful, however, to introduce the notation $\boldsymbol{y}(x) = (y_1(x), y_2(x), \ldots, y_n(x))$ to denote the set of n dependent variables; generally, we use bold characters to denote sets of variables. There is still only one independent variable, so the functional is

$$S[\boldsymbol{y}] = \int_a^b dx\, F(x, \boldsymbol{y}, \boldsymbol{y}'), \quad \boldsymbol{y}(a) = \boldsymbol{A}, \quad \boldsymbol{y}(b) = \boldsymbol{B}, \tag{3.64}$$

where

$$\boldsymbol{y}' = (y_1'(x), y_2'(x), \ldots, y_n'(x)),$$
$$\boldsymbol{A} = (A_1, A_2, \ldots, A_n) \text{ and } \boldsymbol{B} = (B_1, B_2, \ldots, B_n). \tag{3.65}$$

If $\boldsymbol{y}(x)$ and $\boldsymbol{y}(x) + \epsilon \boldsymbol{g}(x)$ are admissible functions, so that $\boldsymbol{g}(a) = \boldsymbol{g}(b) = 0$, we again consider the difference

$$\delta = S[\boldsymbol{y} + \epsilon \boldsymbol{g}] - S[\boldsymbol{y}] = \epsilon \Delta S[\boldsymbol{y}, \boldsymbol{g}] + O(\epsilon^2). \tag{3.66}$$

The Gâteaux derivative, ΔS, is given by the relation

$$\Delta S = \frac{d}{d\epsilon} S[\boldsymbol{y} + \epsilon \boldsymbol{g}] \bigg|_{\epsilon=0}$$
$$= \int_a^b dx\, \frac{d}{d\epsilon} F(x, \boldsymbol{y} + \epsilon \boldsymbol{g}, \boldsymbol{y}' + \epsilon \boldsymbol{g}') \bigg|_{\epsilon=0}, \tag{3.67}$$

and for $\boldsymbol{y}$ to be a stationary path, this must be zero for all suitable $\boldsymbol{g}(x)$. Using the chain rule we have

$$\frac{d}{d\epsilon}F(x, \boldsymbol{y} + \epsilon \boldsymbol{g}, \boldsymbol{y}' + \epsilon \boldsymbol{g}')\bigg|_{\epsilon=0} = \sum_{k=1}^{n}\left(g_k \frac{\partial F}{\partial y_k} + g'_k \frac{\partial F}{\partial y'_k}\right), \tag{3.68}$$

hence

$$\Delta S = \sum_{k=1}^{n} \int_a^b dx \left(g_k \frac{\partial F}{\partial y_k} + g'_k \frac{\partial F}{\partial y'_k}\right). \tag{3.69}$$

Now integrate by parts to cast this in the form

$$\Delta S = \sum_{k=1}^{n} \left[g_k \frac{\partial F}{\partial y'_k}\right]_a^b - \sum_{k=1}^{n} \int_a^b dx \left(\frac{d}{dx}\left(\frac{\partial F}{\partial y'_k}\right) - \frac{\partial F}{\partial y_k}\right) g_k. \tag{3.70}$$

But, since $\boldsymbol{g}(a) = \boldsymbol{g}(b) = 0$, the boundary term vanishes. Further, since $\Delta S = 0$ for *all* suitable $\boldsymbol{g}(x)$, by the same reasoning used when $n = 2$, we obtain the set of n coupled equations

$$\frac{d}{dx}\left(\frac{\partial F}{\partial y'_k}\right) - \frac{\partial F}{\partial y_k} = 0, \quad y_k(a) = A_k, \ y_k(b) = B_k, \ k = 1, 2, \ldots, n. \tag{3.71}$$

This set of n coupled equations is usually nonlinear and difficult to solve. There is, however, one circumstance when the solution is relatively simple; this is when the integrand of the functional $S[\boldsymbol{y}]$ is a quadratic form in both $\boldsymbol{y}$ and $\boldsymbol{y}'$, as in equation (3.72) in the following exercise, which is an important example because it describes small oscillations about an equilibrium position of an n-dimensional dynamical system.

Exercise 3.10

(a) If A and B are real, symmetric, positive definite, $n \times n$ matrices so $A_{ij} = A_{ji}$ and $B_{ij} = B_{ji}$ for $1 \le i, j \le n$, consider the functional

$$S[\boldsymbol{y}] = \int_a^b dx \sum_{i=1}^{n}\sum_{j=1}^{n}\left(y'_i A_{ij} y'_j - y_i B_{ij} y_j\right), \tag{3.72}$$

with the integrand being quadratic in $\boldsymbol{y}$ and $\boldsymbol{y}'$. Show that the n Euler–Lagrange equations are the set of coupled linear equations

$$\sum_{j=1}^{n}\left(A_{kj}\frac{d^2 y_j}{dx^2} + B_{kj} y_j\right) = 0, \quad 1 \le k \le n.$$

(b) Show that if we interpret $\boldsymbol{y}$ as an n-dimensional column vector and its transpose $\boldsymbol{y}^\top$ as a row vector, the functional can be written in the equivalent matrix form

$$S[\boldsymbol{y}] = \int_a^b dx \left(\boldsymbol{y}'^\top A \boldsymbol{y}' - \boldsymbol{y}^\top B \boldsymbol{y}\right),$$

and the Euler–Lagrange equations can be written in the matrix form

$$A\frac{d^2 \boldsymbol{y}}{dx^2} + B\boldsymbol{y} = 0$$

and that, on multiplication by A^{-1}, this can also be written in the form

$$\frac{d^2 \boldsymbol{y}}{dx^2} + A^{-1} B \boldsymbol{y} = 0. \tag{3.73}$$

This exercise is optional and requires knowledge of matrices.

It can be shown that a real symmetric matrix has real eigenvalues: a positive definite matrix is any matrix having only positive eigenvalues.

Whilst not part of this course, we now very briefly outline the theory which shows how the solutions of this equation can be found and how these solutions behave. The matrices A and B have, by definition, only positive eigenvalues, it can therefore be shown that $A^{-1}B$ has non-negative eigenvalues ω_k^2, $k = 1, 2, \ldots, n$ and that there are a set of n orthogonal eigenvectors $\boldsymbol{z}_k$,

$k=1,2,\ldots,n$, possibly complex, which diagonalise $A^{-1}B$. Then if we express $\boldsymbol{y}$ as a linear combination of the $\boldsymbol{z}_k$,

$$\boldsymbol{y} = \sum_{k=1}^{n} a_k(x)\boldsymbol{z}_k, \tag{3.74}$$

it can be shown that

$$\frac{d^2 a_j}{dx^2} + \omega_j^2 a_j = 0, \quad j = 1, 2, \ldots, n. \tag{3.75}$$

These equations for $a_j(x)$ are trivially solved and the constants of integration determined by the initial value of $\boldsymbol{y}$. Even without solving these equations, it is seen that in general the vector $\boldsymbol{y}$ comprises a sum of n periodic components with the frequencies ω_j, $j = 1, 2, \ldots, n$. We emphasise that these results are not an assessed part of the course.

3.4 Changing dependent variables

In this section we consider the effect of changing the dependent variables. A simple example of such a transformation was dealt with in Exercise 3.8, where it was shown how a linear transformation uncoupled the Euler–Lagrange equations. In general the aim of changing variables is to simplify the Euler–Lagrange equations, but it is usually easier to make the transformation to the functional rather than to the Euler–Lagrange equations. The main use of this method is in Newtonian dynamics.

Before explaining the general theory we deal with a specific example, which highlights all salient points. The functional is

$$S[y_1, y_2] = \int_a^b dx \left[\tfrac{1}{2}\left(y_1'^2 + y_2'^2\right) - V(r)\right], \quad r = \sqrt{y_1^2 + y_2^2}, \tag{3.76}$$

where $V(r)$ is any suitable function. This functional occurs frequently because, as will be seen in the next chapter, it arises when describing the motion of a particle moving in a force depending only on the distance from a fixed point. It is special because it depends only upon the combinations $y_1'^2 + y_2'^2$ and $y_1^2 + y_2^2$, which suggests that changing to polar coordinates may lead to simplification.

First, however, we need to be clear about the difference between the present example and that of Section 3.2.2, where polar coordinates were also used. Now we have two dependent variables, y_1 and y_2, representing the Cartesian coordinates of the particle's position, as shown in Figure 3.2; we shall use polar coordinates, r and θ, to replace these coordinates, but the independent variable, x, is unchanged. In Section 3.2.2 we had *one* dependent variable, y, and the polar coordinates were used to change both independent and dependent variables.

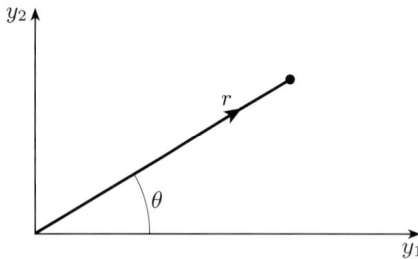

Figure 3.2 Diagram showing the relation between the Cartesian and polar coordinates (y_1, y_2) and (r, θ), respectively

The relations between (y_1, y_2) and (r, θ) are $y_1 = r\cos\theta$ and $y_2 = r\sin\theta$, so that $y_1^2 + y_2^2 = r^2$ and, on using the product rule,

$$\frac{dy_1}{dx} = \frac{dr}{dx}\cos\theta - r\frac{d\theta}{dx}\sin\theta \quad \text{and} \quad \frac{dy_2}{dx} = \frac{dr}{dx}\sin\theta + r\frac{d\theta}{dx}\cos\theta. \qquad (3.77)$$

Squaring and adding these equations gives

$$y_1'^2 + y_2'^2 = r'^2 + r^2\theta'^2. \qquad (3.78)$$

Hence the functional becomes

$$S[r, \theta] = \int_a^b dx\, \left[\tfrac{1}{2}r'^2 + \tfrac{1}{2}r^2\theta'^2 - V(r)\right]. \qquad (3.79)$$

Exercise 3.11

(a) Show that the Euler–Lagrange equations for the functional (3.76) are

$$\frac{d^2 y_1}{dx^2} + V'(r)\frac{y_1}{r} = 0 \quad \text{and} \quad \frac{d^2 y_2}{dx^2} + V'(r)\frac{y_2}{r} = 0. \qquad (3.80)$$

(b) Show that the Euler–Lagrange equations for the functional (3.79) can be written in the form

$$\frac{d^2 r}{dx^2} - \frac{L^2}{r^3} + V'(r) = 0 \quad \text{and} \quad \frac{d\theta}{dx} = \frac{L}{r^2}, \qquad (3.81)$$

where L is a constant. Note that the equation for r does not depend upon θ which is obtained from $r(x)$ by a single integration, so these equations are easier to solve. This exercise shows how useful changing the dependent variables can be, provided the correct choice is made.

Exercise 3.12

This is an algebraically complicated problem which aims to show directly that it is easier to transform the functional rather than directly attack the Euler–Lagrange equations. You may omit this exercise, but it is instructive and will help you understand why the theory of the next chapter is so useful.

(a) By finding expressions for y_1'' and y_2'' in terms of the derivatives of r and θ, show that equations (3.80) become

$$\left(r'' - r\theta'^2\right)\cos\theta - (r\theta'' + 2r'\theta')\sin\theta + V'(r)\cos\theta = 0,$$
$$\left(r'' - r\theta'^2\right)\sin\theta + (r\theta'' + 2r'\theta')\cos\theta + V'(r)\sin\theta = 0.$$

Hint: multiply the first equation by $\cos\theta$, the second by $\sin\theta$, add the results and use the identity $\cos^2\theta + \sin^2\theta = 1$. The second equation is obtained by using a similar trick.

(b) By rearranging these equations, show that

$$r'' - r\theta'^2 + V'(r) = 0 \quad \text{and} \quad \frac{d}{dx}\left(r^2\theta'\right) = 0.$$

3.4 Changing dependent variables

The general theory is not much more complicated that these examples. Suppose that $\boldsymbol{y} = (y_1, y_2, \ldots, y_n)$ and $\boldsymbol{z} = (z_1, z_2, \ldots, z_n)$ are two sets of dependent variables related by the equations

$$y_k = \psi_k(\boldsymbol{z}), \quad k = 1, 2, \ldots, n, \tag{3.82}$$

where we assume, in order to slightly simplify the analysis, that each of the ψ_k is not explicitly dependent upon x. The chain rule gives

$$\frac{dy_k}{dx} = \sum_{i=1}^{n} \frac{\partial \psi_k}{\partial z_i} \frac{dz_i}{dx}, \tag{3.83}$$

showing that each of the y'_k depends linearly upon the z'_i. It is necessary to assume that the transformation between the derivatives is invertible and it may be shown that the condition for this is that the determinant of the matrix with elements $\partial \psi_k / \partial z_i$ is non-zero at all points of the region, which is also the condition for the transformation between $\boldsymbol{y}$ and $\boldsymbol{z}$ to be invertible.

Under this transformation, the functional

$$S[\boldsymbol{y}] = \int_a^b dx\, F(x, \boldsymbol{y}, \boldsymbol{y}') \quad \text{becomes} \quad S[\boldsymbol{z}] = \int_a^b dx\, G(x, \boldsymbol{z}, \boldsymbol{z}'), \tag{3.84}$$

where

$$G(x, \boldsymbol{z}, \boldsymbol{z}') =$$
$$F\left(x, \psi_1(\boldsymbol{z}), \psi_2(\boldsymbol{z}), \ldots, \psi_n(\boldsymbol{z}), \sum_{i=1}^{n} \frac{\partial \psi_1}{\partial z_i} z'_i, \ldots, \sum_{i=1}^{n} \frac{\partial \psi_n}{\partial z_i} z'_i\right); \tag{3.85}$$

that is, $G(x, \boldsymbol{z}, \boldsymbol{z}')$ is obtained from $F(x, \boldsymbol{y}, \boldsymbol{y}')$ simply by replacing $\boldsymbol{y}$ and $\boldsymbol{y}'$. In practice, of course, the transformation (3.82) is chosen to ensure that $G(x, \boldsymbol{z}, \boldsymbol{z}')$ is simpler than $F(x, \boldsymbol{y}, \boldsymbol{y}')$.

Exercise 3.13

Show that under the transformation to cylindrical polar coordinates,

$$y_1 = \rho \cos \phi, \quad y_2 = \rho \sin \phi, \quad y_3 = z,$$

the functional

$$S[y_1, y_2, y_3] = \int_a^b dx\, \left[\tfrac{1}{2}\left(y_1'^{\,2} + y_2'^{\,2} + y_3'^{\,2}\right) - V(\rho)\right], \quad \rho = \sqrt{y_1^2 + y_2^2},$$

becomes

$$S[\rho, \phi, z] = \int_a^b dx\, \left[\tfrac{1}{2}\left(\rho'^{\,2} + \rho^2 \phi'^{\,2} + z'^{\,2}\right) - V(\rho)\right].$$

Find the Euler–Lagrange equations and show that those for ρ and z are uncoupled.

3.5 Symmetries (Optional)

In previous examples, for instance the minimal surface area of Section 2.5, and the brachistochrone of Section 2.7, we solved Euler's equation by using the fact that if the integrand, $F(y, y')$ does not depend explicitly upon x, that is, $\partial F/\partial x = 0$ then

$$\frac{d}{dx}\left(y'\frac{\partial F}{\partial y'} - F\right) = y'(x)\left(\frac{d}{dx}\left(\frac{\partial F}{\partial y'}\right) - \frac{\partial F}{\partial y}\right). \tag{3.86}$$

This result is derived in Exercise 2.6 (page 51) and is important because it reduces the second-order Euler equation to the simpler first-order equation

$$y'\frac{\partial F}{\partial y'} - F = \text{constant}. \tag{3.87}$$

The proof of this result given in the solution to Exercise 2.6 is algebraic and relied only on the fact that $\partial F/\partial x = 0$. In the following optional subsection we re-derive this result using the equivalent but more fundamental notion that the integrand $F(y, y')$ is invariant under translations in x: this is a more fruitful method because it is more readily generalised to other types of transformations; for instance in three-dimensional problems the integrand of the functional may be invariant under all rotations, or just rotations about a given axis. The general theory is described in Section 3.5.2.

The algebra of the following analysis is fairly complicated and requires careful thought at each stage: this material cannot be read and understood quickly and for this reason, this section is optional. However, we encourage you to attempt to understand it because this theory is important for the understanding of advanced dynamics and also, it is a very elegant piece of mathematics. This material also illustrates the important role that symmetries play in simplifying the mathematical description of nature, as was noted in Block I, Chapter 5, in the discussion of vibrating membranes.

3.5.1 Invariance under translations

We develop the general method by applying it to the example treated in Exercise 2.6 (page 51), in which there is a single dependent variable, x and where the integrand does not depend explicitly upon x, that is $\partial F/\partial x = 0$, so can be written as

$$S[y] = \int_a^b dx\, F(y, y'), \quad y(a) = A,\ y(b) = B. \tag{3.88}$$

The admissible functions, $y(x)$, describe curves C, between the points (a, A) and (b, B), in the two-dimensional space with axes Oxy, so that a point, P, on the curve has coordinates $(x, y(x))$, as shown in Figure 3.3.

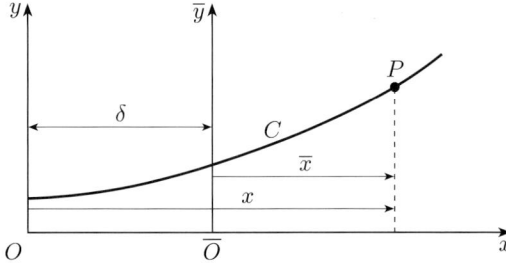

Figure 3.3 Diagram showing the two coordinate systems Oxy and $\overline{O}\overline{x}\overline{y}$, connected by a translation along the x-axis by a distance δ.

3.5 Symmetries (Optional)

Another coordinate system could be $\overline{O}\,\overline{x}\,\overline{y}$, where $\overline{x} = x - \delta$ and $\overline{y} = y$, with the origin, $\overline{O}$, of this system at $x = \delta$, $y = 0$ in the original coordinate system, that is a translation a distance δ along the x-axis. In this coordinate system the curve C has the equation $\overline{y} = \overline{y}(\overline{x})$, so the coordinates of a point P are $(\overline{x}, \overline{y}(\overline{x}))$ and these are related to coordinates in Oxy, $(x, y(x))$, by

$$\overline{x} = x - \delta \quad \text{and} \quad \overline{y}(\overline{x}) = y(x) \quad \text{or} \quad \overline{y}(\overline{x}) = y(\overline{x} + \delta), \tag{3.89}$$

the latter equation defining the function $\overline{y}$; differentiation, using the chain rule, gives

$$\frac{d\overline{y}}{d\overline{x}} = \frac{dy}{dx}\frac{dx}{d\overline{x}} = \frac{dy}{dx}, \quad \text{that is,} \quad \overline{y}'(\overline{x}) = y'(x). \tag{3.90}$$

The functional (3.88) can be computed in either coordinate system and, for reasons that will soon become apparent, we consider the integral, $T[\overline{y}]$, in the $\overline{O}\,\overline{x}\,\overline{y}$ representation over a *limited*, but arbitrary, range $\overline{c} \leq \overline{x} \leq \overline{d}$,

$$T[\overline{y}] = \int_{\overline{c}}^{\overline{d}} du\, F(\overline{y}(u), \overline{y}'(u)) \quad \text{where} \quad \overline{y}'(u) = \frac{d\overline{y}}{du}, \tag{3.91}$$

$\overline{c} = c - \delta$, $\overline{d} = d - \delta$ and $a < c < d < b$. The integrand of T depends on u only through the function $\overline{y}(u)$: at each value of u the integrand has the same value as the integrand of $S[y]$ at the equivalent point, $x = u + \delta$.

Now follows the crucial step: because F does not depend explicitly upon the independent variable it follows that

$$\int_{\overline{c}}^{\overline{d}} du\, F(\overline{y}(u), \overline{y}'(u)) = \int_{c}^{d} dx\, F(y(x), y'(x)), \quad \text{where } x = u + \delta, \tag{3.92}$$

and this is true for *all* δ.

Now consider small values of δ and expand to $O(\delta)$, first writing the integral in the form

$$T = \int_{c-\delta}^{d-\delta} du\, F(\overline{y}(u), \overline{y}'(u))$$

$$= \int_{c}^{d} du\, F(\overline{y}, \overline{y}') + \int_{c-\delta}^{c} du\, F(\overline{y}, \overline{y}') - \int_{d-\delta}^{d} du\, F(\overline{y}, \overline{y}'). \tag{3.93}$$

This expression can be expanded to first-order in δ by using the results

$$\int_{z-\delta}^{z} du\, g(u) = g(z)\,\delta + O(\delta^2), \tag{3.94}$$

and

$$\overline{y}(u) = y(u + \delta) = y(u) + y'(u)\,\delta + O(\delta^2), \tag{3.95}$$

and also

$$\overline{y}'(u) = y'(u) + y''(u)\,\delta + O(\delta^2). \tag{3.96}$$

Now consider the three terms of equation (3.93) separately. To first-order, the first term is

$$T_1 = \int_{c}^{d} du\, F(\overline{y}, \overline{y}') = \int_{c}^{d} du\, F(y + y'\delta, y' + y''\delta) + O(\delta^2). \tag{3.97}$$

Using the Taylor expansion

$$F(y + y'\delta, y' + y''\delta) = F(y, y') + \frac{\partial F}{\partial y} y'\delta + \frac{\partial F}{\partial y'} y''\delta + O(\delta^2), \tag{3.98}$$

where all derivatives are evaluated with $\delta = 0$, this becomes

$$T_1 = \int_{c}^{d} du\, \left[F(y, y') + \delta\frac{\partial F}{\partial y} y' + \delta\frac{\partial F}{\partial y'} y'' \right] + O(\delta^2). \tag{3.99}$$

The second and third terms of equation (3.93) have the same form,

$$\int_{z-\delta}^{z} du\, F(\overline{y}, \overline{y}') = F(\overline{y}(z), \overline{y}'(z))\delta + O(\delta^2)$$
$$= F(y(z), y'(z))\delta + O(\delta^2), \tag{3.100}$$

hence

$$\int_{c-\delta}^{c} du\, F(\overline{y}, \overline{y}') - \int_{d-\delta}^{d} du\, F(\overline{y}, \overline{y}') = -\delta\left[F(y, y')\right]_{c}^{d} + O(\delta^2). \tag{3.101}$$

On combining these relations the expression (3.93) for T becomes

$$T = \int_{c}^{d} du\, \left[F(y, y') + \delta\frac{\partial F}{\partial y}y' + \delta\frac{\partial F}{\partial y'}y''\right]$$
$$- \delta\left[F(y, y')\right]_{c}^{d} + O(\delta^2). \tag{3.102}$$

Because of equation (3.92) this gives

$$0 = \delta\int_{c}^{d} du\, \left(y'\frac{\partial F}{\partial y} + y''\frac{\partial F}{\partial y'}\right) - \delta\left[F(y, y')\right]_{c}^{d} + O(\delta^2). \tag{3.103}$$

Now integrate the second integral by parts,

$$\int_{c}^{d} du\, y''\frac{\partial F}{\partial y'} = \left[y'\frac{\partial F}{\partial y'}\right]_{c}^{d} - \int_{c}^{d} du\, y'\frac{d}{du}\left(\frac{\partial F}{\partial y'}\right). \tag{3.104}$$

Substituting this into (3.103) gives

$$0 = \delta\left[y'\frac{\partial F}{\partial y'} - F\right]_{c}^{d} - \delta\int_{c}^{d} du\, y'\left[\frac{d}{du}\left(\frac{\partial F}{\partial y'}\right) - \frac{\partial F}{\partial y}\right] + O(\delta^2). \tag{3.105}$$

But $y(u)$ is a solution of the Euler–Lagrange equation, so the remaining integrand is identically zero, and hence

$$\left.\left(F - y'\frac{\partial F}{\partial y'}\right)\right|_{x=c} = \left.\left(F - y'\frac{\partial F}{\partial y'}\right)\right|_{x=d}. \tag{3.106}$$

Finally, we observe that c and d are arbitrary and hence, for any x in the interval $a < x < b$, the function

$$F(y, y') - y'\frac{\partial}{\partial y'}F(y, y') = \text{constant}. \tag{3.107}$$

Because the function on the left-hand side is continuous the equality is true in the interval $a \leq x \leq b$. This relation is always true if the integrand of the functional does not depend explicitly upon x, that is, $\partial F/\partial x = 0$. It relates $y'(x)$ to $y(x)$ and by rearranging it we obtain one (or more if, for instance, a square root is involved) first-order equation for the unknown function $y(x)$, which is generally easier to deal with than the second-order Euler–Lagrange equation. Of course, by differentiating equation (3.107) with respect to x the Euler–Lagrange equation is regained, as shown in Exercise 2.6 (page 51).

The function $F - y'\partial F/\partial y'$ is named a *first integral* of the Euler–Lagrange equation, this name being suggestive of it being derived by integrating the original second-order equation once to give a first-order equation. For the same reason in dynamics, quantities that are conserved, for instance energy, linear and angular momentum, are also named first integrals, *integrals of the motion* or *constants of the motion*, and we shall see in the next chapter that these dynamical quantities have exactly the same mathematical origin as the first integral defined in equation (3.107).

3.5 Symmetries (Optional)

As an example of a functional that is not invariant under translations of the independent variable, consider

$$J[y] = \int_a^b dx\, x\, y'(x)^2, \quad y(a) = A,\ y(b) = B. \tag{3.108}$$

It is instructive to go through the above proof to see where and how it breaks down. In this case equation (3.92) becomes

$$\overline{J}[\overline{y}] = \int_{\overline{c}}^{\overline{d}} du\, u\, \overline{y}'(u)^2 = \int_c^d dx\, (x-\delta)\, y'(x)^2, \quad x = u + \delta$$

$$= J[y] - \delta \int_c^d dx\, y'(x)^2 \neq J[y]. \tag{3.109}$$

This proof of equation (3.107) may seem a lot more elaborate than that given in Exercise 2.6 (page 51). However, there are circumstances when the algebra required to use the former method is too unwieldy to be useful, and then the present method is far more transparent. An example of such a problem is given in Exercise 3.33. The following exercise is long and fairly difficult, but it revises all the steps of the preceding analysis.

Exercise 3.14

In this exercise we consider a functional,

$$S[y] = \int_a^b dx\, F(x, y, y') \tag{3.110}$$

that remains unchanged when the dependent variable is changed to $\overline{x}$ where $x = (1+\delta)\overline{x}$, for some constant δ. This represents a change of scale in x, rather than a translation, as considered in the text.

This means that for any c and d, where $a \le c < d \le b$, equation (3.92) is replaced by

$$T = \int_{\overline{c}}^{\overline{d}} du\, F(u, \overline{y}(u), \overline{y}'(u)) = \int_c^d dx\, F(x, y(x), y'(x)), \tag{3.111}$$

where $\overline{y}(\overline{x})$ and $y(x)$ are related by the identity $\overline{y}(\overline{x}) = y(x)$, with $x = (1+\delta)\overline{x}$.

(a) Show that

$$\frac{d}{d\overline{x}}\overline{y}(\overline{x}) = (1+\delta)\frac{dy}{dx} \quad \text{that is} \quad \overline{y}'(\overline{x}) = (1+\delta)y'(x).$$

(b) Show that, to first-order in δ,

$$T = \int_{c-c\delta}^{d-d\delta} du\, F(u, \overline{y}(u), \overline{y}'(u)) + O(\delta^2)$$

$$= \int_c^d du\, F(u, \overline{y}, \overline{y}') + \int_{c-c\delta}^{c} du\, F(u, \overline{y}, \overline{y}')$$

$$- \int_{d-d\delta}^d du\, F(u, \overline{y}, \overline{y}') + O(\delta^2), \tag{3.112}$$

and that

$$\overline{y}(u) = y(u) + y'(u)u\delta + O(\delta^2)$$

and

$$\overline{y}'(u) = y'(u) + \delta\left(y'(u) + u\, y''(u)\right) + O(\delta^2).$$

(c) By expanding the right-hand side of equation (3.112), show that, to first-order, this equation becomes

$$0 = \int_c^d du\, \left[y'\frac{\partial F}{\partial y}u + (y' + uy'')\frac{\partial F}{\partial y'}\right] - [uF(u,y,y')]_c^d.$$

(d) Use integration by parts to show that

$$\int_c^d du\, y'' u \frac{\partial F}{\partial y'} = \left[y' u \frac{\partial F}{\partial y'}\right]_c^d - \int_c^d du\, y' \frac{d}{du}\left(u \frac{\partial F}{\partial y'}\right)$$

and hence that, if $y(u)$ is a stationary path of S, the relation derived in part (c) becomes

$$0 = \left[u\left(y'\frac{\partial F}{\partial y'} - F\right)\right]_{u=c}^d \quad \text{for all } c \text{ and } d.$$

Deduce that for a functional satisfying the given invariance, the first integral is

$$xF - xy'\frac{\partial F}{\partial y'} = \text{constant} \quad \text{for all } a \leq x \leq b.$$

(e) If $F = x^\alpha y'^\beta$, for some constants α and $\beta \neq 1$, show that the first integral derived in part (d) reduces to $y'(x) = Ax^{-\gamma}$ for $\gamma = (1+\alpha)/\beta$ and some constant A. Show that the solutions of this equation are

$$y(x) = \begin{cases} A\ln x + B, & \gamma = 1, \\ \dfrac{A}{1-\gamma}x^{1-\gamma} + B, & \gamma \neq 1. \end{cases}$$

Explain why only the first of these solutions, $\gamma = 1$, is a stationary path of a functional invariant under a scale transformation.

Later, when we discuss the variational formulation of Newtonian dynamics, it will be seen that the equivalent of equation (3.107) becomes the conservation of energy in those circumstances when the forces are conservative and are independent of the time; that is, in Newtonian mechanics energy conservation is a consequence of the invariance of the equations of motion under translations in time. Similarly, we shall see that invariance under translations in space gives rise to conservation of linear momentum, and invariance under rotations in space give rise to conservation of angular momentum.

3.5.2 Noether's theorem

In this section we extend the previous analysis to deal with functionals having several dependent variables. The analysis is a straightforward generalisation of that presented above but takes time to absorb, so we recommend that you do not try to understand it properly until finishing Chapter 4, by which time the significance of Noether's theorem will be clearer. For this first reading, try to understand the fundamental ideas and try to avoid getting lost in algebraic details. That is, you should try to understand the definition of an invariant functional, the meaning of Noether's theorem, rather than the proof, and should be able to do Exercises 3.16-3.18.

There are two ingredients to Noether's theorem:

(i) functionals that are invariant under transformations of both dependent and independent variables;
(ii) families of transformations that depend upon one or more real parameters.

We consider both of these elements in turn in relation to the functional

$$S[\boldsymbol{y}] = \int_a^b dx\, F(x, \boldsymbol{y}, \boldsymbol{y}'), \quad \boldsymbol{y} = (y_1, y_2, \ldots, y_n), \tag{3.113}$$

which has stationary paths defined by the solutions of the Euler–Lagrange equations. Notice that we have not mentioned the boundary conditions: this is because they play no role in the general theorem.

3.5 Symmetries (Optional)

The value of the functional usually depends upon the path taken, which in this section is not always restricted to being stationary. We shall consider the change in the value of the functional when the path is changed according to a given transformation: in particular, we are interested in those transformations which change the path but not the value of the functional.

Consider, for instance, the two functionals

$$S_1[\boldsymbol{y}] = \int_0^1 dx \, (y_1'^2 + y_2'^2) \quad \text{and} \quad S_2[\boldsymbol{y}] = \int_0^1 dx \, (y_1'^2 + y_2'^2) \, y_1. \quad (3.114)$$

A path γ can be defined by the pair of functions $(f(x), g(x))$, $0 \le x \le 1$, and on each γ the functionals have a value.

Consider the transformation

$$\begin{aligned} \overline{y}_1 &= y_1 \cos\alpha - y_2 \sin\alpha, \\ \overline{y}_2 &= y_1 \sin\alpha + y_2 \cos\alpha, \end{aligned} \quad \text{with inverse} \quad \begin{aligned} y_1 &= \overline{y}_1 \cos\alpha + \overline{y}_2 \sin\alpha, \\ y_2 &= -\overline{y}_1 \sin\alpha + \overline{y}_2 \cos\alpha, \end{aligned} \quad (3.115)$$

which can be interpreted as an anti-clockwise rotation in the (y_1, y_2)-plane through an angle α, independent of x. Hence under this transformation the curve γ is rotated bodily to the curve $\overline{\gamma}$, as shown in Figure 3.4.

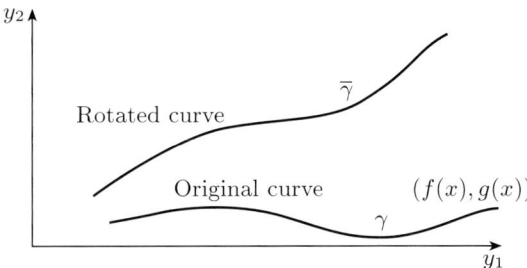

Figure 3.4 Diagram showing the rotation of the curve γ anti-clockwise through an angle α to the curve $\overline{\gamma}$.

The points on $\overline{\gamma}$ are parametrised by $(\overline{f}(x), \overline{g}(x))$, $0 \le x \le 1$, where

$$\overline{f} = f \cos\alpha - g \sin\alpha \quad \text{and} \quad \overline{g} = f \sin\alpha + g \cos\alpha. \quad (3.116)$$

Hence on the path γ, the functional S_1 has the value

$$S_1(\gamma) = \int_0^1 dx \, [f'(x)^2 + g'(x)^2] \quad (3.117)$$

and on the path $\overline{\gamma}$ it has the value

$$S_1(\overline{\gamma}) = \int_0^1 dx \, [\overline{f}'(x)^2 + \overline{g}'(x)^2] = \int_0^1 dx \, [f'(x)^2 + g'(x)^2]. \quad (3.118) \quad \text{See Exercise 3.15.}$$

Hence the functional has the same value on γ and on $\overline{\gamma}$ for all α. That is, $S_1[\boldsymbol{y}]$ is invariant with respect to the rotation defined in Equations (3.115).

On the other hand, the values of S_2 on γ are

$$S_2(\gamma) = \int_0^1 dx \, [f'(x)^2 + g'(x)^2] \, f(x), \quad (3.119)$$

and on $\overline{\gamma}$:

$$\begin{aligned} S_2(\overline{\gamma}) &= \int_0^1 dx \, [\overline{f}'(x)^2 + \overline{g}'(x)^2] \, \overline{f}(x) \\ &= \int_0^1 dx \, [f'(x)^2 + g'(x)^2] \, [f(x) \cos\alpha - g(x) \sin\alpha] \\ &= S_2(\gamma) \cos\alpha - \sin\alpha \int_0^1 dx \, [f'(x)^2 + g'(x)^2] \, g(x). \quad (3.120) \end{aligned}$$

In this case the functional has different values on γ and on $\overline{\gamma}$, unless α is an integer multiple of 2π. That is, $S_2[y]$ is not invariant with respect to the rotation (3.115).

Exercise 3.15

Derive equations (3.118) and (3.120).

The transformation (3.115) does not involve changes to the independent variable x, but the transformation considered in the previous subsection involved only a change in the independent variable, via a translation along the x-axis; see Figure 3.3. In general, it is necessary to deal with a transformation in both dependent and independent variables, which can be written as

$$\overline{x} = \Phi(x, \boldsymbol{y}, \boldsymbol{y}') \quad \text{and} \quad \overline{y}_k = \Psi_k(x, \boldsymbol{y}, \boldsymbol{y}'), \quad k = 1, 2, \ldots, n. \tag{3.121}$$

We assume that these relations can be inverted to give x and $\boldsymbol{y}$ in terms of $\overline{x}$ and $\overline{\boldsymbol{y}}$. For any curve γ, defined by a specific function $\boldsymbol{y} = \boldsymbol{f}(x)$, $a \leq x \leq b$, this transformation moves γ to another curve $\overline{\gamma}$ defined by the equation $\overline{\boldsymbol{y}} = \overline{\boldsymbol{f}}(\overline{x})$.

Definition: The functional (3.113) is said to be *invariant* under the transformation (3.121) if

$$\overline{G} = G, \tag{3.122}$$

where

$$\overline{G} = \int_{\overline{c}}^{\overline{d}} d\overline{x}\, F\left(\overline{x}, \overline{\boldsymbol{y}}, \frac{d\overline{\boldsymbol{y}}}{d\overline{x}}\right) \quad \text{and} \quad G = \int_{c}^{d} dx\, F\left(x, \boldsymbol{y}, \frac{d\boldsymbol{y}}{dx}\right), \tag{3.123}$$

for all c and d satisfying $a \leq c < d \leq b$, with $\overline{c}$ and $\overline{d}$ given by equation (3.121) by putting x equal to c and d, respectively.

The meaning of the equality $\overline{G} = G$ is easiest to understand if $\overline{x} = x$. Then the functions $\boldsymbol{y}(x)$ and $\overline{\boldsymbol{y}}(x)$ define two curves, γ and $\overline{\gamma}$ in an n-dimensional space, each parametrised by the independent variable x. The functional G is the integral of $F(x, \boldsymbol{y}, \boldsymbol{y}')$ along γ and $\overline{G}$ is the integral of the *same* function along $\overline{\gamma}$.

By contrast, in the case $x \neq \overline{x}$ the only difference is that the parametrisation along γ and $\overline{\gamma}$ is changed. In practice, the usual change to the independent variable, x, is the uniform shift $\overline{x} = x + \delta$, where δ is independent of x, $\boldsymbol{y}$ and $\boldsymbol{y}'$; this is the example dealt with in the previous subsection.

A one-parameter family of transformations is the set of transformations

$$\overline{x} = \Phi(x, \boldsymbol{y}, \boldsymbol{y}'; \delta) \quad \text{and} \quad \overline{y}_k = \Psi_k(x, \boldsymbol{y}, \boldsymbol{y}'; \delta), \quad k = 1, 2, \ldots, n, \tag{3.124}$$

depending upon the parameter δ, which reduces to the identity when $\delta = 0$, that is,

$$x = \Phi(x, \boldsymbol{y}, \boldsymbol{y}'; 0) \quad \text{and} \quad y_k = \Psi_k(x, \boldsymbol{y}, \boldsymbol{y}'; 0), \quad k = 1, 2, \ldots, n, \tag{3.125}$$

and where Φ and all the Ψ_k have continuous first derivatives in all variables, including δ. This last condition ensures that the transformation is invertible in the neighbourhood of $\delta = 0$. An example of a one-parameter transformation is defined by equations (3.115), which becomes the identity when $\alpha = 0$.

3.5 Symmetries (Optional)

Exercise 3.16

Which of the following is a one-parameter family of transformations?

(a) $\overline{x} = x - y\delta$, $\quad \overline{y} = y + x\delta$.

(b) $\overline{x} = x\cosh\delta - y\sinh\delta$, $\quad \overline{y} = -y\cosh\delta + y\sinh\delta$.

(c) $\overline{x} = x$ and $\overline{\boldsymbol{y}} = \boldsymbol{y}\exp(A\delta)$, where A is a square non-singular $n \times n$ matrix, and the exponential of a matrix is defined by the infinite series

$$\exp(A\delta) = \sum_{k=0}^{\infty} \frac{\delta^k A^k}{k!} = 1 + A\delta + \tfrac{1}{2}A^2\delta^2 + \tfrac{1}{6}A^3\delta^3 + \cdots. \quad (3.126)$$

Families of transformations are very common and are often generated by solutions of differential equations, as illustrated by the following exercise.

Exercise 3.17

(a) Show that the solution of the equation

$$\frac{dy}{dt} = y(1-y), \quad 0 \le y(0) \le 1, \quad (3.127)$$

is

$$y = \Psi(z,t) = \frac{ze^t}{1 + (e^t - 1)z}, \quad z = y(0). \quad (3.128)$$

(b) Show that this defines a one-parameter family of transformations, $y = \Psi(z,t)$, with parameter t.

Exercise 3.18

Show that the functional

$$S[\boldsymbol{y}] = \int_a^b dx\, \left(y_1'^{\,2} - y_2'^{\,2}\right) \quad (3.129)$$

is invariant under the transformation

$$\overline{y}_1 = y_1\cosh\delta + y_2\sinh\delta, \quad \overline{y}_2 = y_1\sinh\delta + y_2\cosh\delta, \quad \overline{x} = x + \delta\,g(x), \quad (3.130)$$

only if $g(x)$ is a constant.

We have finally arrived at the main part of this section, the statement and proof of Noether's theorem. The theorem is far easier to understand than its proof, so after reading the theorem we suggest that you do Exercises 3.20 to 3.22, leaving the actual proof until you have time to understand it. The theorem was published in 1918 by Emmy Noether (1882–1935), a German mathematician, considered to be one of the most creative abstract algebraists of modern times. The theorem was derived for certain variational principles, and has important applications to physics, especially relativity and quantum mechanics, besides systematising many of the known results of classical dynamics.

The theorem deals with arbitrarily small changes in the coordinates, so in equation (3.124) we assume $|\delta| \ll 1$ and write the transformation as

$$\overline{x} = x + \phi(x,\boldsymbol{y},\boldsymbol{y}')\delta + O(\delta^2), \quad \phi = \left.\frac{\partial\Phi}{\partial\delta}\right|_{\delta=0},$$
$$\overline{y}_k = y_k + \psi_k(x,\boldsymbol{y},\boldsymbol{y}')\delta + O(\delta^2), \quad \psi_k = \left.\frac{\partial\Psi_k}{\partial\delta}\right|_{\delta=0}, \quad (3.131)$$

where we have used the $\delta = 0$ limit defined after equation (3.124). In all subsequent analysis, second-order terms in the parameter, here δ, are ignored.

Exercise 3.19

Show that, to first-order in α, the rotation defined by equations (3.115) becomes

$$\overline{y}_1 = y_1 + \psi_1 \alpha, \quad \psi_1 = -y_2,$$
$$\overline{y}_2 = y_2 + \psi_2 \alpha, \quad \psi_2 = y_1.$$

Noether's theorem

If the functional

$$S[\boldsymbol{y}] = \int_c^d dx\, F(x, \boldsymbol{y}, \boldsymbol{y}') \qquad (3.132)$$

is invariant under the family of transformations (3.131), for arbitrary c and d, then

$$\sum_{k=1}^n \frac{\partial F}{\partial y'_k} \psi_k + \left(F - \sum_{k=1}^n y'_k \frac{\partial F}{\partial y'_k} \right) \phi = \text{constant}. \qquad (3.133)$$

along each stationary path of $S[\boldsymbol{y}]$.

The function defined on the left-hand side of equation 3.133 is often named a *first integral* of the Euler–Lagrange equations. Equation (3.133) is the significant result and is the required generalisation of the simpler result first derived in Exercise 2.6 (page 51) and used in Section 2.5 to find the minimum surface of revolution, and in Section 2.7 for the brachistochrone problem.

In the one-dimensional case, $n = 1$, and when $\overline{y} = y$ ($\psi = 0$) and $\phi = 1$, equation (3.133) reduces to the result derived in the previous section, equation (3.107) (page 114). The proof of Noether's theorem is given after the following exercises.

Exercise 3.20

Use the fact that the functional

$$S[\boldsymbol{y}] = \int_a^b dx\, \left(y_1'^2 + y_2'^2 \right)$$

is invariant under the rotation defined by equations (3.115), and the result derived in Exercise 3.19, to show that a first integral is

$$y_1 y_2' - y_2 y_1' = \text{constant}.$$

3.5 Symmetries (Optional)

Exercise 3.21

Show that the functional

$$S[\mathbf{y}] = \int_a^b dx\, \left(y_1'^2 + y_2'^2\right)$$

is invariant under each of the following three transformations only if $g(x)$ is a constant.

(a) $\overline{y}_1 = y_1 + g(x)\,\delta$, $\quad \overline{y}_2 = y_2$, $\quad \overline{x} = x$.
(b) $\overline{y}_1 = y_1$, $\quad \overline{y}_2 = y_2 + g(x)\,\delta$, $\quad \overline{x} = x$.
(c) $\overline{y}_1 = y_1$, $\quad \overline{y}_2 = y_2$, $\quad \overline{x} = x + g(x)\,\delta$.

In the case $g(x) = 1$, show that these three transformations lead to the following first integrals, respectively.

(a) $y_1' = \text{constant}$ (b) $y_2' = \text{constant}$ (c) $y_1'^2 + y_2'^2 = \text{constant}$

Exercise 3.22

Show that the functional

$$S[\mathbf{y}] = \int_a^b dx\, \left[\tfrac{1}{2}\left(y_1'^2 + y_2'^2\right) + V(y_1 - y_2)\right],$$

where $V(z)$ is a differentiable function, is invariant under the transformation

$$\overline{y}_1 = y_1 + \delta g(x), \quad \overline{y}_2 = y_2 + \delta g(x), \quad \overline{x} = x$$

only if $g(x)$ is a constant, and that in this circumstance a first integral is

$$y_1' + y_2' = \text{constant}.$$

Exercise 3.23

Show that for the scale transformation $x = (1+\delta)\overline{x}$ considered in Exercise 3.14, $\phi = -x$, and that in this case the first integral given in equation (3.133) is the same as that derived in Exercise 3.14.

Proof of Noether's theorem

Noether's theorem is proved by substituting the transformation (3.131) into the functional (3.132) and expanding to first-order in δ. The algebra is messy, so we proceed in two stages.

First, we assume that $\overline{x} = x$, that is, $\phi = 0$, which simplifies the algebra. It is easiest to start with the transformed functional

$$\overline{T} = \int_c^d dx\, F\left(x, \overline{\mathbf{y}}, \frac{d\overline{\mathbf{y}}}{dx}\right) \quad (\text{since } \overline{x} = x). \tag{3.134}$$

Now substitute for $\overline{\mathbf{y}}$ and $\overline{\mathbf{y}}'$ and expand to first-order in δ, to obtain

$$\overline{T} = \int_c^d dx\, F\left(x, \mathbf{y} + \delta\boldsymbol{\psi}, \mathbf{y}' + \delta\frac{d\boldsymbol{\psi}}{dx}\right)$$

$$= \int_c^d dx\, F(x, \mathbf{y}, \mathbf{y}') + \delta \int_c^d dx \sum_{k=1}^n \left(\frac{\partial F}{\partial y_k}\psi_k + \frac{\partial F}{\partial y_k'}\frac{d\psi_k}{dx}\right). \tag{3.135}$$

But the first term is merely the untransformed functional which equals the transformed functional – because it is invariant under the transformation. Also, using integration by parts

$$\int_c^d dx\, \frac{\partial F}{\partial y_k'}\frac{d\psi_k}{dx} = \left[\psi_k \frac{\partial F}{\partial y_k'}\right]_c^d - \int_c^d dx\, \psi_k \frac{d}{dx}\left(\frac{\partial F}{\partial y_k'}\right) \tag{3.136}$$

and hence, by substituting this result into equation (3.135) we obtain

$$0 = \delta \sum_{k=1}^{n} \left[\psi_k \frac{\partial F}{\partial y'_k} \right]_c^d + \delta \int_c^d dx \sum_{k=1}^{n} \psi_k \left\{ \frac{\partial F}{\partial y_k} - \frac{d}{dx}\left(\frac{\partial F}{\partial y'_k}\right) \right\}. \tag{3.137}$$

The term in curly brackets is, by virtue of the Euler–Lagrange equations, zero on a stationary path. It follows that

$$\sum_{k=1}^{n} \psi_k \frac{\partial F}{\partial y'_k}\bigg|_{x=d} = \sum_{k=1}^{n} \psi_k \frac{\partial F}{\partial y'_k}\bigg|_{x=c}, \tag{3.138}$$

and since c and d are arbitrary, we obtain equation (3.133), with $\phi = 0$.

In the second stage of this analysis we consider the general case, $\phi \neq 0$, which proceeds similarly but is algebraically more complicated. As before we start with the transformed functional, which is now

$$\overline{T} = \int_{\overline{c}}^{\overline{d}} d\overline{x}\, F\left(\overline{x}, \overline{y}, \frac{d\overline{y}}{d\overline{x}}\right), \tag{3.139}$$

where $\overline{d} = d + \delta\phi(d)$, $\overline{c} = c + \delta\phi(c)$, with $\phi(c)$ denoting $\phi(c, \mathbf{y}(c), \mathbf{y}'(c))$ and similarly for $\phi(d)$. Now we have to change the integration variable and limits, besides expanding F. First consider the derivative $d\overline{\mathbf{y}}/d\overline{x}$; using equations (3.131) and the chain rule,

$$\frac{d\overline{\mathbf{y}}}{d\overline{x}} = \left(\frac{d\mathbf{y}}{dx} + \delta \frac{d\boldsymbol{\psi}}{dx}\right)\frac{dx}{d\overline{x}}, \quad \text{but} \quad \frac{d\overline{x}}{dx} = 1 + \delta\frac{d\phi}{dx} \tag{3.140}$$

and so, to first-order in δ, we have

$$\frac{d\overline{\mathbf{y}}}{d\overline{x}} = \left(\frac{d\mathbf{y}}{dx} + \delta\frac{d\boldsymbol{\psi}}{dx}\right)\left(1 - \delta\frac{d\phi}{dx}\right) = \frac{d\mathbf{y}}{dx} + \delta\left(\frac{d\boldsymbol{\psi}}{dx} - \frac{d\mathbf{y}}{dx}\frac{d\phi}{dx}\right). \tag{3.141}$$

Thus the integral becomes

$$\overline{T} = \int_c^d dx\, \frac{d\overline{x}}{dx} F\left[x + \delta\phi, \mathbf{y} + \delta\boldsymbol{\psi}, \frac{d\mathbf{y}}{dx} + \delta\left(\frac{d\boldsymbol{\psi}}{dx} - \frac{d\mathbf{y}}{dx}\frac{d\phi}{dx}\right)\right]. \tag{3.142}$$

Now expand to first-order in δ and use the fact that the functional is invariant. After some algebra we find that

$$0 = \delta \int_c^d dx \left[\left(F - \sum_{k=1}^n \frac{\partial F}{\partial y'_k}\frac{dy_k}{dx}\right)\frac{d\phi}{dx} + \phi\frac{\partial F}{\partial x} \right. \\ \left. + \sum_{k=1}^n \left(\psi_k \frac{\partial F}{\partial y_k} + \frac{\partial F}{\partial y'_k}\frac{d\psi_k}{dx}\right)\right]. \tag{3.143}$$

Now integrate those terms containing the total derivatives of ϕ and ψ_k by parts to cast this equation into the form

$$0 = \delta \left[\left(F - \sum_{k=1}^n \frac{\partial F}{\partial y'_k}\frac{dy_k}{dx}\right)\phi + \sum_{k=1}^n \psi_k \frac{\partial F}{\partial y'_k}\right]_c^d \\ + \delta \int_c^d dx\, \phi \left\{\frac{\partial F}{\partial x} - \frac{d}{dx}\left(F - \sum_{k=1}^n \frac{\partial F}{\partial y'_k}\frac{dy_k}{dx}\right)\right\} \\ + \delta \int_c^d dx \sum_{k=1}^n \psi_k \left[\frac{\partial F}{\partial y_k} - \frac{d}{dx}\left(\frac{\partial F}{\partial y'_k}\right)\right]. \tag{3.144}$$

Notice that if $\phi = 0$ this is the same as equation (3.137). Finally, we need to show that on stationary paths the integrals are zero. The second integral is clearly zero, by virtue of the Euler–Lagrange equations. On expanding the integrand of the first integral the term in curly brackets becomes

$$\frac{\partial F}{\partial x} - \left[\frac{\partial F}{\partial x} + \sum_{k=1}^{n}\left(\frac{\partial F}{\partial y_k}y'_k + \frac{\partial F}{\partial y'_k}y''_k\right)\right] + \sum_{k=1}^{n}\frac{\partial F}{\partial y'_k}y''_k$$
$$+ \sum_{k=1}^{n} y'_k \frac{d}{dx}\left(\frac{\partial F}{\partial y'_k}\right). \quad (3.145)$$

Using the Euler–Lagrange equations to modify the last term, it is seen that this expression is zero. Hence, because c and d are arbitrary, we have shown that the function

$$\left(F - \sum_{k=1}^{n}\frac{\partial F}{\partial y'_k}y'_k\right)\phi + \sum_{k=1}^{n}\psi_k\frac{\partial F}{\partial y'_k}, \quad (3.146)$$

where $\boldsymbol{y}(x)$ is evaluated along a stationary path, is independent of x, and is a constant.

3.6 Further Exercises

Exercise 3.24

Using the functional

$$S[y] = \int_a^b dx\,\left(y'^{\,2} - \omega^2 y^2\right)$$

and the change of variable $z = x^{1/c}$, show that the differential equation $y'' + \omega^2 y = 0$ is transformed into

$$z\frac{d^2 y}{dz^2} + (1-c)\frac{dy}{dz} + c^2\omega^2 z^{2c-1} y = 0.$$

Exercise 3.25

Show that the Euler–Lagrange equations for the functional

$$S[y_1, y_2] = \int_a^b dx\, F(y'_1, y'_2),$$

which depends only upon the first derivatives of y_1 and y_2, are

$$\frac{\partial^2 F}{\partial y'^{\,2}_1}y''_1 + \frac{\partial^2 F}{\partial y'_1 \partial y'_2}y''_2 = 0, \qquad \frac{\partial^2 F}{\partial y'_1 \partial y'_2}y''_1 + \frac{\partial^2 F}{\partial y'^{\,2}_2}y''_2 = 0.$$

Deduce that, provided the determinant

$$d = \begin{vmatrix} \dfrac{\partial^2 F}{\partial y'^{\,2}_1} & \dfrac{\partial^2 F}{\partial y'_1 \partial y'_2} \\ \dfrac{\partial^2 F}{\partial y'_1 \partial y'_2} & \dfrac{\partial^2 F}{\partial y'^{\,2}_2} \end{vmatrix}$$

is non-zero, the stationary paths are the straight lines

$$y_1(x) = Ax + B, \quad y_2(x) = Cx + D,$$

where A, B, C and D are constants. Describe the solution if $d = 0$.

What is the equivalent condition to $d \neq 0$ if there is only one dependent variable?

Exercise 3.26

If $\Phi(x, y_1, y_2)$ is any twice-differentiable function, show that the functionals

$$S_1[y_1, y_2] = \int_a^b dx\, F(x, y_1, y_2, y_1', y_2')$$

and

$$S_2[y_1, y_2] = \int_a^b dx\, [F(x, y_1, y_2, y_1', y_2') + \Psi(x, y_1, y_2, y_1', y_2')],$$

where

$$\Psi = \frac{\partial \Phi}{\partial x} + \frac{\partial \Phi}{\partial y_1} y_1' + \frac{\partial \Phi}{\partial y_2} y_2',$$

lead to the same Euler–Lagrange equation.

Note that this is the direct generalisation of the result derived in Exercise 2.28 (page 71).

Exercise 3.27

Consider the two functionals

$$S_1[y_1, y_2] = \int_a^b dx\, \left[\tfrac{1}{2}\left(y_1'^2 + y_2'^2\right) + g_1(x) y_1' + g_2(x) y_2' - V(x, y_1, y_2)\right]$$

and

$$S_2[y_1, y_2] = \int_a^b dx\, \left[\tfrac{1}{2}\left(y_1'^2 + y_2'^2\right) - \overline{V}(x, y_1, y_2)\right],$$

where

$$\overline{V} = V + g_1'(x) y_1 + g_2'(x) y_2.$$

Use the result proved in the previous exercise to show that they have identical Euler–Lagrange equations.

Exercise 3.28

(a) Show that the Euler–Lagrange equation of the functional

$$S[y] = \int_0^\infty dx\, e^{-x} \sqrt{y - e^x y'}$$

is

$$\frac{d^2 y}{dx^2} - (3e^{-x} - 1)\frac{dy}{dx} + 2e^{-2x} y = 0. \tag{3.147}$$

(b) Show that the change of variables $u = e^{-x}$, with inverse $x = -\ln u$, transforms this functional to

$$S[Y] = \int_0^1 du\, \sqrt{Y(u) + Y'(u)}, \quad Y(u) = y(-\ln u),$$

and that the Euler–Lagrange equation for this functional is

$$\frac{d^2 Y}{du^2} + 3\frac{dY}{du} + 2Y = 0. \tag{3.148}$$

(c) By making the substitution $x = -\ln u$, show directly that equation (3.148) transforms into equation (3.147).

Exercise 3.29

Show that the stationary paths of the functional

$$S[y, z] = \int_0^{\pi/2} dx \, \left(y'^2 + z'^2 + 2yz\right),$$

with the boundary conditions $y(0) = 0$, $z(0) = 0$, $y(\pi/2) = 3/2$ and $z(\pi/2) = 1/2$, satisfy the equations

$$\frac{d^2y}{dx^2} - z = 0, \quad \frac{d^2z}{dx^2} - y = 0.$$

Show that the solution of these equations is

$$y(x) = \frac{\sinh x}{\sinh(\pi/2)} + \tfrac{1}{2}\sin x, \quad z(x) = \frac{\sinh x}{\sinh(\pi/2)} - \tfrac{1}{2}\sin x.$$

Exercise 3.30

Show that if $y = G(z)$, where $G(z)$ is differentiable, the functional

$$S[y] = \int_a^b dx \, F(x, y, y')$$

transforms to

$$S[z] = \int_a^b dx \, F(x, G(z), G'(z)z'),$$

with associated Euler–Lagrange equation

$$\frac{d}{dx}\left(G'(z)\frac{\partial F}{\partial y'}\right) - \frac{\partial F}{\partial y}G'(z) - \frac{\partial F}{\partial y'}G''(z)\,z' = 0.$$

Exercise 3.31

Consider the functional

$$S[y] = \int_1^2 dx \, \frac{y'^2}{x^2}, \quad y(1) = A, \quad y(2) = B,$$

where A and B are positive and different.

(a) Using the fact that $dy/dx = \dfrac{1}{dx/dy}$, show that if y becomes the independent variable the functional becomes

$$S[x] = \int_A^B dy \, \frac{1}{x^2 x'}, \quad x(A) = 1, \quad x(B) = 2,$$

where $x' = dx/dy$.

(b) Show that the Euler–Lagrange equation for the functional $S[x]$ is

$$\frac{d}{dy}\left(\frac{1}{x^2 x'^2}\right) - \frac{2}{x^3 x'} = 0,$$

which simplifies to

$$\frac{d^2x}{dy^2} + \frac{2}{x}\left(\frac{dx}{dy}\right)^2 = 0.$$

(c) By writing this last equation in the form

$$\frac{1}{x^2}\frac{d}{dy}\left(x^2 \frac{dx}{dy}\right) = 0, \quad x(A) = 1, \quad x(B) = 2,$$

and integrating twice, show that the required solution is

$$x^3 = \frac{7y + B - 8A}{B - A}.$$

3.7 Harder Exercises

Exercise 3.32

The ordinary Bessel function, denoted by $J_n(x)$, may be defined to be proportional to the solution of the second-order differential equation

$$x^2 \frac{d^2 y}{dx^2} + x\frac{dy}{dx} + (x^2 - n^2)y = 0, \quad n = 0, 1, 2, \ldots, \qquad (3.149)$$

which behaves as $(x/2)^n$ near the origin.

See Block I, Chapter 5, Subsection 5.2.5.

(a) Show that equation (3.149) is the Euler–Lagrange equation of the functional

$$F[y] = \int_0^X dx \left[xy'(x)^2 - \left(x - \frac{n^2}{x}\right) y(x)^2 \right], \quad X > 0,$$

where the admissible functions have continuous second derivatives for $0 \le x \le X$. Note that the derivation of equation (3.149) from this functional requires care because there is no boundary condition at $x = X$. This omission can be rectified using the fact that the differential equation is linear.

(b) Define a new independent variable u by the equation $x = f(u)$, where $f(u)$ is invertible, and set $w(u) = y(f(u))$, to cast this functional into the form

$$F[w] = \int_{u_0}^{u_1} du \left[\frac{f(u)}{f'(u)} w'(u)^2 - \left(f(u) f'(u) - n^2 \frac{f'(u)}{f(u)} \right) w(u)^2 \right],$$

where $f(u_0) = 0$ and $f(u_1) = X$.

(c) Hence show that if $f(u) = e^u$, $w(u)$ satisfies the equation

$$\frac{d^2 w}{du^2} + \left(e^{2u} - n^2 \right) w = 0, \qquad (3.150)$$

and deduce that a solution of equation (3.150) is $w(u) = J_n(e^u)$.

Exercise 3.33

Show that the Euler–Lagrange equation of the functional, first considered in Exercise 2.33 (page 73)

$$S[y] = \int_a^b dx\, F(x, y, y', y''),$$

with the boundary conditions

$$y(a) = A_1, \quad y'(a) = A_2, \quad y(b) = B_1, \quad y'(b) = B_2,$$

has the first integral

$$\frac{d}{dx}\left(\frac{\partial F}{\partial y''}\right) - \frac{\partial F}{\partial y'} = \text{constant}$$

if the integrand does not depend explicitly upon $y(x)$, and the first integral

$$y'' \frac{\partial F}{\partial y''} - \left(\frac{d}{dx}\left(\frac{\partial F}{\partial y''}\right) - \frac{\partial F}{\partial y'} \right) y' - F = \text{constant}$$

if the integrand does not depend explicitly upon x.

[Hint: the second part of this question is most easily done using the theory described in Section 3.5.1.]

Solutions to Exercises in Chapter 3

Solution 3.1

(a) If $x = u^2$ the chain rule gives

$$\frac{dy}{dx} = \frac{dy}{du}\frac{du}{dx} = \frac{dy}{du}\frac{1}{dx/du} = \frac{1}{2u}\frac{dY}{du},$$

where $Y(u) = y(u^2)$, so the functional becomes

$$S[Y] = 2\int_0^U du\, u\left(\frac{1}{4u^2}\left(\frac{dY}{du}\right)^2 - \omega^2 Y^2\right), \quad \text{where} \quad u = x^{1/2},$$

$$= \frac{1}{2}\int_0^U du\left(\frac{1}{u}Y'^2 - 4u\omega^2 Y^2\right).$$

The Euler–Lagrange equation is now

$$\frac{d}{du}\left(\frac{Y'}{u}\right) + 4u\omega^2 Y = 0,$$

which expands to $uY'' - Y' + 4u^3\omega^2 Y = 0$. With $\omega = 1$ this gives equation (3.1), with $a = 2$.

(b) The second derivative is obtained using the chain rule again,

$$\frac{d^2 y}{dx^2} = \frac{d}{du}\left(\frac{1}{2u}\frac{dY}{du}\right)\frac{1}{dx/du} = \frac{1}{2u}\frac{d}{du}\left(\frac{1}{2u}\frac{dY}{du}\right)$$

$$= \frac{1}{2u}\left(\frac{1}{2u}\frac{d^2Y}{du^2} - \frac{1}{2u^2}\frac{dY}{du}\right) = \frac{1}{4u^3}\left(u\frac{d^2Y}{du^2} - \frac{dY}{du}\right).$$

Hence $y'' + \omega^2 y = 0$ becomes

$$\frac{1}{4u^3}\left(u\frac{d^2Y}{du^2} - \frac{dY}{du}\right) + \omega^2 Y = 0,$$

giving the same Euler–Lagrange equation for $Y(u)$, as before.

Solution 3.2

(a) Here the integrand is $F = y'^2$ so $\partial F/\partial y' = 2y'$ and the Euler–Lagrange equation is $y'' = 0$; alternatively apply the first integral, equation (2.33) (page 49), to give $y'^2 = c$.

(b) If $y = G(z)$ the chain rule gives

$$\frac{dy}{dx} = G'(z)\frac{dz}{dx},$$

and the functional becomes

$$S[z] = \int_a^b dx\, G'(z)^2 z'^2.$$

Now the integrand is $F = G'(z)^2 z'^2$, giving

$$\frac{\partial F}{\partial z'} = 2G'(z)^2 z' \quad \text{and} \quad \frac{\partial F}{\partial z} = 2G''(z)\, G'(z)\, z'^2,$$

so the Euler–Lagrange equation becomes

$$\frac{d}{dx}\left(G'(z)^2 z'\right) - G''(z)\, G'(z)\, z'^2 = 0.$$

But

$$\frac{d}{dx}\left(G'(z)^2 z'\right) = G'(z)^2 z'' + 2G''(z)\, G'(z)\, z'^2,$$

and hence the Euler–Lagrange equation becomes

$$G'(z)^2\, z'' + G''(z)G'(z)z'^2 = 0,$$

which is the result quoted.

Now make the same change of variables in the original Euler–Lagrange equation $y'' = 0$. The chain rule gives

$$\frac{dy}{dx} = G'(z)\frac{dz}{dx} \quad \text{and} \quad \frac{d^2y}{dx^2} = G'(z)\frac{d^2z}{dx^2} + G''(z)\left(\frac{dz}{dz}\right)^2,$$

which leads to the same equation as derived above, provided $G'(z) \neq 0$.

Solution 3.3

(a) With the polar coordinates $x = r\cos\theta$, $y = r\sin\theta$, the derivative $y'(x)$ is given by equation (3.25) (page 100), hence

$$1 + y'(x)^2 = \frac{r'^2 + r^2}{(r'\cos\theta - r\sin\theta)^2}.$$

Since we are assuming that $y'(x)$ is bounded, from the definition of admissible functions, $r'\cos\theta - r\sin\theta \neq 0$ on the curve. Here, for simplicity, we assume $r'\cos\theta - r\sin\theta > 0$: in the opposite case the analysis changes slightly, but the final result is the same. The functional becomes

$$S[r] = \int_{\theta_a}^{\theta_b} d\theta \, \frac{dx}{d\theta} \frac{r\sqrt{r'^2 + r^2}}{r'\cos\theta - r\sin\theta},$$

where $\tan\theta_a = y(a)/a$ and $\tan\theta_b = y(b)/b$. But by equation (3.22),

$$\frac{dx}{d\theta} = r'\cos\theta - r\sin\theta, \quad \text{hence} \quad S[r] = \int_{\theta_a}^{\theta_b} d\theta \, r\sqrt{r^2 + r'^2}.$$

If $F = r\sqrt{r^2 + r'^2}$, its derivatives are

$$\frac{\partial F}{\partial r} = \frac{r'^2 + 2r^2}{\sqrt{r^2 + r'^2}}, \quad \frac{\partial F}{\partial r'} = \frac{rr'}{\sqrt{r^2 + r'^2}},$$

and the Euler–Lagrange equation is

$$\frac{d}{d\theta}\left(\frac{rr'}{\sqrt{r^2 + r'^2}}\right) - \frac{r'^2 + 2r^2}{\sqrt{r^2 + r'^2}} = 0.$$

Expanding this gives

$$\frac{rr'' + r'^2}{\sqrt{r^2 + r'^2}} - \frac{rr'(rr' + r'r'')}{(r^2 + r'^2)^{3/2}} - \frac{2r^2 + r'^2}{\sqrt{r^2 + r'^2}} = 0,$$

and this reduces to $r^3 r'' - 3r^2 r'^2 - 2r^4 = 0$, which on dividing by r^3, gives the quoted equation.

(b) In order to simplify this, consider the first two derivatives of $1/r^\alpha$,

$$\frac{d}{d\theta}\left(\frac{1}{r^\alpha}\right) = -\alpha\frac{r'}{r^{\alpha+1}} \quad \text{and} \quad \frac{d^2}{d\theta^2}\left(\frac{1}{r^\alpha}\right) = -\alpha\frac{r''}{r^{\alpha+1}} + \alpha(\alpha+1)\frac{r'^2}{r^{\alpha+2}}$$

and hence

$$r''(\theta) = (\alpha+1)\frac{r'^2}{r} - \frac{r^{\alpha+1}}{\alpha}\frac{d^2}{d\theta^2}\left(\frac{1}{r^\alpha}\right).$$

Thus if we set $\alpha = 2$ and substitute for $r''(\theta)$, our equation becomes

$$-\frac{r^3}{2}\frac{d^2}{d\theta^2}\left(\frac{1}{r^2}\right) - 2r = 0, \quad \text{or} \quad \frac{d^2 z}{d\theta^2} + 4z = 0, \quad \text{where} \quad z = \frac{1}{r^2}.$$

The general solution of the equation for z is $z = r^{-2} = A\cos 2\theta + B\sin 2\theta$, but since

$$\sin 2\theta = 2\sin\theta\cos\theta = \frac{2xy}{r^2} \quad \text{and} \quad \cos 2\theta = \cos^2\theta - \sin^2\theta = \frac{x^2 - y^2}{r^2},$$

this becomes $A(x^2 - y^2) + 2Bxy = 1$.

Solution 3.4

The functional is
$$S[x] = \int_A^B dy\, \frac{dx}{dy} F\left(\frac{1}{dx/dy}\right) = \int_A^B dy\, x'(y) F(1/x'(y)),$$
which is the required result.

Solution 3.5

Applying the boundary conditions gives, for $y_1(x)$,
$$0 = A + C \quad \text{and} \quad 1 = B + C\cosh(\pi/2) + D\sinh(\pi/2),$$
and for $y_2(x)$,
$$0 = -A + C \quad \text{and} \quad -1 = -B + C\cosh(\pi/2) + D\sinh(\pi/2).$$
Since $A = C$ and $A + C = 0$, we have $A = C = 0$. Then the equations for B and D become
$$B + D\sinh(\pi/2) = 1 \quad \text{and} \quad -B + D\sinh(\pi/2) = -1.$$
Adding these two solutions gives $2D\sinh(\pi/2) = 0$, so $D = 0$ and hence $B = 1$. Thus the required solution is $y_1(x) = \sin x$ and $y_2(x) = -\sin x$.

Solution 3.6

The conditions at $x = 0$ give $y_1(0) = A + C = 0$ and $y_2(0) = -A + C = 0$, so $A = C = 0$. The conditions at $x = \pi$ give $y_1(\pi) = D\sinh\pi = \alpha$ and $y_2(\pi) = D\sinh\pi = \beta$. If $\alpha \neq \beta$, there is no solution. If $\alpha = \beta$, the solution is
$$y_1(x) = B\sin x + \alpha \frac{\sinh x}{\sinh\pi}, \quad y_2(x) = -B\sin x + \alpha \frac{\sinh x}{\sinh\pi} \quad \text{for all } B.$$

Solution 3.7

In this case $F = y_1'^2 + y_2'^2 + y_1' y_2'$, so
$$\frac{\partial F}{\partial y_1} = \frac{\partial F}{\partial y_2} = 0, \quad \frac{\partial F}{\partial y_1'} = 2y_1' + y_2' \quad \text{and} \quad \frac{\partial F}{\partial y_2'} = 2y_2' + y_1'.$$
Hence the two Euler–Lagrange equations are
$$\frac{d}{dx}(2y_1' + y_2') = 0 \quad \text{and} \quad \frac{d}{dx}(2y_2' + y_1') = 0.$$
These may be integrated directly to give $2y_1' + y_2' = a_1$ and $2y_2' + y_1' = a_2$, where a_1 and a_2 are constants. Now integrate again to obtain
$$2y_1 + y_2 = a_1 x + b_1 \quad \text{and} \quad 2y_2 + y_1 = a_2 x + b_2.$$
But the boundary conditions at $x = 0$ give $b_1 = 1$ and $b_2 = 2$; at $x = 1$ we have $4 = a_1 + b_1$ and $5 = a_2 + b_2$; hence $a_1 = 3$ and $a_2 = 3$, and the solutions are $2y_1 + y_2 = 3x + 1$ and $2y_2 + y_1 = 3x + 2$. Multiplying the first equation by 2 and subtracting the second now gives $y_1 = x$ and $y_2 = x + 1$.

Solution 3.8

We can write the functional in the form
$$S[y_1, y_2] = \int_0^1 dx\, \left[\left(y_1' + \tfrac{1}{2}y_2'\right)^2 + \tfrac{3}{4}y_2'^2\right].$$
Thus, on defining $z_1 = y_1 + y_2/2$, this becomes
$$S[z_1, y_2] = \int_0^1 dx\, \left(z_1'^2 + \tfrac{3}{4}y_2'^2\right), \quad z_1(0) = \tfrac{1}{2},\ z_1(1) = 2,\ y_2(0) = 1,\ y_2(1) = 2.$$
The associated Euler–Lagrange equations are simply $z_1'' = 0$ and $y_2'' = 0$, which can be integrated directly to yield $z_1 = A_1 x + B_1$ and $y_2 = A_2 x + B_2$. The boundary

conditions for $z_1(x)$ give $B_1 = 1/2$ and $A_1 + B_1 = 2$, so $A_1 = 3/2$. For $y_2(x)$ we find $B_2 = 1$ and $A_2 + B_2 = 2$, so $A_2 = 1$. Hence the solution is

$$z_1 = \tfrac{3}{2}x + \tfrac{1}{2} \quad \text{and} \quad y_2 = x + 1, \quad \text{so} \quad y_1(x) = z_1(x) - \tfrac{1}{2}y_2(x) = x.$$

Solution 3.9

Since

$$S[y_1, y_2] = \int_a^b dx\, F(x, y_1, y_2, y_1', y_2'),$$

from the definition (3.51) we require the derivative

$$\Delta(\epsilon) = \frac{d}{d\epsilon} \int_a^b dx\, F(x, y_1 + \epsilon g_1, y_2 + \epsilon g_2, y_1' + \epsilon g_1', y_2' + \epsilon g_2')$$

$$= \int_a^b dx\, \frac{d}{d\epsilon} F(x, y_1 + \epsilon g_1, y_2 + \epsilon g_2, y_1' + \epsilon g_1', y_2' + \epsilon g_2').$$

But, the chain rule gives

$$\frac{d}{d\epsilon} F(x, y_1 + \epsilon g_1, y_2 + \epsilon g_2, y_1' + \epsilon g_1', y_2' + \epsilon g_2') = \frac{\partial F}{\partial y_1} g_1 + \frac{\partial F}{\partial y_2} g_2 + \frac{\partial F}{\partial y_1'} g_1' + \frac{\partial F}{\partial y_2'} g_2'.$$

Now set $\epsilon = 0$ to obtain equation (3.50), since $\Delta S = \Delta(0)$.

Solution 3.10

(a) In this example $F(\mathbf{y}, \mathbf{y}') = \sum_{i=1}^n \sum_{j=1}^n \left(y_i' A_{ij} y_j' - y_i B_{ij} y_j \right)$ giving

$$\frac{\partial F}{\partial y_k'} = \sum_{j=1}^n A_{kj} y_j' + \sum_{i=1}^n y_i' A_{ik},$$

which, because $A_{ij} = A_{ji}$, gives

$$\frac{\partial F}{\partial y_k'} = 2\sum_{j=1}^n A_{kj} \frac{dy_j}{dx}, \quad 1 \le k \le n.$$

Similarly,

$$\frac{\partial F}{\partial y_k} = -2\sum_{j=1}^n B_{kj} y_j, \quad 1 \le k \le n,$$

which gives the quoted Euler–Lagrange equations.

(b) Using the standard rules for matrix multiplication, we see that

$$\mathbf{y}'^\top A \mathbf{y}' = \sum_{i=1}^n \sum_{j=1}^n y_i' A_{ij} y_j', \quad \mathbf{y}^\top B \mathbf{y} = \sum_{i=1}^n \sum_{j=1}^n y_i B_{ij} y_j,$$

which gives the required functional. Similarly,

$$(B\mathbf{y})_k = \sum_{j=1}^n B_{kj} y_j \quad \text{and} \quad \left(A\frac{d^2\mathbf{y}}{dx^2}\right)_k = \sum_{j=1}^n A_{kj} \frac{d^2 y_j}{dx^2}, \quad 1 \le k \le n,$$

so the Euler–Lagrange equation is $A\mathbf{y}'' + B\mathbf{y} = 0$, and assuming that A is non-singular, so that A^{-1} exists (as is the case for the positive-definite matrix A) we obtain the given equation by multiplying by A^{-1}.

Solution 3.11

(a) If
$$S[y_1, y_2] = \int_a^b dx \left[\tfrac{1}{2}\left(y_1'^{\,2} + y_2'^{\,2}\right) - V(r)\right], \quad r = \sqrt{y_1^2 + y_2^2},$$

we have
$$\frac{\partial F}{\partial y_k'} = y_k', \quad \frac{\partial F}{\partial y_k} = -\frac{dV}{dr}\frac{\partial r}{\partial y_k} \quad \text{and} \quad r\frac{\partial r}{\partial y_k} = y_k, \quad k = 1, 2.$$

Hence the Euler–Lagrange equations are
$$\frac{d^2 y_k}{dx^2} + \frac{dV}{dr}\frac{y_k}{r} = 0, \quad k = 1, 2.$$

(b) For the functional
$$S[r, \theta] = \int_a^b dx \left[\tfrac{1}{2} r'^{\,2} + \tfrac{1}{2} r^2 \theta'^{\,2} - V(r)\right],$$

we have $\partial F/\partial \theta' = r^2 \theta'$ and $\partial F/\partial \theta = 0$. The Euler–Lagrange equation for θ is
$$\frac{d}{dx}\left(r^2 \frac{d\theta}{dx}\right) = 0, \quad \text{giving} \quad \theta' = \frac{L}{r^2},$$

for some constant L. Also $\partial F/\partial r' = r'$ and $\partial F/\partial r = r\theta'^{\,2} - V'(r)$, so the Euler–Lagrange equation for r is
$$\frac{d^2 r}{dx^2} + V'(r) - r\theta'^{\,2} = 0.$$

Substituting for $\theta' = L r^{-2}$ gives the required equation.

Solution 3.12

(a) We need to find the second derivatives of y_1 and y_2 in terms of the derivatives of r and θ. Since the equations relating y_1 and y_2 to r and θ are products, this is most easily achieved using Leibniz's product rule: for second derivatives this is
$$\frac{d^2}{dx^2} f(x) g(x) = f'' g + 2 f' g' + f g''.$$

Hence for $y_1 = r\cos\theta$ we have
$$\frac{d^2 y_1}{dx^2} = \frac{d^2 r}{dx^2}\cos\theta + 2\frac{dr}{dx}\frac{d}{dx}\cos\theta + r\frac{d^2}{dx^2}\cos\theta.$$

But
$$\frac{d}{dx}\cos\theta = -\frac{d\theta}{dx}\sin\theta, \quad \frac{d^2}{dx^2}\cos\theta = -\frac{d^2\theta}{dx^2}\sin\theta - \left(\frac{d\theta}{dx}\right)^2 \cos\theta,$$

hence
$$\frac{d^2 y_1}{dx^2} = \frac{d^2 r}{dx^2}\cos\theta - 2\frac{dr}{dx}\frac{d\theta}{dx}\sin\theta - r\frac{d^2\theta}{dx^2}\sin\theta - r\left(\frac{d\theta}{dx}\right)^2 \cos\theta$$
$$= \left(r'' - r\theta'^{\,2}\right)\cos\theta - \left(r\theta'' + 2r'\theta'\right)\sin\theta. \tag{3.151}$$

Similarly, for $y_2 = r\sin\theta$ we have
$$\frac{d^2 y_2}{dx^2} = \frac{d^2 r}{dx^2}\sin\theta + 2\frac{dr}{dx}\frac{d}{dx}\sin\theta + r\frac{d^2}{dx^2}\sin\theta.$$

But
$$\frac{d}{dx}\sin\theta = \frac{d\theta}{dx}\cos\theta, \quad \frac{d^2}{dx^2}\sin\theta = \frac{d^2\theta}{dx^2}\cos\theta - \left(\frac{d\theta}{dx}\right)^2 \sin\theta,$$

hence
$$\frac{d^2 y_2}{dx^2} = \frac{d^2 r}{dx^2}\sin\theta + 2\frac{dr}{dx}\frac{d\theta}{dx}\cos\theta + r\frac{d^2\theta}{dx^2}\cos\theta - r\left(\frac{d\theta}{dx}\right)^2 \sin\theta$$
$$= \left(r'' - r\theta'^{\,2}\right)\sin\theta + \left(r\theta'' + 2r'\theta'\right)\cos\theta. \tag{3.152}$$

The equations (3.80) are therefore
$$\left(r'' - r\theta'^2\right)\cos\theta - (r\theta'' + 2r'\theta')\sin\theta + V'(r)\cos\theta = 0,$$
$$\left(r'' - r\theta'^2\right)\sin\theta + (r\theta'' + 2r'\theta')\cos\theta + V'(r)\sin\theta = 0.$$

(b) Multiplying the first of these by $\cos\theta$, the second by $\sin\theta$ and adding, gives
$$r'' - r\theta'^2 + V'(r) = 0.$$

Multiplying the first of these by $\sin\theta$, the second by $\cos\theta$ and subtracting, gives
$$r\theta'' + 2r'\theta' = 0, \quad \text{but} \quad \frac{1}{r}\frac{d}{dx}\left(r^2\theta'\right) = r\theta'' + 2r'\theta',$$
which gives the second equation.

Solution 3.13

Differentiating the given expressions for y_1, y_2 and y_3 with respect to x gives
$$y_1' = \rho'\cos\phi - \rho\phi'\sin\phi, \quad y_2' = \rho'\sin\phi + \rho\phi'\cos\phi, \quad y_3' = z',$$
so $y_1'^2 + y_2'^2 + y_3'^2 = \rho'^2 + \rho^2\phi'^2 + z'^2$. Hence the functional becomes
$$S[\rho, \phi, z] = \int_a^b dx \left[\tfrac{1}{2}\left(\rho'^2 + \rho^2\phi'^2 + z'^2\right) - V(\rho)\right].$$

The three Euler–Lagrange equations for ρ, ϕ and z are, respectively,
$$\frac{d^2\rho}{dx^2} - \rho\left(\frac{d\phi}{dx}\right)^2 + V'(\rho) = 0,$$
$$\frac{d}{dx}\left(\rho^2 \frac{d\phi}{dx}\right) = 0,$$
$$\frac{d^2 z}{dx^2} = 0.$$

The second of these equations gives $\phi' = L\rho^{-2}$, where L is a constant, and hence the first becomes $\rho'' - L^2\rho^{-3} + V'(\rho) = 0$. Hence in this coordinate system the Euler–Lagrange equations for ρ and z are uncoupled.

Solution 3.14

(a) Since $\overline{y}(\overline{x}) = y(x)$ the chain rule gives
$$\frac{d}{d\overline{x}}\overline{y}(\overline{x}) = \frac{d}{dx}y(x)\frac{dx}{d\overline{x}} = (1+\delta)\, y'(x).$$

(b) We have
$$\overline{d} = \frac{d}{1+\delta} = d\left(1 - \delta + O(\delta^2)\right) = d - d\delta + O(\delta^2),$$
with a similar expression for $\overline{c}$. Hence, to first-order the expression for T is
$$T = \int_{\overline{c}}^{\overline{d}} du\, F(u, \overline{y}(u), \overline{y}'(u)) = \int_{c-c\delta}^{d-d\delta} du\, F(u, \overline{y}(u), \overline{y}'(u)) + O(\delta^2).$$

Splitting the integration into three integrations over the intervals $(c - c\delta, c)$, (c, d) and $(d - d\delta, d)$ gives
$$T = \int_{c-c\delta}^{d-d\delta} du\, F(u, \overline{y}, \overline{y}') + O(\delta^2)$$
$$= \int_c^d du\, F(u, \overline{y}, \overline{y}') + \int_{c-c\delta}^c du\, F(u, \overline{y}, \overline{y}') - \int_{d-d\delta}^d du\, F(u, \overline{y}, \overline{y}') + O(\delta^2).$$

Also, by definition, we have, $\overline{y}(\overline{x}) = (x)$ so $\overline{y}(x/(1+\delta)) = y(x)$ giving
$$\overline{y}(u) = y(u + u\delta) = y(u) + y'(u)u\delta + O(\delta^2).$$

Differentiation with respect to u then gives
$$\overline{y}'(u) = y'(u) + \delta \frac{d}{du}(u\, y'(u)) + O(\delta^2) = y'(u) + \delta\left(y'(u) + u\, y''(u)\right) + O(\delta^2).$$

(c) With these expressions the functional T may be expanded. The integrand becomes
$$F(u, y + y'\delta, y' + (y' + uy'')\,\delta) = F + \left(uy'\frac{\partial F}{\partial y} + (y' + uy'')\frac{\partial F}{\partial y'}\right)\delta + O(\delta^2),$$
where F and its derivatives are evaluated at (u, y, y').

The boundary terms are
$$\int_{c-c\delta}^{c} du\, F = c\, F(c, y(c), y'(c))\,\delta + O(\delta^2)$$
and
$$\int_{d-d\delta}^{d} du\, F = d\, F(d, y(d), y'(d))\,\delta + O(\delta^2).$$

Hence on using equation (3.111), that is the expression of the fact that the functional is invariant under the transformation, to order δ we obtain
$$0 = \int_{c}^{d} du\, \left[uy'\frac{\partial F}{\partial y} + (y' + uy'')\frac{\partial F}{\partial y'}\right] - [uF]_{c}^{d} + O(\delta). \qquad (3.153)$$

(d) Integration by parts gives
$$\int_{c}^{d} du\, y''\left(u\frac{\partial F}{\partial y'}\right) = \left[y'\, u\frac{\partial F}{\partial y'}\right]_{c}^{d} - \int_{c}^{d} du\, y'\frac{d}{du}\left(u\frac{\partial F}{\partial y'}\right).$$

Thus, equation (3.153) becomes
$$0 = \left[u\left(y'\frac{\partial F}{\partial y'} - F\right)\right]_{c}^{d} + \int_{c}^{d} du\, \left[uy'\frac{\partial F}{\partial y} - uy'\frac{d}{du}\left(\frac{\partial F}{\partial y'}\right)\right] + O(\delta).$$

If y is a solution of the Euler–Lagrange equation, then $\dfrac{d}{du}\left(\dfrac{\partial F}{\partial y'}\right) = \dfrac{\partial F}{\partial y}$, so the integral is zero, hence
$$\left[u\left(y'\frac{\partial F}{\partial y'} - F\right)\right]_{c}^{d} = 0 \quad \text{for all } c \text{ and } d.$$

The required result follows.

(e) If $F = x^{\alpha} y'^{\beta}$, the first integral is, for some constant c,
$$(\beta - 1)x^{\alpha+1} y'^{\beta} = c \quad \text{or} \quad \frac{dy}{dx} = \frac{A}{x^{\gamma}}, \quad \gamma = \frac{1 + \alpha}{\beta}.$$

If $\gamma = 1$ the solution is $y = A \ln x + B$; otherwise it is $y(x) = Ax^{1-\gamma}/(1 - \gamma) + B$.

It is necessary that the functional
$$S = \int_{c}^{d} dx\, x^{\alpha} y'^{\beta}$$
be invariant under the scaling $x = (1 + \delta)\overline{x}$. If we set $z = \ln x$ and $\overline{z} = \ln \overline{x}$, we see that $z = \overline{z} + \ln(1 + \delta)$, so with this independent variable the scale transformation becomes a translation, which is the case dealt with in the text. Transforming the independent variable to z gives
$$S = \int_{\overline{c}}^{\overline{d}} dz\, e^{(1 + \alpha - \beta)z}\left(\frac{dy}{dz}\right)^{\beta} \quad \text{since} \quad \frac{dy}{dx} = \frac{dy}{dz}\frac{dz}{dx} = \frac{1}{e^z}\frac{dy}{dz}.$$

In order for the functonal to be translationally invariant this must not depend explicitly upon z and hence we need $1 + \alpha = \beta$, that is $\gamma = 1$.

Solution 3.15

Express the integrand of the left-hand integral in terms of $f(x)$ and $g(x)$ using the given relations. We have

$$\overline{f}'(x)^2 = f'(x)^2 \cos^2\alpha + g'(x)^2 \sin^2\alpha - 2f'(x)g'(x)\sin\alpha\cos\alpha,$$
$$\overline{g}'(x)^2 = f'(x)^2 \sin^2\alpha + g'(x)^2 \cos^2\alpha + 2f'(x)g'(x)\sin\alpha\cos\alpha.$$

Since $\cos^2\alpha + \sin^2\alpha = 1$ this gives

$$\overline{f}'(x)^2 + \overline{g}'(x)^2 = f'(x)^2 + g'(x)^2$$

and hence the two results.

Solution 3.16

All the given transformations depend continuously upon only one parameter, δ. Transformations (a) and (c) reduce to the identity when $\delta = 0$, but (b) gives $(\overline{x}, \overline{y}) = (x, -y)$. Hence (a) and (c) define a one-parameter family of transformations, but (b) does not.

Solution 3.17

(a) Separating variables puts the equation in the form

$$\int_z^y du \, \frac{1}{u(1-u)} = \int_0^t dt, \quad z = y(0).$$

The integrand of the left-hand side may be decomposed into partial fractions

$$\frac{1}{u(1-u)} = \frac{1}{u} + \frac{1}{1-u},$$

so integration gives

$$\left[\ln\left(\frac{u}{1-u}\right)\right]_z^y = t$$

and rearranging gives

$$y = \Psi(z, t) = \frac{ze^t}{1 + (e^t - 1)z}.$$

(b) There is one parameter, t, and at $t = 0$, we have $y = z$, which is clearly correct because this is just the initial condition.

Solution 3.18

In this case

$$\overline{T} = \int_{\overline{c}}^{\overline{d}} d\overline{x} \left[\left(\frac{d\overline{y}_1}{d\overline{x}}\right)^2 - \left(\frac{d\overline{y}_2}{d\overline{x}}\right)^2\right].$$

But

$$\frac{d\overline{y}_1}{d\overline{x}} = \left(\frac{dy_1}{dx}\cosh\delta + \frac{dy_2}{dx}\sinh\delta\right)\frac{dx}{d\overline{x}} \quad \text{and} \quad \frac{d\overline{y}_2}{d\overline{x}} = \left(\frac{dy_1}{dx}\sinh\delta + \frac{dy_2}{dx}\cosh\delta\right)\frac{dx}{d\overline{x}},$$

so

$$\left(\frac{d\overline{y}_1}{d\overline{x}}\right)^2 - \left(\frac{d\overline{y}_2}{d\overline{x}}\right)^2 = \left(\frac{dx}{d\overline{x}}\right)^2 \left[\left(\frac{dy_1}{dx}\right)^2 - \left(\frac{dy_2}{dx}\right)^2\right].$$

Hence

$$\overline{T} = \int_c^d dx \, \frac{d\overline{x}}{dx}\left(\frac{dx}{d\overline{x}}\right)^2 \left[\left(\frac{dy_1}{dx}\right)^2 - \left(\frac{dy_2}{dx}\right)^2\right]$$

$$= \int_c^d dx \, \frac{1}{1 + \delta g'(x)} \left[\left(\frac{dy_1}{dx}\right)^2 - \left(\frac{dy_2}{dx}\right)^2\right].$$

Hence $\overline{T} = T$ only if $g'(x) = 0$, that is, if $g(x)$ is a constant.

Solution 3.19

The Taylor expansions of the trigonometric functions are,

$$\cos\alpha = 1 - \frac{\alpha^2}{2!} + \frac{\alpha^4}{4!} + \cdots + \frac{(-1)^n \alpha^{2n}}{(2n)!} + \cdots = 1 + O(\alpha^2),$$

$$\sin\alpha = \alpha - \frac{\alpha^3}{3!} + \frac{\alpha^5}{5!} + \cdots + \frac{(-1)^n \alpha^{2n+1}}{(2n+1)!} + \cdots = \alpha + O(\alpha^3).$$

Hence, to $O(\alpha)$, from transformation (3.115) we have $\bar{y}_1 = y_1 - \alpha y_2$, $\bar{y}_2 = \alpha y_1 + y_2$, which can be written in the form $\bar{y}_1 = y_1 + \alpha \psi_1$, $\bar{y}_2 = y_2 + \alpha \psi_2$, where $\psi_1 = -y_2$ and $\psi_2 = y_1$.

Solution 3.20

In this case $\phi = 0$, $\psi_1 = -y_2$ and $\psi_2 = y_1$ so equation (3.133) becomes

$$-\frac{\partial F}{\partial y_1'} y_2 + \frac{\partial F}{\partial y_2'} y_1 = 2y_1 y_2' - 2y_2 y_1' = \text{constant}.$$

Solution 3.21

Consider the general transformation

$$\bar{y}_1 = y_1 + \delta g_1(x), \quad \bar{y}_2 = y_2 + \delta g_2(x), \quad \bar{x} = x + \delta g_3(x),$$

so that

$$\frac{d\bar{y}_1}{d\bar{x}} = \left(\frac{dy_1}{dx} + \delta g_1'(x)\right)\frac{dx}{d\bar{x}}, \quad \frac{d\bar{y}_2}{d\bar{x}} = \left(\frac{dy_2}{dx} + \delta g_2'(x)\right)\frac{dx}{d\bar{x}},$$

and the functional

$$\bar{T} = \int_{\bar{c}}^{\bar{d}} d\bar{x} \left(\left(\frac{d\bar{y}_1}{d\bar{x}}\right)^2 + \left(\frac{d\bar{y}_2}{d\bar{x}}\right)^2\right)$$

becomes, to first-order in δ,

$$\bar{T} = \int_c^d dx \, \frac{1}{1+\delta g_3'(x)} \left[\left(\frac{dy_1}{dx}\right)^2 + \left(\frac{dy_2}{dx}\right)^2 + 2\delta \left(g_1'(x)\frac{dy_1}{dx} + g_2'(x)\frac{dy_2}{dx}\right)\right].$$

Now consider the three specific cases:

(a) If $g_1 = g$ and $g_2 = g_3 = 0$, then

$$\bar{T} = T + 2\delta \int_c^d dx \, g'(x) \frac{dy_1}{dx}.$$

(b) If $g_2 = g$ and $g_1 = g_3 = 0$, then

$$\bar{T} = T + 2\delta \int_c^d dx \, g'(x) \frac{dy_2}{dx}.$$

(c) If $g_3 = g$ and $g_1 = g_2 = 0$, then

$$\bar{T} = \int_c^d dx \, \frac{1}{1+\delta g'(x)} \left[\left(\frac{dy_1}{dx}\right)^2 + \left(\frac{dy_2}{dx}\right)^2\right].$$

In all cases $\bar{T} = T$ only if $g'(x) = 0$, that is, if $g(x)$ is a constant, which we set to unity in the following. The resulting first integrals are found as follows:

(a) $\psi_1 = 1$, $\psi_2 = \phi = 0$, hence $\frac{\partial F}{\partial y_1'} = \text{constant}$, giving $y_1' = \text{constant}$.

(b) $\psi_2 = 1$, $\psi_1 = \phi = 0$, hence $\frac{\partial F}{\partial y_2'} = \text{constant}$, giving $y_2' = \text{constant}$.

(c) $\phi = 1$, $\psi_1 = \psi_2 = 0$, hence $F - y_1' \frac{\partial F}{\partial y_1'} - y_2' \frac{\partial F}{\partial y_2'} = \text{constant}$, giving $y_1'^2 + y_2'^2 = \text{constant}$.

In this example the first integral arising from the invariance with respect to translations in x can also be derived from the two first integrals due to the invariance under the translations in y_1 and y_2. Thus not all symmetries lead to new first integrals.

Solution 3.22

In this case, since $\overline{x} = x$, we have
$$\overline{T} = \int_c^d dx \left[\frac{1}{2}\left(\left(\frac{d\overline{y}_1}{dx}\right)^2 + \left(\frac{d\overline{y}_2}{dx}\right)^2 \right) + V(\overline{y}_1 - \overline{y}_2) \right].$$

Replacing $\overline{y}_k$ by $y_k + \delta g(x)$ and expanding to first-order in δ gives
$$\overline{T} = \int_c^d dx \left[\frac{1}{2}\left(\left(\frac{dy_1}{dx}\right)^2 + \left(\frac{dy_2}{dx}\right)^2 \right) + \delta g'(x)\left(\frac{dy_1}{dx} + \frac{dy_2}{dx}\right) + V(y_1 - y_2) \right],$$
$$= T + \delta \int_c^d dx \left(\frac{dy_1}{dx} + \frac{dy_2}{dx} \right) g'(x).$$

Thus $\overline{T} = T$ only if $g = $ constant. In this case equation (3.133) becomes, on setting $g = 1$ so that $\psi_1 = \psi_2 = 1$,
$$\frac{\partial F}{\partial y'_1} + \frac{\partial F}{\partial y'_2} = \text{constant}, \quad \text{that is,} \quad y'_1 + y'_2 = \text{constant}.$$

Solution 3.23

In this example $\overline{x} = \dfrac{x}{1+\delta} = x - \delta x + O(\delta^2)$ and $\overline{y} = y$. From equation (3.131) $\phi = -x$, and from equation (3.133)
$$x\left(F - y'\frac{\partial F}{\partial y'}\right) = \text{constant}.$$

Solution 3.24

If $x = z^c$,
$$\frac{dy}{dx} = \frac{dy}{dz}\frac{dz}{dx} = \frac{1}{cz^{c-1}}\frac{dy}{dz},$$
so the functional becomes
$$S[y] = c\int_{\overline{a}}^{\overline{b}} dz\, z^{c-1}\left(\frac{1}{c^2 z^{2c-2}} y'^2 - \omega^2 y^2 \right) = \frac{1}{c}\int_{\overline{a}}^{\overline{b}} dz \left(\frac{y'^2}{z^{c-1}} - c^2\omega^2 z^{c-1} y^2 \right)$$
where $\overline{a} = a^{1/c}$ and $\overline{b} = b^{1/c}$. The factor $1/c$, external to the integral, is ignored in the following analysis.

The Euler–Lagrange equation for this functional is derived from the relations
$$\frac{\partial F}{\partial y'} = \frac{2y'}{z^{c-1}} \quad \text{and} \quad \frac{\partial F}{\partial y} = -2c^2\omega^2 z^{c-1} y,$$
and is
$$\frac{d}{dz}\left(\frac{2y'}{z^{c-1}} \right) + 2c^2\omega^2 z^{c-1} y = 0,$$
which expands to
$$\frac{1}{z^{c-1}}\frac{d^2 y}{dz^2} - \frac{c-1}{z^c}\frac{dy}{dz} + c^2\omega^2 z^{c-1} y = 0.$$

Multiply by z^c to obtain the required equation. This is the transformed form of the differential equation $y'' + \omega^2 y = 0$, because the latter is the Euler–Lagrange equation corresponding to the original functional.

Solution 3.25

The functional depends only upon y_1' and y_2', so the Euler–Lagrange equations are

$$\frac{d}{dx}\left(\frac{\partial F}{\partial y_1'}\right) = 0 \quad \text{and} \quad \frac{d}{dx}\left(\frac{\partial F}{\partial y_2'}\right) = 0.$$

Expanding these equations gives

$$\frac{\partial^2 F}{\partial y_1'^2}y_1'' + \frac{\partial^2 F}{\partial y_1' \partial y_2'}y_2'' = 0 \quad \text{and} \quad \frac{\partial^2 F}{\partial y_1' \partial y_2'}y_1'' + \frac{\partial^2 F}{\partial y_2'^2}y_2'' = 0.$$

If $d \neq 0$, the only solution of these equations is $y_1'' = 0$ and $y_2'' = 0$, and integration gives the family of straight lines $y_1(x) = Ax + B$ and $y_2(x) = Cx + D$.

If $d = 0$, the two linear equations for y_1'' and y_2'' are the same. In this case we use the original equation, which integrates to $F_{y_1'}(y_1', y_2') = c_1$, where c_1 is a constant. The other equation, $F_{y_2'}(y_1', y_2') = c_2$, is the same because $d = 0$. Now there is insufficient information to find $y_1(x)$ and $y_2(x)$: all we know is that their derivatives are related.

Consider the simple example,

$$S[y_1, y_2] = \int_a^b dx\, (y_1' + y_2')^2$$

for which $d = 0$, and the Euler–Lagrange equations are both $y_1'' + y_2'' = 0$, so that $y_1 + y_2 = Ax + B$, for some constants A and B: that is, only the sum is known. Another way of seeing this is to introduce two new variables $u = y_1 + y_2$ and $v = y_1 - y_2$ so the functional, $S = \int_a^b dx\, u'^2$, depends only upon u and $u = Ax + B$, but $v(x)$ is undetermined.

If there is only one dependent variable, the Euler–Lagrange equation is

$$\frac{d}{dx}\left(\frac{\partial F}{\partial y'}\right) = 0, \quad \text{or} \quad \frac{\partial^2 F}{\partial y'^2}y'' = 0.$$

If $\partial F/\partial y' = $ constant, there is no unique solution for $y(x)$; thus the equivalent of $d \neq 0$ is that $\partial F/\partial y'$ is not a constant.

Solution 3.26

Observe that

$$\frac{d\Phi}{dx} = \frac{\partial \Phi}{\partial x} + \frac{\partial \Phi}{\partial y_1}y_1' + \frac{\partial \Phi}{\partial y_2}y_2',$$

so

$$S_2[y_1, y_2] = \int_a^b dx\, \left[F(x, y_1, y_2, y_1', y_2') + \frac{d\Phi}{dx}\right]$$
$$= S_1[y_1, y_2] + [\Phi(x, y_1, y_2)]_{x=a}^b.$$

The boundary term is independent of the path, so the stationary paths of the two functionals, S_1 and S_2, are identical and have the same Euler–Lagrange equations.

Solution 3.27

Replace each of the terms $g_k(x)y_k'$, $k = 1, 2$, in the original functional by

$$g_k(x)y_k' = \frac{d}{dx}(g_k(x)y_k) - g_k'y_k, \quad k = 1, 2,$$

to obtain S_2, with the addition of the terms $(g_k(x)y_k(x))'$ in the integrand. Then apply the result of Exercise 3.26, with $\Phi = -g_1 y_1 - g_2 y_2$.

Solution 3.28

(a) The required derivatives of the integrand are
$$\frac{\partial F}{\partial y} = \frac{e^{-x}}{2\sqrt{y-e^x y'}}, \quad \frac{\partial F}{\partial y'} = -\frac{1}{2\sqrt{y-e^x y'}}.$$
The Euler–Lagrange equation is therefore
$$\frac{d}{dx}\left(\frac{1}{2\sqrt{y-e^x y'}}\right) + \frac{e^{-x}}{2\sqrt{y-e^x y'}} = 0.$$
On expanding this expression we obtain
$$\frac{e^x y'' + e^x y' - y'}{4(y-e^x y')^{3/2}} + \frac{e^{-x}}{2\sqrt{y-e^x y'}} = 0.$$
Rearranging this gives $y'' - (3e^{-x} - 1)y' + 2e^{-2x}y = 0$.

(b) If $u = e^{-x}$, $x = -\ln u$, and $Y(u) = y(x(u))$, we have, using the chain rule
$$\frac{dy}{dx} = \frac{dY}{du}\frac{du}{dx} = -e^{-x}\frac{dY}{du}, \quad \text{hence} \quad e^x \frac{dy}{dx} = -\frac{dY}{du} = -Y'.$$
When $x = 0$, $u = 1$ and when $x = \infty$, $u = 0$, so for any function $f(x)$
$$\int_0^\infty dx\, f(x) = \int_1^0 du\, \frac{dx}{du} f(x(u)) = \int_0^1 du\, \frac{1}{u} f(-\ln u).$$
Thus the functional becomes
$$S = \int_0^1 du\, \sqrt{Y(u) + Y'(u)}.$$
In this case $\mathcal{F} = \sqrt{Y + Y'}$ and $\partial\mathcal{F}/\partial Y = \partial\mathcal{F}/\partial Y' = \frac{1}{2}(Y + Y')^{-1/2}$, and the Euler–Lagrange equation is
$$\frac{d}{du}\left(\frac{1}{2\sqrt{Y+Y'}}\right) - \frac{1}{2\sqrt{Y+Y'}} = 0 \quad \text{or} \quad Y'' + 3Y' + 2Y = 0.$$

(c) If $x = -\ln u$,
$$\frac{dY}{du} = \frac{dy}{dx}\frac{dx}{du} = -e^x \frac{dy}{dx}$$
and similarly
$$\frac{d^2 Y}{du^2} = e^x \frac{d}{dx}\left(e^x \frac{dy}{dx}\right),$$
so equation (3.148) for $y(x) = Y(u(x))$ becomes $y'' - (3e^{-x} - 1)y' + 2e^{-2x}y = 0$.

Solution 3.29

If $F = y'^2 + z'^2 + 2yz$, then
$$\frac{\partial F}{\partial y'} = 2y', \quad \frac{\partial F}{\partial y} = 2z \quad \text{and} \quad \frac{\partial F}{\partial z'} = 2z', \quad \frac{\partial F}{\partial z} = 2y$$
so the Euler–Lagrange equations are
$$\frac{d^2 y}{dx^2} - z = 0 \quad \text{and} \quad \frac{d^2 z}{dx^2} - y = 0.$$
Putting $z = y''$ into the second equation gives $d^4 y/dx^4 - y = 0$. Now put $y = \alpha \exp(\lambda x)$, where α and λ are constants, to obtain $\lambda^4 = 1$, that is $\lambda = \pm 1$ and $\pm i$. Thus the general solution is
$$y = A\cos x + B\sin x + C\cosh x + D\sinh x,$$
and
$$z = y'' = -A\cos x - B\sin x + C\cosh x + D\sinh x.$$

The boundary conditions at $x = 0$ give $A + C = 0$ and $C - A = 0$, hence $A = C = 0$.
The boundary conditions at $x = \pi/2$ then give
$$\tfrac{3}{2} = B + D\sinh(\pi/2) \quad \text{and} \quad \tfrac{1}{2} = -B + D\sinh(\pi/2).$$

Subtracting and adding these equations gives $B = 1/2$, and $2D\sinh(\pi/2) = 2$. Hence
$$y = \frac{\sinh x}{\sinh(\pi/2)} + \tfrac{1}{2}\sin x \quad \text{and} \quad z = \frac{\sinh x}{\sinh(\pi/2)} - \tfrac{1}{2}\sin x.$$

Solution 3.30

Using the expression for y' derived at the begining of the solution of Exercise 3.2(b), the functional becomes
$$S[y] = \int_a^b dx\, F(x, G(z), G'(z)z'), \quad \text{where} \quad z' = \frac{dz}{dx}.$$

Since
$$\frac{\partial F}{\partial z'} = \frac{\partial F}{\partial y'}\frac{\partial y'}{\partial z'} = G'(z)\frac{\partial F}{\partial y'} \quad \text{and} \quad \frac{\partial F}{\partial z} = \frac{\partial F}{\partial y}\frac{\partial y}{\partial z} + \frac{\partial F}{\partial y'}\frac{\partial y'}{\partial z} = \frac{\partial F}{\partial y}G'(z) + \frac{\partial F}{\partial y'}G''(z)z',$$

the Euler–Lagrange equation becomes
$$\frac{d}{dx}\left(G'(z)\frac{\partial F}{\partial y'}\right) - \frac{\partial F}{\partial y}G'(z) - \frac{\partial F}{\partial y'}G''(z)\,z' = 0.$$

Solution 3.31

(a) The functional is
$$S[x] = \int_A^B dy\, \frac{dx}{dy}\frac{1}{x^2}\frac{1}{x'(y)^2} = \int_A^B dy\, \frac{1}{x^2 x'(y)}, \quad x(A) = 1,\ x(B) = 2.$$

(b) Putting $F = 1/(x^2 x')$, we have
$$\frac{\partial F}{\partial x'} = -\frac{1}{(xx')^2} \quad \text{and} \quad \frac{\partial F}{\partial x} = -\frac{2}{x^3 x'},$$

so the Euler–Lagrange equation is
$$\frac{d}{dy}\left(\frac{1}{(xx')^2}\right) - \frac{2}{x^3 x'} = 0.$$

But
$$\frac{d}{dy}\left(\frac{1}{(xx')^2}\right) = -\frac{2x''}{x^2 x'^3} - \frac{2}{x^3 x'},$$

which gives the required equation.

(c) Integrating once gives $x^2 x' = c$ for some constant c. Integrating again gives $x^3/3 = cy + d$. The boundary conditions give
$$\tfrac{1}{3} = cA + d \quad \text{and} \quad \tfrac{8}{3} = cB + d,$$

giving
$$x^3 = \frac{7y + B - 8A}{B - A}.$$

Solution 3.32

(a) This problem is slightly different from all previous examples because there are no boundary conditions prescribed: this means that some care is needed. If $y(x)$ is a stationary path and $y(x) + \epsilon g(x)$ a varied path, the standard analysis gives

$$\Delta F = 2 \int_0^X dx \left[xy'(x)g'(x) - \left(x - \frac{n^2}{x} \right) y(x)g(x) \right],$$

and integration by parts gives

$$\Delta F = 2 \left[xy'(x)g(x) \right]_0^X - 2 \int_0^X dx\, g(x) \left[\frac{d}{dx}(xy') + \left(x - \frac{n^2}{x} \right) y(x) \right].$$

The boundary term vanishes at $x = 0$ because $y'(x)$ is finite here. However it does not vanish at $x = X$. We solve this problem by imposing the boundary condition $y(X) = Y \neq 0$ on the admissible functions, so now $g(X) = 0$ and the boundary term vanishes. Hence the Euler–Lagrange equation is

$$\frac{d}{dx}(xy') + \left(x - \frac{n^2}{x} \right) y(x) = 0,$$

or

$$x^2 \frac{d^2 y}{dx^2} + x \frac{dy}{dx} + \left(x^2 - n^2 \right) y = 0.$$

This is a linear equation; the solution is therefore undetermined to within a multiplicative constant, so the value of Y, provided it is not zero, is irrelevant.

(b) If $x = f(u)$, then the chain rule gives

$$\frac{dy}{dx} = \frac{dy}{du}\frac{du}{dx} = \frac{w'(u)}{f'(u)},$$

and the functional becomes

$$F[w] = \int_{u_0}^{u_1} du\, f'(u) \left[\frac{f(u)}{f'(u)^2} w'(u)^2 - \left(f(u) - \frac{n^2}{f(u)} \right) w(u)^2 \right]$$

$$= \int_{u_0}^{u_1} du \left[\frac{f(u)}{f'(u)} w'(u)^2 - \left(f(u)f'(u) - \frac{f'(u)}{f(u)} n^2 \right) w(u)^2 \right],$$

where $f(u_0) = 0$ and $f(u_1) = X$.

(c) If $f = e^u$ this functional becomes

$$F[w] = \int_{-\infty}^{u_1} du \left[w'(u)^2 - \left(e^{2u} - n^2 \right) w(u)^2 \right], \quad \text{where } u_1 = \ln X.$$

The Euler–Lagrange equation for this functional is $w''(u) + \left(e^{2u} - n^2 \right) w = 0$, and this has a solution $w(u) = J_n(e^u)$.

Solution 3.33

In the first case, if $\partial F / \partial y = 0$ the differential equation derived in Exercise 2.33 can be written as

$$\frac{d}{dx}\left(\frac{d}{dx}\left(\frac{\partial F}{\partial y''} \right) - \frac{\partial F}{\partial y'} \right) = 0, \quad \text{hence} \quad \frac{d}{dx}\left(\frac{\partial F}{\partial y''} \right) - \frac{\partial F}{\partial y'} = \text{constant}.$$

In the second case, if $\partial F / \partial x = 0$ it is easiest to use a version of the analysis described in Section 3.5.1. Equation (3.93) becomes

$$T = \int_{c-\delta}^{d-\delta} du\, F(\overline{y}(u), \overline{y}'(u), \overline{y}''(u))$$

$$= \int_c^d du\, F(\overline{y}, \overline{y}', \overline{y}'') + \int_{c-\delta}^c du\, F(\overline{y}, \overline{y}', \overline{y}'') - \int_{d-\delta}^d du\, F(\overline{y}, \overline{y}', \overline{y}'').$$

On putting $u = v + \delta$ and expanding to first-order in δ, we obtain

$$T = \int_c^d dv \left[F(y, y', y'') + \delta \frac{\partial F}{\partial y} y' + \delta \frac{\partial F}{\partial y'} y'' + \delta \frac{\partial F}{\partial y''} y''' \right] - \delta \left[F(y, y', y'') \right]_c^d + O(\delta^2),$$

and because G is invariant under x-translations, this gives

$$0 = \int_c^d dv \left[\frac{\partial F}{\partial y} y' + \frac{\partial F}{\partial y'} y'' + \frac{\partial F}{\partial y''} y''' \right] - \left[F(y, y', y'') \right]_c^d + O(\delta), \tag{3.154}$$

which is the equivalent of equation (3.103). Now integrate by parts:

$$\int_c^d dv\, y'' \frac{\partial F}{\partial y'} = \left[y' \frac{\partial F}{\partial y'} \right]_c^d - \int_c^d dv\, y' \frac{d}{dv}\left(\frac{\partial F}{\partial y'} \right),$$

$$\int_c^d dv\, y''' \frac{\partial F}{\partial y''} = \left[y'' \frac{\partial F}{\partial y''} - y' \frac{d}{dv}\left(\frac{\partial F}{\partial y''} \right) \right]_c^d + \int_c^d dv\, y' \frac{d^2}{dv^2}\left(\frac{\partial F}{\partial y''} \right).$$

Hence equation (3.154) becomes

$$\left[y'' \frac{\partial F}{\partial y''} - y' \frac{d}{dv}\left(\frac{\partial F}{\partial y''} \right) + y' \frac{\partial F}{\partial y'} - F \right]_c^d + \int_c^d dv\, y' \left(\frac{\partial F}{\partial y} - \frac{d}{dv}\left(\frac{\partial F}{\partial y'} \right) + \frac{d^2}{dv^2}\left(\frac{\partial F}{\partial y''} \right) \right) = O(\delta).$$

But the integrand is zero and the end points c and d are arbitrary, hence

$$y'' \frac{\partial F}{\partial y''} - \left(\frac{d}{dx}\left(\frac{\partial F}{\partial y''} \right) - \frac{\partial F}{\partial y'} \right) y' - F = \text{constant}.$$

CHAPTER 4
Newtonian dynamics

4.1 Introduction

The principal aims of this chapter are to formulate Newton's equations of motion as a variational principle, to derive from this Lagrange's form of these equations, and to show how these may be used to simplify the derivation of the equations of motion for certain types of mechanical systems.

Joseph-Louis Lagrange, 1736–1813.

You may be wondering why Newton's laws, as formulated in previous courses, need reformulating: there are several reasons and, in no particular order of significance, the main ones are listed below.

(1) A variational formulation of the equations of motion is more elegant and leads to powerful methods of finding and understanding solutions of the equations of motion; see point (3).

(2) The application of Newton's equations to even quite simple systems is awkward and tedious. This difficulty was overcome by Lagrange who, building upon the work of James Bernoulli and D'Alembert, reformulated Newton's equations of motion in terms of *generalised coordinates* – a term that will be defined in Section 4.3.1.

James Bernoulli, 1654–1705.

Jean le Rond d'Alembert, 1717–1783, was an illegitimate child, abandoned on the steps of the church of St Jean le Rond, from which his forenames came. This church, originally adjacent to the north wall of Notre Dame Cathedral, was demolished between 1742 and 1765.

The complete transformation to the modern formulation was made by the Irish mathematician Hamilton in 1827 with the introduction of the Hamiltonian function – although an important element of modern physics this is not part of the present course. He also described solutions of Newton's equations in geometric terms using what is now named *Hamilton's principle*, from which Lagrange's equations of motion can be derived. The main aspects of Hamilton's principle (which does not involve Hamilton's function) will be described in Section 4.4.

It is worth noting here that towards the end of the 19th century, as physics progressed to the study of atomic particles, the classical mechanics of Newton was found to fail. Then Hamilton's ideas were important in the development of the new quantum mechanics that describes the dynamics of atoms and molecules.

Sir William Rowan Hamilton, 1805–1865.

(3) By expressing Lagrange's equations of motion as a variational principle, the underlying mathematical structure of the equations of motion is clarified. This leads to the development of techniques and methods that help understand the behaviour of solutions and also help with their computation.

Perhaps the most dramatic example of this is Poincaré's discovery of chaos in 1898, made by extending the geometric ideas first formulated by Hamilton.

Jules Henri Poincaré, 1854–1912.

(4) A variational principle formulation of the equations of motion is independent of any particular coordinate system, because of the invariance of the Euler–Lagrange equations discussed in Section 3.2. This feature provides a route to powerful methods of approximation, to the more general theories of special and general relativity, and is helpful in the formulation of the wave equations that describe electromagnetic radiation. These topics are beyond the scope of this course, though a wave equation will be derived in Subsection 4.5.4, in order to show how the first and last parts of this course are connected.

In Section 4.2 we quote Newton's laws and derive Newton's equations of motion for three typical situations: these three examples are chosen to illustrate some of the technical problems that occur, and which are overcome by Lagrange's formulation, described in Section 4.4. The material contained in Section 4.2 covers the same ground as previous level 2 courses: it is *not* assessed in this course, but you need to be familiar with the ideas in order to understand later sections. In Section 4.3 we discuss Newton's laws and introduce the important ideas of generalised coordinates and kinetic and potential energy, all of which are essential to Lagrange's formulation of Newton's equations of motion.

Lagrange's equations and Hamilton's principle are described in Section 4.4. In the final section before the end of chapter exercises we apply Lagrange's method to a variety of problems, ending with the derivation of the wave equation, which extends the theory to dynamical systems of infinitely many interacting particles and relates the theory of this section to that developed in Block 1, Chapter 1.

The assessment questions based on this chapter will be on material contained in Sections 4.4 and 4.5, though an understanding of this work is possible only if you are familiar with the ideas discussed in earlier sections. None of the historical material is assessed, and may be omitted.

4.1.1 On nomenclature

Newtonian dynamics describes the motion of physical objects. In order to obtain manageable equations, we idealise these physical objects by mathematical constructs, and also make various kinds of approximations. The link between the mathematics and the physical world is by experiment, often overlooked but one of the crucial methods of validating our approximations.

The primary concepts of Newtonian dynamics are *mass*, *length* and *time*; here we assume the intuitive notion of these; you should be aware, however, that these concepts are more difficult to understand than first appearances suggest.

The basic object of dynamics is a *particle*, a mathematical object with mass, but no volume or structure. Particles are used to approximate a wide variety of physical objects, for example, electrons and stars. It is a useful approximation where the physical size of objects is small by comparison with the distance between them, and where their internal structure is assumed to play an insignificant role in their relative motion.

An example is the relative motion of the Earth and the Sun, which for many purposes is accurately described using two particles, because their radii, respectively 4×10^3 and 4×10^5 miles, are far less than the mean distance between them, 9.3×10^7 miles, and over this distance the physical structure and slight departures from spherical symmetry of each does not significantly affect the relative motion, except over very long times. If, on the other hand, the rotation of the Earth is of interest, then it needs to

4.1 Introduction

be treated as a body with structure. As usual the approximation required depends upon the questions being asked, but it is often difficult to know a priori which approximations should be made.

The next object of dynamics is the *rigid body*, which is used to approximate a physical object that does not deform significantly under the forces present; for instance a steel ball bearing can often be treated as rigid. The distinction between a rigid body and a particle is that the former has an orientation in space, so it can rotate and have angular momentum; the spinning Earth is an example. In this chapter we do not deal with rigid bodies, nevertheless we shall use the terms body and particle freely.

The collection of objects being considered is often referred to as *the system* and its mathematical description is named a *dynamical system*: sometimes the distinction between these entities is blurred. The motion of a dynamical system is described by differential equations, generally referred to as *the equations of motion*.

A number of other common English adjectives are used in a more precise manner than in common parlance, but such precision is essential for our simplified approximations of reality. Those used freely in this chapter are listed below.

Light: For example, 'a light rod' – this means that the mathematical object has zero mass, and is useful for approximating a physical object with a mass sufficiently small that we can assume it to have a negligible effect on the motion. The opposite of light is *heavy*, which means that the mass cannot be ignored.

Inextensible: This term is used to describe rods or strings, for instance – it means that the length of the mathematical object does not change and is used when the forces involved do not significantly change the dimensions of the physical object.

Rigid: For example, a 'rigid rod' – it means that the mathematical object does not bend and is used when the physical object bends little. A synonym is *stiff*. A rigid body can be constructed by joining two or more particles with light, inextensible, stiff rods. A rigid body need not be inextensible, but it must not bend.

Smooth: This is a particularly useful term meaning *without friction*; synonyms are *free* and *frictionless*. The opposite of smooth is *rough*, and for such surfaces friction cannot be ignored.

Dot notation: In this chapter we shall use the notation introduced by Newton whereby the first and second derivatives of a function of time, $f(t)$, are represented by $\dot{f}$ and $\ddot{f}$ respectively, that is, $\dot{f} = df/dt$ and $\ddot{f} = d^2f/dt^2$.

4.2 Newton's laws

In this section we use Newton's laws to derive the equations of motion for some simple systems. It is assumed that you are familiar with Newton's laws, and have used them to derive equations for some simple dynamical systems. The conventional statements of Newton's three laws are as follows.

Law I   Every body continues in its state of rest, or moves with constant velocity in a straight line, unless a force is applied to it.

Law II   The change in motion is proportional to the force acting and takes place in the direction of the straight line along which the force acts.

Law III   For every action (that is force) there is always an equal and opposite reaction: or, the forces two bodies exert on each other are always equal and opposite directions.

You may be familiar with a different version of the second law in which "change in motion" is replaced by "rate of change of velocity". Newton used the former because he experimented with impulses; see Exercise 4.30 (page 185). The version given here describes his observations more appropriately and includes continuously varying forces.

These laws provide an accurate description of a wide variety of physical phenomena. The mathematical expression of the second law is normally called *Newton's equation of motion*. These laws are, however, awkward to apply in all but the simplest situations, and the mathematical structure of the resulting differential equations is not usually clear. The Lagrangian formulation of these principles overcomes these problems.

In the following three subsections we apply Newton's laws to a variety of examples in order to demonstrate some of these problems. We start with some quite simple examples and end by deriving complicated equations of motion for the double pendulum. It must be emphasised that this material is not assessed, but provides essential background that you should understand. Moreover, we shall use the examples provided here to illustrate the more general discussion that follows in Section 4.3, and which leads to the statement of Hamilton's principle in Section 4.4. Although this material is not assessed we advise you to do as many of the exercises as time permits.

4.2.1 Particle moving on a parabolic wire

In this example a bead is constrained to slide under gravity on a smooth rigid wire in the shape of the parabola $y = ax^2/2$, $a > 0$, where the y-axis is vertically upwards, as shown in Figure 4.1.

4.2 Newton's laws

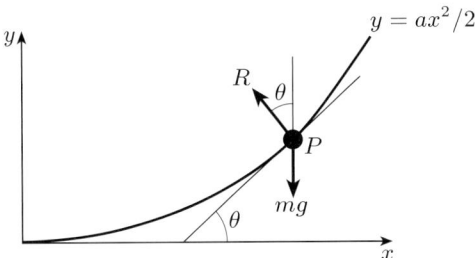

Figure 4.1 Diagram showing a bead of mass m at a point P on a wire in the shape of the parabola, $y = ax^2/2$, for $x \geq 0$. The tangent through P makes an angle θ with the x-axis.

In the first instance we shall suppose the wire to be stationary, then we shall allow it to move bodily in the y-direction in a prescribed manner. In both cases the position of the bead can be defined by one coordinate, and the system is said to have one degree of freedom. Here we shall take this to be the x-coordinate, but others could be used; for instance we could use the y-coordinate or the distance along the curve from any given point.

In order to use Newton's equations directly we introduce a force of constraint which, in this case, is a reaction normal to the tangent at P and of magnitude R, which is necessary to keep the bead on the wire. The reaction is perpendicular to the wire because we assume that there is no friction, so there is no force between the bead and the wire along the tangent. If the gradient of the wire at P is $\tan\theta$, as shown in Figure 4.1, then the force of reaction makes an angle θ to the vertical through P. Thus the x- and y-components of Newton's equations of motion are, respectively,

$$m\ddot{x} = -R\sin\theta \quad \text{and} \quad m\ddot{y} = R\cos\theta - mg. \tag{4.1}$$

Now we need to eliminate the force of reaction; this is achieved by multiplying the first equation by $\cos\theta$, the second by $\sin\theta$ and adding to give

$$m(\ddot{x}\cos\theta + \ddot{y}\sin\theta) = -mg\sin\theta. \tag{4.2}$$

Now divide by $m\cos\theta$ to give

$$\ddot{x} + \ddot{y}\tan\theta = -g\tan\theta. \tag{4.3}$$

But x and y are related because the bead is constrained to the wire, that is, $y = ax^2/2$, and the chain rule gives

$$\dot{y} = ax\dot{x} \quad \text{and} \quad \ddot{y} = ax\ddot{x} + a\dot{x}^2. \tag{4.4}$$

Also we have $\tan\theta = y'(x) = ax$. Thus equation (4.3) becomes

$$\ddot{x}\left(1 + a^2x^2\right) + a^2x\dot{x}^2 + agx = 0. \tag{4.5}$$

Later, we shall see that this is Lagrange's equation, and we shall learn how it may be derived more simply without introducing the reaction force.

Exercise 4.1

By differentiating the quantity

$$E = \tfrac{1}{2}m\dot{x}^2\left(1 + a^2 x^2\right) + \tfrac{1}{2}mgax^2 \tag{4.6}$$

with respect to t and using the equation of motion (4.5), show that E is a positive constant. Hence show that the equation of motion can be integrated once to yield the first order equation

$$\frac{dx}{dt} = \pm\sqrt{\frac{2E - mgax^2}{m(1 + a^2 x^2)}} \quad \text{where} \quad |x| \le \sqrt{\frac{2E}{mga}}.$$

We shall see later that the quantity E, defined in equation (4.6), is the total energy of this system, and is a constant on all solutions of the equations of motion, with a value that depends upon the initial conditions.

A vertically moving wire

Now suppose that the wire is moving vertically with given displacement $\gamma(t)$ from a fixed point at time t. We can represent this motion by allowing the coordinate system Oxy to move, so that the height of the x-axis from a fixed horizontal line is $\gamma(t)$. Then the acceleration in the y-direction relative to this line is $\ddot{y} + \ddot{\gamma}$, so this replaces $\ddot{y}$ in equation (4.1), which then becomes

$$m\ddot{x} = -R\sin\theta \quad \text{and} \quad m\ddot{y} = R\cos\theta - m(g + \ddot{\gamma}). \tag{4.7}$$

Thus the effect of moving the wire is the same as modifying the gravitational force. On using the constraint, $y = ax^2/2$, we obtain, the equation of motion,

$$\ddot{x}\left(1 + a^2 x^2\right) + a^2 x \dot{x}^2 + a\left(g + \ddot{\gamma}\right)x = 0. \tag{4.8}$$

If the wire is moving uniformly, that is $\dot{\gamma} = $ constant, then this equation is identical to the original equation. Otherwise the energy, E, defined by equation (4.6), is not a constant, as shown in Exercise 4.2; it can be shown that there is no equivalent constant quantity.

This example may seem rather artificial; however, it illustrates a very common circumstance. For instance, if we ride in an aircraft through turbulent air or in a vehicle over rough ground, the coordinate system Oxy is fixed in the vehicle and the additional 'force' – the term $ma\ddot{\gamma}$ in equation (4.8) – is the force felt by a passenger of mass m.

Exercise 4.2

Show that when the wire is moving, the quantity E, defined in equation (4.6), varies as follows

$$\frac{dE}{dt} = -\tfrac{1}{2}ma\ddot{\gamma}\frac{d}{dt}\left(x^2\right).$$

In this example the elimination of the reaction R was achieved relatively painlessly. In other circumstances the equivalent analysis is more difficult and the purpose of Lagrange's theory, developed later, is to make this process easier.

4.2.2 The simple pendulum

The simple pendulum – also known as the vertical pendulum – is one of the simplest dynamical systems and has attracted the attention of numerous philosophers, although Galileo discovered practically all of its major properties. Until the early 20th century it provided the primary method of accurately measuring time. Even today the dynamics of the pendulum is important in the study of more complicated nonlinear systems and in some aspects of molecular dynamics. Some of these aspects of the pendulum are discussed further in Subsection 4.5.1.

Our idealised pendulum comprises a heavy particle, P, of mass m, firmly attached to a light, stiff, inextensible rod of length l. The other end of the rod is attached to a fixed point, O, about which it can swing freely, that is, without friction and air resistance. Such a system is shown schematically in Figure 4.2.

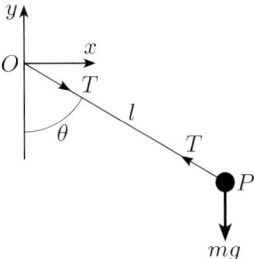

Figure 4.2 Diagram showing the idealised pendulum of length l, with smooth hinge at O and particle of mass m at P. The coordinate axes are shown with Oy vertically upwards.

We shall assume that the rod OP can rotate about the pivot O, so there are no constraints on the value of θ, the angle between OP and the downward vertical. A bicycle wheel with a suitable weight attached to a spoke and with the axle held firmly is an example of a mathematically equivalent system. The configuration of this system is specified uniquely by the single variable θ, and we expect Newton's equation of motion to depend only upon this angle and its derivatives. In this example the configurations corresponding to $\theta + 2n\pi$, for integer values of n, are physically equivalent, so we expect the equations of motion to depend periodically upon θ.

The simplest approach is to use the conservation of energy. However, this method cannot be used for more complicated systems, so here we apply Newton's laws directly using a widely applicable method.

There are two forces acting upon the mass P:

(a) the external downward force of gravity, with magnitude mg;

(b) the tension, of magnitude T, which acts along the line OP and constrains the distance OP to be constant; the tension also acts on the hinge at O, but a frictionless hinge is unaffected by the tension, and the tension has no effect on the motion of the mass, other than to constrain it to move on a circle.

The mass P moves in the plane of the paper. We define a coordinate system with origin at the point of support, O, with the y-axis vertically upwards and the x-axis as shown in Figure 4.2 and describe the position of P using Cartesian coordinates, (x, y), which are

$$x = l \sin\theta \quad \text{and} \quad y = -l \cos\theta. \tag{4.9}$$

The force acting on P is the vector sum of the tension, having Cartesian coordinates $(-T\sin\theta, T\cos\theta)$, and the gravitational force $(0, -mg)$. Hence the two components of Newton's equations of motion are

$$m\ddot{x} = -T\sin\theta \quad \text{and} \quad m\ddot{y} = T\cos\theta - mg. \tag{4.10}$$

Thus we have two equations and three unknowns, x, y and T. However, because the length OP is fixed, x and y are related with $x^2 + y^2 = l^2$. The tension T may be eliminated from equations (4.10) by multiplying the equation for $\ddot{x}$ by $\cos\theta$, the equation for $\ddot{y}$ by $\sin\theta$ and adding the resulting equations, to obtain

$$m\left(\ddot{x}\cos\theta + \ddot{y}\sin\theta\right) = -mg\sin\theta. \tag{4.11}$$

But, from the geometric relations of equation (4.9) we have, on differentiating each twice,

$$\begin{aligned}\dot{x} &= l\dot{\theta}\cos\theta, \quad \ddot{x} = l\left(\ddot{\theta}\cos\theta - \dot{\theta}^2\sin\theta\right), \\ \dot{y} &= l\dot{\theta}\sin\theta, \quad \ddot{y} = l\left(\ddot{\theta}\sin\theta + \dot{\theta}^2\cos\theta\right),\end{aligned} \tag{4.12}$$

so $\ddot{x}\cos\theta + \ddot{y}\sin\theta = l\ddot{\theta}$, and equation (4.11) becomes

$$\frac{d^2\theta}{dt^2} = -\frac{g}{l}\sin\theta. \tag{4.13}$$

This is the required equation for θ: it is independent of the mass m and depends upon l and g only via the ratio g/l. Notice that, as suggested above, this equation of motion depends periodically upon θ. It can be shown that the solutions of this differential equation cannot be expressed as finite combinations of elementary functions.

You should note that the main objective of the analysis between equations (4.9) and (4.13) is the removal of the tension T. In Section 4.5 we shall see how to derive the equations of motion without introducing the tension.

In this chapter, we shall not usually concern ourselves with the solutions of the equations of motion, but this problem is a little special so we note that if $|\theta|$ is small, $\sin\theta \simeq \theta$ and an approximation to the equation of motion is

$$\ddot{\theta} + \omega^2\theta = 0, \quad \omega = \sqrt{\frac{g}{l}}, \tag{4.14}$$

which has the general solution $\theta(t) = A\sin(\omega t + \alpha)$, where the amplitude, A, and the phase, α, are constants, depending upon the initial conditions.

Note that $\theta(t) = A\sin(\omega t + \alpha)$ is an exact solution of an approximate equation, and it is not a priori obvious that it is also an approximate solution of the exact equation. In this case the solution of the exact equation (4.13) is known so we know that for sufficiently small times, the solution of the approximate equation is close to this exact solution.

Thus the motion is periodic with period $2\pi/\omega$, which is independent of the amplitude, so the system is said to be *isochronous*, the property that makes the pendulum so useful as a time-keeper. The isochrony of the pendulum was first explicitly stated by Galileo (in a letter dated 1602), who at the same time noted that the period was independent of the mass. That the period is proportional to the square root of the length, l, was later discovered by Galileo, and published in 1632. If the amplitude of the motion increases too much, the approximation $\sin\theta \simeq \theta$ breaks down and the period of the motion then depends upon the amplitude, see Exercise 4.29 (page 184).

A moving point of support

We now consider a variant of the simple pendulum in which the hinge is not rigidly attached to a stationary point, but is fixed to a heavy particle, of mass M, that can slide smoothly along a straight horizontal rail, as shown in Figure 4.3. In this example we use a coordinate system fixed relative to the rail, with the y-axis vertically upwards and Ox along the rail. The resulting analysis is easier than if we used a moving origin at A.

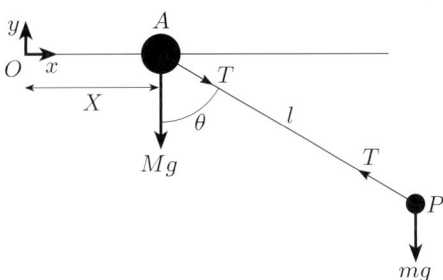

Figure 4.3 Schematic diagram of a simple pendulum with its hinge attached, at A, to a particle of mass M that can slide freely on a horizontal rail along the x-axis. As in Figure 4.2, the rod AP is rigid, light, inextensible with length l, has a particle of mass m firmly attached at P and can swing freely at A.

The reason for treating this system is to illustrate how to analyse a system that requires two variables to define its configuration. As before, there is the angle θ between the pendulum and the downward vertical: in addition there is the distance, X, between the particle at A and a given stationary point, O, on the rail. The tension in the pendulum now acts on A and causes it to accelerate.

Using the coordinate system Oxy shown in the figure, we see that the Cartesian coordinates of P are $(X + l\sin\theta, -l\cos\theta)$, as in equation (4.9), but with the distance X added to the x-coordinate. The equation of motion of the mass at A is

$$M\ddot{X} = T\sin\theta, \tag{4.15}$$

and the equations of motion for the mass at P are

$$m\frac{d^2x}{dt^2} = -T\sin\theta \quad \text{and} \quad m\frac{d^2y}{dt^2} = T\cos\theta - mg. \tag{4.16}$$

These are identical to the equations of the fixed pendulum, equations (4.10); however, this motion is affected by the motion of the mass at A through the tension T. Comparing the equations for x and X, we see that $M\ddot{X} + m\ddot{x} = 0$, that is $M\dot{X} + m\dot{x} =$ constant; the momentum in the x-direction is constant, which is just Newton's first law applied to the centre of mass of the system – we expand upon these ideas later, so do not be concerned if they are unfamiliar.

We now need to eliminate the tension from equations (4.16), but because $x = X + l\sin\theta$, the analysis is not as simple as before. Equation (4.11) is still valid, but now differentiation gives

$$\dot{x} = \dot{X} + l\dot{\theta}\cos\theta, \quad \ddot{x} = \ddot{X} + l\left(\ddot{\theta}\cos\theta - \dot{\theta}^2\sin\theta\right), \tag{4.17}$$

$$\dot{y} = l\dot{\theta}\sin\theta, \quad \ddot{y} = l\left(\ddot{\theta}\sin\theta + \dot{\theta}^2\cos\theta\right), \tag{4.18}$$

so $\ddot{x}\cos\theta + \ddot{y}\sin\theta = \ddot{X}\cos\theta + l\ddot{\theta}$ and the equation of motion (4.13) is replaced by

$$\ddot{X}\cos\theta + l\ddot{\theta} = -g\sin\theta. \tag{4.19}$$

Now we use the relation $m\ddot{x} = -M\ddot{X}$ with second of equations (4.17) to obtain

$$\ddot{X} = -\mu l\left(\ddot{\theta}\cos\theta - \dot{\theta}^2\sin\theta\right), \quad \mu = \frac{m}{M+m}. \tag{4.20}$$

By substituting this into equation (4.19) and rearranging we obtain

$$\left(1 - \mu\cos^2\theta\right)l\ddot{\theta} + \mu l\dot{\theta}^2\sin\theta\cos\theta + g\sin\theta = 0. \tag{4.21}$$

Notice that the equation for θ does not depend upon the other variable, X: there is a good reason for this which will be discussed later, in Section 4.5.1.

The equation of motion (4.21) looks quite complicated. In fact, this equation can be reduced to a simpler first-order equation, but we leave this simplification until Section 4.5.1, specifically equation (4.107) (page 186).

Exercise 4.3

Show that if M is sufficiently large, so that μ can be approximated by zero, equation (4.21) reduces to the original pendulum equation, equation (4.13), and the equation for X, becomes $\ddot{X} = 0$. Explain the meaning of the latter equation.

Exercise 4.4

Show that the quantity

$$E = \tfrac{1}{2}(M+m)\dot{X}^2 + \tfrac{1}{2}ml^2\dot{\theta}^2 + ml\dot{\theta}\dot{X}\cos\theta - mgl\cos\theta$$

is a constant of the motion. [Hint: compute $\dot{E}$ and use the equations of motion, 4.19 and (4.20), to show that this is zero.]

4.2.3 The double pendulum

The double pendulum was studied extensively in the 18th century by Daniel Bernoulli and Euler, with three published works between 1738 and 1741, and by Johann Bernoulli in 1742: but only for the small amplitude oscillations were the correct equations of motion obtained. Thus, half a century after the publication of Newton's laws, their application was not routine.

The double pendulum is shown schematically in Figure 4.4: it comprises a simple pendulum suspended from another simple pendulum by a frictionless hinge. Both pendulums move in the same plane.

Daniel Bernoulli, 1700–1782.
Leonhard Euler, 1707–1783.
Johann Bernoulli, 1667–1748.

A discussion of the history and evolution of dynamics during this period is given in *The evolution of dynamics: vibration theory from 1687 to 1742* by J. T. Cannon and S. Dostrovsky (Springer, 1981).

4.2 Newton's laws

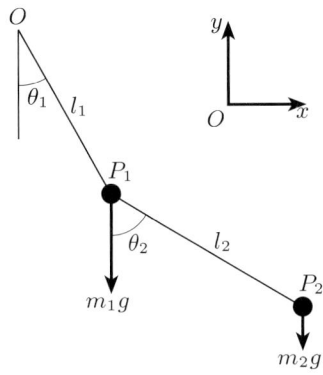

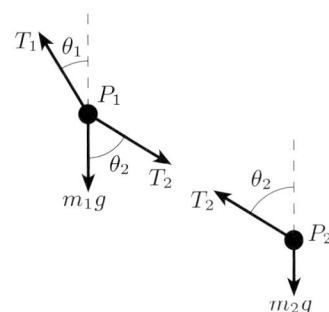

Figure 4.4 The double pendulum.

Figure 4.5 Diagram showing the forces acting on each particle of the double pendulum.

In this system the rods OP_1 and P_1P_2 are rigid, light and inextensible. The masses m_1 and m_2 are fixed at P_1 and P_2 and the two hinges at O and P_1 are smooth. Further, we assume that neither θ_1 nor θ_2 is constrained so both may take all values.

The configuration of this system is given uniquely by the values of the angles θ_1 and θ_2, and the system is said to have two degrees of freedom. We expect Newton's equations to involve only these angles and their derivatives. Further, the configurations corresponding to $(\theta_1 + 2n_1\pi, \theta_2 + 2n_2\pi)$, where n_1 and n_2 are integers, are physically identical, so the equations of motion must be 2π-periodic in these angles.

The derivation of the equations of motion that follows is complicated. Our aim is to show directly the difficulties that can arise when Newton's equations are applied to relatively simple mechanical problems, with the hope that you will appreciate the efforts made by Lagrange to provide a more sophisticated, yet simpler method. You will not be assessed on this material, but please read it and note the opportunities for error.

Equation of motion for P_2

As with the simple pendulum we first need to find the Cartesian coordinates of P_1 and P_2 and, as in that case, we use a reference frame Oxy with the origin at O and the y-axis vertically upwards, as shown in Figure 4.4. The coordinates of P_1 are given by equation (4.9), with a suitable notation change:

$$x_1 = l_1 \sin\theta_1 \quad \text{and} \quad y_1 = -l_1 \cos\theta_1. \tag{4.22}$$

The coordinates of P_2 with respect to P_1 are similar, $(l_2 \sin\theta_2, -l_2 \cos\theta_2)$, so the coordinates with respect to O are

$$\begin{aligned} x_2 &= x_1 + l_2 \sin\theta_2 = l_1 \sin\theta_1 + l_2 \sin\theta_2, \\ y_2 &= y_1 - l_2 \cos\theta_2 = -l_1 \cos\theta_1 - l_2 \cos\theta_2. \end{aligned} \tag{4.23}$$

The equations of motion for P_2 are similar to those derived for the simple pendulum, equation (4.10), because the forces on this mass are similar, compare Figure 4.5 and 4.2. These equations are

$$m_2 \frac{d^2 x_2}{dt^2} = -T_2 \sin\theta_2 \quad \text{and} \quad m_2 \frac{d^2 y_2}{dt^2} = T_2 \cos\theta_2 - m_2 g. \tag{4.24}$$

As before, we need to eliminate T_2 from these equations. This is achieved by multiplying the first equation by $\cos\theta_2$, the second by $\sin\theta_2$ and adding the results to obtain

$$m_2\left(\ddot{x}_2\cos\theta_2 + \ddot{y}_2\sin\theta_2\right) = -m_2 g\sin\theta_2, \tag{4.25}$$

which is the equivalent of equation (4.11). The next step, however, is more difficult because both x_2 and y_2 depend upon θ_1 besides θ_2. We require the accelerations $\ddot{x}_2$ and $\ddot{y}_2$ to be expressed in terms of the angles: for this we first differentiate equations (4.23) with respect to t, to obtain

$$\dot{x}_2 = l_1\dot{\theta}_1\cos\theta_1 + l_2\dot{\theta}_2\cos\theta_2, \quad \dot{y}_2 = l_1\dot{\theta}_1\sin\theta_1 + l_2\dot{\theta}_2\sin\theta_2. \tag{4.26}$$

A second differentiation gives

$$\begin{aligned}\ddot{x}_2 &= l_1\left(\ddot{\theta}_1\cos\theta_1 - \dot{\theta}_1^2\sin\theta_1\right) + l_2\left(\ddot{\theta}_2\cos\theta_2 - \dot{\theta}_2^2\sin\theta_2\right), \\ \ddot{y}_2 &= l_1\left(\ddot{\theta}_1\sin\theta_1 + \dot{\theta}_1^2\cos\theta_1\right) + l_2\left(\ddot{\theta}_2\sin\theta_2 + \dot{\theta}_2^2\cos\theta_2\right).\end{aligned} \tag{4.27}$$

From these, we obtain

$$\ddot{x}_2\cos\theta_2 + \ddot{y}_2\sin\theta_2 = l_1\left(\ddot{\theta}_1\cos(\theta_1 - \theta_2) - \dot{\theta}_1^2\sin(\theta_1 - \theta_2)\right) + l_2\ddot{\theta}_2, \tag{4.28}$$

where the identities

$$\begin{aligned}\cos(u-v) &= \cos u\cos v + \sin u\sin v, \\ \sin(u-v) &= \sin u\cos v - \cos u\sin v\end{aligned} \tag{4.29}$$

have been used.

Hence equation (4.25) becomes

$$l_1\ddot{\theta}_1\cos(\theta_1 - \theta_2) - l_1\dot{\theta}_1^2\sin(\theta_1 - \theta_2) + l_2\ddot{\theta}_2 = -g\sin\theta_2, \tag{4.30}$$

which is one of the required equations of motion. Notice that this equation is independent of the masses and that if the rod OP_1 is held stationary ($\dot{\theta}_1 = 0$), this equation becomes equation (4.13), with suitable changes of variables, as should be the case.

Equation of motion for P_1

The force on P_1 involves the tension T_2, so we first need to use equations (4.24) to express T_2 in terms of the angles. This is achieved by multiplying the first of equation (4.24) by $\sin\theta_2$, the second by $\cos\theta_2$ and subtracting to obtain

$$m_2\left(\ddot{x}_2\sin\theta_2 - \ddot{y}_2\cos\theta_2\right) = -T_2 + m_2 g\cos\theta_2. \tag{4.31}$$

Now use equations (4.27) again to express the left-hand side of this equation in the form

$$\ddot{x}_2\sin\theta_2 - \ddot{y}_2\cos\theta_2 = -l_1\ddot{\theta}_1\sin(\theta_1 - \theta_2) - l_1\dot{\theta}_1^2\cos(\theta_1 - \theta_2) - l_2\dot{\theta}_2^2, \tag{4.32}$$

hence

$$T_2 = m_2\left(g\cos\theta_2 + l_1\ddot{\theta}_1\sin(\theta_1 - \theta_2) + l_1\dot{\theta}_1^2\cos(\theta_1 - \theta_2) + l_2\dot{\theta}_2^2\right). \tag{4.33}$$

The force on P_1 involves both tensions, T_1 and T_2, and the gravitational force, so its equation of motion is more complicated than that of P_2. The x- and y-components of the force acting on P_1 are

$$\begin{aligned}F_{1x} &= -T_1\sin\theta_1 + T_2\sin\theta_2, \\ F_{1y} &= T_1\cos\theta_1 - T_2\cos\theta_2 - m_1 g,\end{aligned} \tag{4.34}$$

so the equations of motion are

$$\begin{aligned}m_1\ddot{x}_1 &= -T_1\sin\theta_1 + T_2\sin\theta_2, \\ m_1\ddot{y}_1 &= T_1\cos\theta_1 - T_2\cos\theta_2 - m_1 g.\end{aligned} \tag{4.35}$$

Since T_2 is known, equation (4.33), we need only eliminate T_1 from these equations to find the equation of motion. As usual this is achieved by multiplying the first equation by $\cos\theta_1$, the second by $\sin\theta_1$ and adding, to obtain,

$$m_1\left(\ddot{x}_1\cos\theta_1 + \ddot{y}_1\sin\theta_1\right) = -T_2\sin(\theta_1-\theta_2) - m_1 g\sin\theta_1. \quad (4.36)$$

But, as in the analysis following equation (4.12) we have $\ddot{x}_1\cos\theta_1 + \ddot{y}_1\sin\theta_1 = l_1\ddot{\theta}_1$, so this equation of motion becomes

$$m_1 l_1 \ddot{\theta}_1 = -T_2\sin(\theta_1-\theta_2) - m_1 g\sin\theta_1. \quad (4.37)$$

We can now use the expression (4.33) for T_2 to simplify this equation. Note, however, that the algebra is complicated and, if you wish, ignore it and go straight to the final result, equation (4.40).

On substituting for T_2 we obtain

$$m_1 l_1 \ddot{\theta}_1 = -m_1 g\sin\theta_1 - m_2\sin\theta_{12}\left(g\cos\theta_2 + l_1\ddot{\theta}_1\sin\theta_{12} + l_1\dot{\theta}_1^2\cos\theta_{12} + l_2\dot{\theta}_2^2\right), \quad (4.38)$$

where $\theta_{12} = \theta_1 - \theta_2$. In this expression the term $\sin^2\theta_{12}$ occurs and we replace it by $1 - \cos^2\theta_{12}$, then rearrange the equation to

$$(m_1+m_2) l_1 \ddot{\theta}_1 = -g\left(m_1\sin\theta_1 + m_2\sin\theta_{12}\cos\theta_2\right)$$
$$- m_2\left(\left[-l_1\ddot{\theta}_1\cos\theta_{12} + l_1\dot{\theta}_1^2\sin\theta_{12}\right]\cos\theta_{12} + l_2\dot{\theta}_2^2\sin\theta_{12}\right). \quad (4.39)$$

Now use equation (4.30) to replace the term in square brackets to obtain

$$l_1\ddot{\theta}_1 + g\sin\theta_1 + \frac{m_2 l_2}{m_1+m_2}\left(\ddot{\theta}_2\cos(\theta_1-\theta_2) + \dot{\theta}_2^2\sin(\theta_1-\theta_2)\right) = 0. \quad (4.40)$$

This and equation (4.30) are the final equations of motion for the double pendulum. They are a pair of coupled, nonlinear, second-order differential equations for θ_1 and θ_2, the solutions of which cannot be expressed as a finite combination of known functions. Notice that the masses m_1 and m_2 occur only as the ratio $m_2/(m_1+m_2)$. A far simpler method of deriving these equations is given in Section 4.5.2.

4.3 A discussion of Newton's laws

In the previous section Newton's laws were quoted and then applied to some representative problems. Although sometimes a daunting task, particularly the last example, these applications were relatively straightforward, provided panic is avoided – but it helps that the collective wisdom accumulated over the past four centuries ensures that the correct answers are known. For novel systems of any complexity the derivation of the equations of motion is not easy and there are many opportunities for error. Lagrange's method, described in the next section, makes this task easier.

Newton's laws were presented together in *Principia* I (1687), but the ideas had been evolving for some time before. Indeed, Newton stated that the first two laws were known to Galileo and they were stated in *Horologium Oscillatorium*, published by Huygens in 1673. The third law is new: Newton

Sir Isaac Newton, 1642–1727.

Galileo Galilei, 1564–1642.

Christiaan Huygens, 1629–1695.

applied this specifically to collisions, and it is related to momentum conservation, previously described by Wallis, Wren and Huygens in submissions to the Royal Society in 1668.

John Wallis, 1616–1703.
Christopher Wren, 1632–1723.

Newton's laws are not self-evident. Indeed, from the time of Aristotle to the 17th century, when Galileo stated the first law (published in 1612), the contrary was believed. The Aristotelian description of motion is complicated and includes the view that the natural state of a body is one of rest, which follows from the notion that everything must be moved by something, a proposition that is in direct contradiction to the modern idea of relative motion; Galileo's analysis, discussed briefly in Section 4.5.1, changed the description to something very close to the modern view. Nevertheless, even today, for the Earth-bound, it takes some effort of thought to overcome the Aristotelian view. This is why the first law was so significant, and why it is stated separately even if redundant in modern physics.

Aristotle, 384–322 BC.

After 1687 a great deal of effort was expended in applying Newton's laws, in attempts to understand to which physical systems they were applicable, in trying to find solutions to various problems, particularly those related to the solar system and in trying to understand the nature of these solutions. The first problem was to derive Kepler's laws from first principles. These three laws were published by Kepler between 1609 and 1619. The first two, published in 1609, are: (i) Planets describe ellipses round the Sun, the Sun being at one focus; (ii) the line joining the Sun to any planet sweeps out equal areas in equal times. The third law, published in 1619, states that the squares of the periods of the planets are proportional to the cubes of the major axes of their orbits. The first two laws were deduced from careful observations of the orbit of Mars. The theoretical origin of Kepler's third law is discussed in Exercise 4.53. Amongst many other works, Kepler also suggested that tides were caused by the attraction of the Moon.

Johannes Kepler, 1571–1630.

The next major triumph was the accurate prediction of the return date of what we now call Halley's comet. Comets have been observed since ancient times as isolated events. In the West, the oldest record is a Babylonian inscription dating from 1140 BC, but in China records date from 1600 BC, with an observation of Halley's comet in 467 BC. However, the notion of comets returning had not been considered seriously much before the time of Newton. Tycho Brahe carefully observed the 1577 comet and concluded that it was in an approximately circular orbit round the Sun; hence giving credence to the view that comets might be part of the solar system. During the 17th century various authors suggested that comets move on parabolic orbits, and Newton used observations of the great comet of 1680 (now thought to have a period of about 8800 years) to compute its orbit which he estimated to be a parabola with the Sun as its focus. Halley then attempted to compute the orbits of 24 comets for which sufficiently accurate data was known and found that among these 24 examples three, those of 1531, 1607 and 1682, seemed almost identical and that the observations were separated by the almost equal intervals of 76 and 75 years. Halley concluded that these orbits were actually ellipses corresponding to the same comet, which would return in 1758.

The word comet is from the Greek for 'long-haired one', a reference to the comet's tail.

Tycho Brahe, 1546–1601.

Edmond Halley, 1656–1742.

The precise calculation of the return date was not easy because the orbit is perturbed by Jupiter and Saturn. In 1748 three French mathematicians Clairaut, Lalande and Lepaute sequestered themselves to calculate the return date of the comet, last seen in 1682. Lalande wrote: "During the six months we calculated from morning to night, sometimes even at meals; the consequence of which was, that I contracted an illness which changed my constitution for the rest of my life." The difficulty in accounting for the effects of Jupiter and Saturn was exacerbated by the fact that probably the

Alexis Chaude Clairaut, 1713–1765.
Joseph-Jérôme le Français Lalande, 1732–1807.
Nichole-Reine Lepaute, 1723–1788.

4.3 A discussion of Newton's laws

only calculating aids they had were logarithms. These approximate computations of the orbit predicted the date of return to its perihelion to be April 1759; its actual return in March, just within the errors quoted by Clairaut, provided confirmation that Newton's laws were more widely applicable than originally thought.

The perihelion of an orbit is the point of closest approach to the Sun.

The trajectory of Halley's comet has now been computed backwards and all 30 previous passages described in historical documents, over 22 centuries, have been authenticated, including the 1066 passage shown in the Bayeux tapestry.

It took longer, however, to understand the behaviour of the solutions of Newton's equations – over two hundred years after their first publication, in 1898 Poincaré showed that these could behave in the most alarmingly complicated manner, which we now name *chaos*.

The simplicity of Newton's laws is deceptive: they combine definitions with observations from nature, and use partly intuitive concepts together with some unexamined assumptions about the properties of space and time. These ideas are conceptually difficult and it was not until two hundred years after Newton that the Austrian physicist Mach in 1883 analysed some of these problems logically: this type of analysis played a significant role in the development of Einstein's special theory of relativity.

Ernst Mach, 1838–1916.

Albert Einstein, 1879–1955.

The first law implies the existence of isolated bodies and *inertial* coordinate systems: this needs some explanation. Any experiment is performed within an assumed *reference frame*, usually defined by the laboratory in which it is performed. In principle, experiments done in a particular laboratory will allow one to deduce, or at least confirm, physical laws and in particular the laws of mechanics. However, the same experiment performed in different laboratories may produce different results. For instance, a laboratory in a uniformly accelerating space rocket, or fixed to the end of a large rotating beam, would not confirm Law I, without taking into account the acceleration of the laboratory, of which the scientist may not be aware. A simple example of this effect is the motion of a particle on an accelerating wire considered on page 148.

An inertial reference frame is one in which Law I holds; essentially, Law I asserts the existence of an inertial frame. Further, it is assumed that the laws of motion are invariant under transformations between inertial reference frames. The limiting case considered in Exercise 4.3 is an example, because the uniform motion of the pendulum support does not affect the pendulum motion. Thus, two laboratories moving uniformly relative to each other would deduce the same dynamical properties for the pendulum. This is Galileo's relativity principle which states that space and time are absolute; that length, time and mass are independent of the relative (uniform) motion of the observer. A consequence of this principle is that the speed of light depends upon the relative motion of the observer, an inference contradicted by the experiments of Michelson and Morley in 1887. Despite, this, because the speed of light is so large, about $3 \times 10^{10} \mathrm{cm\,s}^{-1}$, or 186,000 $\mathrm{miles\,s}^{-1}$, Galileo's relativity principle is approximately true in many important situations. In an inertial reference frame the second law implies the first.

Albert Abraham Michelson, 1852–1931.

Edward Williams Morley, 1838-1923.

Newton's second law relates the 'change in motion' to an impressed force. Here the term 'motion' means the momentum, a vector quantity we often denote by **p**. Newton used the term 'change' rather than the more modern 'rate of change' because his experimental examples involved impulses; one such example is described after Exercise 4.30 (page 185). For a body of constant mass, its momentum is usually the product of its mass, m, and its velocity, **v**; the main exception being the motion of charged particles in electromagnetic fields, which we shall not consider. The modern version of Newton's second law is, in symbolic form,

$$\frac{d\mathbf{p}}{dt} = \mathbf{F} \quad \text{or, for constant mass,} \quad m\frac{d\mathbf{v}}{dt} = \mathbf{F}, \tag{4.41}$$

where $\mathbf{F}$ is the force acting on the body: the first of these equations applies when the mass varies, as in rocket motion. These equations are known as Newton's equations of motion.

The notion of a force is difficult. First we observe that it is a vector. Second, implicit in this law is the fact that forces add vectorially, a fact that Newton added as a corollary; this result is an empirical law, crucial for our understanding of many particle systems. The first observations leading to this law were those of Galileo in his description of projectile motion, in which he separated horizontal and vertical components of motion.

Forces always arise from interactions between particles, and if we ever found an acceleration without an interaction, we would be in a dreadful mess. It is the interaction which is physically significant and which is responsible for the force. For this reason, when we isolate a body sufficiently from its surroundings, we expect it to move uniformly in an inertial frame. An isolated body is one from which interactions have been eliminated – logically, this means that isolated bodies cannot be seen, because observing an object means interacting with it using, for instance, light; however, light exerts a very small force so we can see only approximately isolated bodies: a deep analysis of this problem leads to Heisenberg's uncertainty principle and to quantum mechanics, which is beyond the scope of this course.

4.3.1 Internal and external forces

In this and the next section we introduce the idea of generalised coordinates, starting with a discussion of N interacting particles. If you wish to have in mind a realistic example, for $N = 3$ think of the Sun–Earth–Jupiter system as a set of 3 interacting particles: another is the Sun–Earth–Moon system.

Any pair of isolated particles will interact through a force acting along the line joining their centres because, by definition, this is the only preferred direction The force on particle i due to particle j is denoted by $\mathbf{F}_{ij}$, and similarly the force on j due to i by $\mathbf{F}_{ji}$. Newton's third law is satisfied only if $\mathbf{F}_{ij} = -\mathbf{F}_{ji}$. Thus for the two particle Sun–Earth system the attractive forces are as shown in Figure 4.6.

Figure 4.6 Schematic diagram of the interaction between the Earth, of mass m_E, and the Sun, with mass m_S, often accurately described by assuming both to be point particles.

Note that in this case the magnitudes of the force on each particle are identical, but the effects on each are quite different, simply because the Sun is far more massive than the earth, $m_\mathrm{S} = 328\,900 m_\mathrm{E}$; a consequence of this is explored in Exercise 4.5.

The interaction between any pair of particles is unaffected by any third particle. The pairwise forces $\mathbf{F}_{ij}$, which usually depends upon the positions

4.3 A discussion of Newton's laws

of particle i and j, are normally named *internal*, for reasons that will become clear later; see equation (4.45).

In addition to these pairwise, internal forces each particle of a system may be affected by *external* forces, due to another system apart from that being considered. Usually external forces are gravitational or electromagnetic in origin. We denote the external force on particle i by $\mathbf{F}_i^{(e)}$. An example showing the difference between internal and external forces is discussed next and from this we shall see that this distinction depends upon the approximation used to describe the system.

Consider the Sun–Earth–Jupiter system, with each body being treated as a single particle, with the aim of understanding the motion of the Earth. It is possible, and a good approximation, to consider this system of three particles in isolation and then the only forces are the internal forces between the three particles: there are no external forces. This, however, gives rise to equations of motion that are very hard to solve.

The force between the Earth and Jupiter is, however, relatively weak by comparison to that between each planet and the Sun, so the simplest first approximation is obtained by ignoring this interaction, and then the Earth and Jupiter orbit the Sun unaffected by each other. This approximation conforms to Kepler's description of the solar system giving slightly elliptic orbits at mean distances of 1 AU and 5.2 AU, respectively, from the Sun, and with periods of one (tropical) year and 11.86 years, respectively.

The weaker interaction between the Earth and Jupiter has an asymmetric effect because Jupiter is so massive, $m_J \simeq 320 m_E$, which means that a better but more complicated approximation, may be obtained by assuming that the motion of Jupiter is unaffected by the Earth, but that the Earth is affected by Jupiter. This asymmetry is represented mathematically by treating the effect of Jupiter as an external force on the Earth, as shown schematically, in Figure 4.7. In this approximation to the motion of the Earth the only internal force is that between the Earth and the Sun, and the resulting equations of motion are easier to understand.

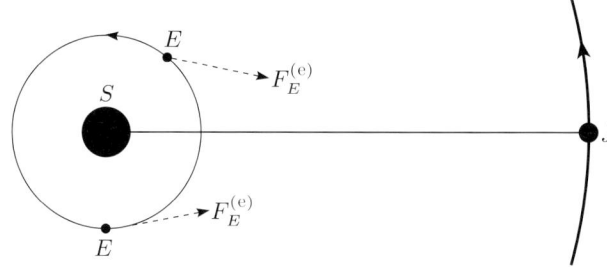

Figure 4.7 Diagram showing how the effect of Jupiter on the motion of the Earth may be represented by a time-dependent external force, $F_E^{(e)}$, with the force on Jupiter due to the Earth being ignored.

Note that at different times of the year this external force is in different directions and has different magnitudes, so depends upon the time.

For N particles each could be affected by the $N-1$ interactions with the other particles. The motion of two interacting particles is well understood and solved in any text book on dynamics. The motion of three or more interacting particles is, however, exceedingly difficult to understand. Indeed, the *three-body problem* remains one of the most celebrated problems in celestial mechanics; it can be simply stated as follows. Three particles move in space under their mutual gravitational attraction; given their initial positions and

An AU is the astronomical unit of length and is defined to be approximately the mean distance between the Earth and the Sun and is 1.496×10^8 km (9.3×10^7 miles); there are 63 240 AU in a light year.

A *tropical year* is defined to be the time between successive vernal equinoxes. An equinox is the time when the Earth's axis of rotation is perpendicular to the line between the Sun and the centre of the Earth. The two yearly equinoxes are at about 21st March (the vernal equinox) and 22nd September (the autumnal equinox). A tropical year is 365.2422 days, with a day defined to be 86 400 seconds.

velocities determine their subsequent motion. The importance of this problem can be judged from the fact that between 1750 and 1900 more than 800 papers on this topic were published and new work is still being produced.

In general, for N interacting particles, where the ith has known mass, m_i, and position vector $\mathbf{r}_i$, with Cartesian coordinates, $\mathbf{r}_i = (x_i, y_i, z_i)$, with respect to a given reference frame, there are $3N$ degrees of freedom. The three Newtonian equations of motion for each of the N particles are, in vector form

$$m_i \frac{d^2 \mathbf{r}_i}{dt^2} = \mathbf{F}_i^{(e)} + \sum_{\substack{j=1 \\ j \neq i}}^{N} \mathbf{F}_{ij}, \quad i = 1, 2, \ldots, N. \tag{4.42}$$

Note that in this equation the $j = i$ term of the sum is omitted because the quantity $\mathbf{F}_{ii}$ is not defined: in some texts, for algebraic convenience, it is defined to be zero. More important you should note that normally all forces depend upon the position vectors. These $3N$ equations are the mathematical statement of Newton's second law. In these differential equations we normally know all the masses and all the forces, which usually depend upon the position vectors in a known manner: the objective is either to solve these equations for $\mathbf{r}_i(t)$ given some initial conditions, that is $\mathbf{r}_i(t_0), \dot{\mathbf{r}}_i(t_0)$, $k = 1, 2, \ldots, N$, or to obtain some qualitative understanding of the solution: for instance, in studies of the solar system we should like to know whether the distance between the Earth and the Sun remains within given limits for all times. In recent years this type of problem has become of public concern with various groups worrying about the distance between the Earth and individual asteroids of the larger variety.

In principle, equations (4.42) may be solved, possibly numerically, for each of the $3N$ variables provided all the forces together with all the initial conditions are known. In practice, except for the simplest of systems, these equations can be solved only by numerical integrations and then, with a given accuracy, only for a finite time, which depends upon the accuracy required: furthermore, it is usually difficult to understand the properties of a system from numerical solutions.

4.3.2 The centre of mass

Before proceeding it is worth noting one of the consequences of Newton's third l law, which manifests itself in the symmetry relation $\mathbf{F}_{ij} = -\mathbf{F}_{ji}$. This involves the *centre of mass* of a system of N particles which is defined by the vector $\overline{\mathbf{R}}$,

$$M\overline{\mathbf{R}} = \sum_{i=1}^{N} m_i \mathbf{r}_i, \quad M = \sum_{i=1}^{N} m_i, \tag{4.43}$$

so that M is the total mass. We now show that the centre of mass of a system of N particles moves like a single particle, of mass M equal to the total mass of the system, acted upon by the sum of all the external forces, and that this motion is independent of the internal forces.

This is shown by differentiating the vector $\overline{\mathbf{R}}$ twice and then using equation (4.42) to obtain

$$M\ddot{\overline{\mathbf{R}}} = \sum_{i=1}^{N} m_i \ddot{\mathbf{r}}_i = \sum_{i=1}^{N} \mathbf{F}_i^{(e)} + \sum_{i=1}^{N} \sum_{\substack{j=1 \\ j \neq i}}^{N} \mathbf{F}_{ij}. \tag{4.44}$$

4.3 A discussion of Newton's laws

But in the second sum, the terms $\mathbf{F}_{rs}$ and $\mathbf{F}_{sr}$ occur, for each r and s, only once and occur as $\mathbf{F}_{rs} + \mathbf{F}_{sr} = 0$; thus the double sum is zero.

Thus the equation of motion for the centre of mass is

$$M\ddot{\mathbf{R}} = \sum_{i=1}^{N} \mathbf{F}_i^{(e)}, \tag{4.45}$$

which is Newton's equation for a particle of mass M at $\overline{\mathbf{R}}$, acted on by the sum of all external forces. This means that the centre of mass motion is unaffected by the forces between the constituent particles, which is why they are named *internal* forces. If the sum of the external forces is zero, then $\ddot{\mathbf{R}} = 0$ and the centre of mass is either stationary or moves with constant velocity: in either case, it is then usually convenient to put the coordinate origin at the centre of mass.

Equation (4.45) also means that a rigid body can often be treated as a single particle with all external forces acting through the centre of mass.

We have already seen an example of equation (4.45). For the simple pendulum hanging from a freely moving mass, the equations of motion being given by equations (4.15) and (4.16) (page 151), it was noted that $M\ddot{X} + m\ddot{x} = 0$, but this is just the x-component of $\ddot{\mathbf{R}}$ because there are no external forces in this direction.

Exercise 4.5

(a) The centre of mass, C, of two particles of mass m_1 and m_2, with position vectors $\mathbf{r}_1$ and $\mathbf{r}_2$, is given by equation (4.43). Show that the vectors from $\mathbf{r}_1$ to C and $\mathbf{r}_2$ to C are, respectively,

$$\overline{\mathbf{r}}_1 = \overline{\mathbf{R}} - \mathbf{r}_1 \qquad \overline{\mathbf{r}}_2 = \overline{\mathbf{R}} - \mathbf{r}_2$$
$$= \frac{m_2}{m_1 + m_2}\mathbf{r}_{12} \quad \text{and} \quad = -\frac{m_1}{m_1 + m_2}\mathbf{r}_{12}$$

where $\mathbf{r}_{12} = \mathbf{r}_2 - \mathbf{r}_1$ is the vector from m_1 to m_2.

(b) The masses of the Earth, m_E, and the Sun, m_S, are related by $m_\mathrm{S} = 328\,900 m_\mathrm{E}$, and the mean distance between their centres is 149.6×10^6 km. Use the result derived in the first part of this exercise to show that the centre of mass of the the Earth–Sun system is a distance of 455 km from the Sun's centre.

(c) The two most massive bodies in the solar system are the Sun and Jupiter, with masses related by $m_\mathrm{S} = 1047.4 m_\mathrm{J}$, and the mean distance between their centres is 778.3×10^6 km. Show that centre of mass of the the Jupiter–Sun system is a distance of 7.43×10^5 km from the Sun's centre. Note that the radius of the Sun is 6.91×10^5 km.

4.3.3 Generalised coordinates

In any description of a dynamical system it is necessary to specify the positions and velocities of the constituent parts, and our objective is to determine how these change with time. The positions are always defined with reference to a coordinate system, usually fixed or moving with a constant velocity. The simplest coordinate system is the Cartesian system with the equations of motion being expressed in terms of the Cartesian coordinates of each particle. Although the simplest system, it rarely results in the simplest equations of motion, so it is almost always necessary to use other coordinate systems. The choice of coordinate system is so important to our understanding of any equations of motion that it is important to be able to formulate these equations in terms of appropriate coordinates.

Generalised coordinates and the associated theory provide this flexibility by enabling the equations of motion to be written in terms of the smallest number of equations. You have already seen such coordinates in action when the equations for the two Cartesian coordinates of the simple pendulum, equations (4.10), became one equation for θ, equation (4.13): in this example θ is the generalised coordinate which automatically allows for the length of the pendulum to be constant.

Another example is the system comprising two particles connected by a light, rigid inextensible rod, of length l; the two position vectors $\mathbf{r}_1$ and $\mathbf{r}_2$ are not independent because they are constrained by the relation $|\mathbf{r}_1 - \mathbf{r}_2| = l$: a hydrogen molecule in which the internuclear distance is assumed fixed is an example, which is a good approximation at low temperatures. A coordinate system describing these two particles needs to take into account the constraint.

Another example is a single particle constrained to move along a curved wire, or on a surface: in these cases the position vector is allowed to move in only one or two dimensions respectively and the coordinate system used needs to reflect this.

Constraints are important because they limit the possible configurations of the system and this simplifies the equations of motion. They are introduced because we need to approximate the system to obtain manageable equations of motion involving only the essential variables. For instance, the solid block (which may be treated as a single particle), of mass M, sliding smoothly on a fixed, horizontal surface, shown on the left in Figure 4.8, is constrained to move in two rather than three dimensions.

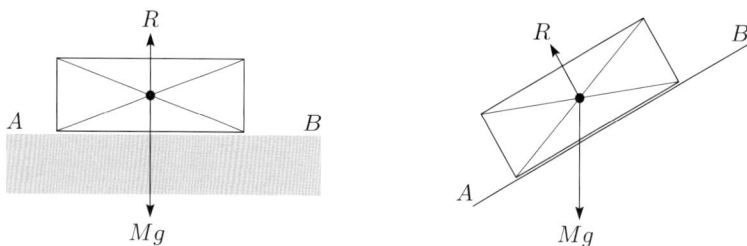

Figure 4.8 Diagram showing the forces on a solid block on a horizontal surface (left) and a tilted surface (right).

Gravity exerts a downward force, Mg, on the block, but on the left the surface does not allow the block to move in this direction, so Newton's first law shows that there is no net vertical force on the block, and the third law implies that there must be an upward force of reaction from the surface to exactly compensate the downward force of gravity. This is denoted by

4.3 A discussion of Newton's laws

R in the figure, and is perpendicular to the surface. This reaction can be rationalised by considering the forces between the constituent molecules of the block and the surface. But for most purposes this is not helpful and Newton's third law cuts through this complex atomic description to a far simpler and usable description, giving $R = Mg$. For a smooth surface the force of reaction is perpendicular to the surface.

Now tip the surface, AB, as shown on the right in the figure. The reaction **R** remains perpendicular to the surface, because the molecular interactions are unaffected by gravity, and is no longer parallel to the gravitational force, so there is a net downward force. Provided **R** remains perpendicular to the surface, the second law shows that the block slides down the plane. This is what happens if there is no friction and then the value of **R** does not affect the motion. In this example the reaction **R** is known as a *force of constraint*, and is similar to that introduced to determine the equations of motion of a bead on a wire, in Section 4.2.1. The reaction **R** is a consequence of the geometric constraint restricting the block to lie on the table. Generally, geometric constraints are associated with forces that impose these constraints: these are known as *forces of constraint*.

Another simple example is the pulley system shown in Figure 4.9.

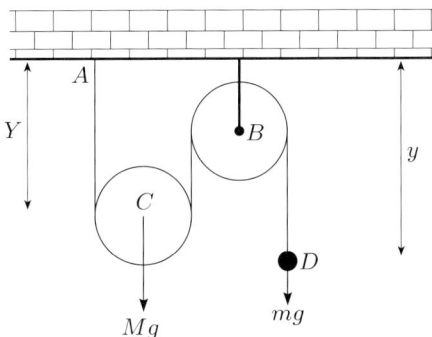

Figure 4.9 A simple pulley system comprising a light inextensible rope fixed to the ceiling at A, a pulley B, also fixed to the ceiling, and a second, pulley C, of mass M, and a particle D of mass m, both free to move in the vertical plane. The rope passes round both pulleys and is attached to the particle D.

In this example we assume that the pulleys rotate with no friction, have no rotational energy and that the rope does not stretch. These assumptions mean that the tension T in the rope, not shown in the diagram, has the same value at all points. The constraint is that the length of the rope is constant, which means that the heights of the particle D and the pulley C are not independent (provided the rope does not break). This system is analysed using Lagrangian methods in Exercise 4.28.

There are different types of constraints. In this course we consider only those that can be expressed in terms of equations like

$$f(\mathbf{r}_1, \mathbf{r}_2, \ldots, \mathbf{r}_N, t) = 0 \tag{4.46}$$

where f is a function of the position vectors and, possibly, the time t. For instance, the block on the left-hand side of Figure 4.8 has the constraint $z = $ constant, where the z-axis is vertically upwards and there is a similar relation when the plane is tipped. For the pulley system shown in Figure 4.9, the two distances between the pulley C and the particle D and the ceiling, Y and y respectively, are related by $2Y + y = $ constant, because the rope is inextensible.

Constraints of the type defined in equation (4.46) are named *holonomic* constraints and a system with such constraints is named a *holonomic system*. Other types of constraints exist and are named *nonholonomic* constraints, but are not considered in this course.

Before describing the properties of holonomic constraints, we list a few examples.

A bead
Constrained to slide on stiff wire with a fixed shape, as in Section 4.2.1, a bead has one degree of freedom because its position is uniquely described by a single variable.

A moving wire
A variant of the previous example is obtained by moving a wire in a prescribed manner. A single variable defines the position of the particle so the system still has one degree of freedom.

Alternatively, we could allow the shape of the curve to change in a prescribed manner: the system still has one degree of freedom; see Exercise 4.48 for an example.

The pendulum
The simple pendulum has one degree of freedom and the double pendulum has two.

Motion on a surface
A point on a two-dimensional surface, stationary or moving in a prescribed manner, requires two variables, so a particle sliding on such a surface has two degrees of freedom.

A block sliding on a plane surface
A block sliding on a plane surface has three degrees of freedom, because the centre of mass of the block requires two coordinates to define its position. A third coordinate is needed to define the orientation of axes fixed in the block relative to axes fixed on the surface.

Child's spinning top
With the point of contact, O, fixed, a child's spinning top has three degrees of freedom because an axis OP, through the point of contact at O, fixed along the symmetry axis of the top, requires two angles to define its orientation, the vector $\mathbf{n}$, and the angle of rotation about this axis, ϕ; see Figure 4.10.

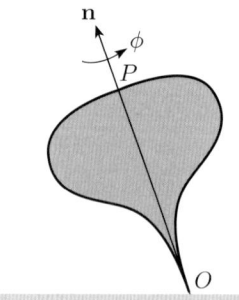

Figure 4.10 Diagram showing the coordinates of a child's top.

If the point of contact is allowed to move around the plane, then two more degrees of freedom are added.

A solid body
A solid body moving in space has six degrees of freedom: there are three coordinates for the position of its centre of mass; and three further angles defining its orientation.

4.3 A discussion of Newton's laws

In general if there are M independent holonomic constraints, defined by the M equations

$$C_j(\mathbf{r}_1, \mathbf{r}_2, \ldots, \mathbf{r}_N, t) = 0, \quad j = 1, 2, \ldots, M, \tag{4.47}$$

then it can be shown that the physically possible positions of the system, that is the values of all $3N$ coordinates (x_i, y_i, z_i), $i = 1, 2, \ldots, N$, can be described by $n = 3N - M$ coordinates q_k, $k = 1, 2, \ldots, n$, so we must have $M \leq 3N - 1$. The number n is important and is called the *number of degrees of freedom* of the system and the n coordinates q_k are called the *generalised coordinates*. For any given system the number of degrees of freedom is unique, but there are infinitely many choices of the generalised coordinates.

Generalised coordinates arising from holonomic constraints satisfy two conditions:

(i) their values completely determine the configuration of the system;
(ii) they may be varied arbitrarily and independently without violating the constraints of the system.

The n generalised coordinates q_k, $k = 1, 2, \ldots, n$, are frequently used to define coordinates in an abstract space which is named the *configuration space*. For instance a single particle moving in three dimensions has for its configuration space the usual three-dimensional Euclidean space. If the particle was confined to the Oxy-plane the configuration space would be this plane: but it might be confined to the surface of a sphere or a torus, in which case the configuration space would be different.

There are two principal methods of dealing with holonomic constraints. One is to introduce the forces of constraint, for example, the reaction on the wire in Section 4.2.1 and the tension in the pendulum in Sections 4.2.2 and 4.2.3. The second method is to formulate the equations of motion in terms of generalised coordinates, so the constraints are built in at the beginning. In general, when applicable, the second method is easiest to apply: the general formalism is described in Section 4.4 and it is applied to some problems in Section 4.5.

Exercise 4.6

For each of the following systems, state the number of degrees of freedom and suggest suitable generalised coordinate(s).

(a) A particle moving on a fixed, stationary circle.

(b) A particle moving perpendicular to a given stationary plane.

(c) A particle moving on the surface of a sphere.

(d) A rigid body rotating about an axis fixed in space.

(e) A particle moving on a circle which is rotating about a diameter in a prescribed manner.

(f) A particle moving on a circle which is free to rotate about a horizontal diameter.

(g) A rigid body (i) moving freely in space, and (ii) with a stationary centre of mass.

(h) A planet moving round a star, fixed in space.

Exercise 4.7

(a) Consider the pulley system illustrated in Figure 4.9. If the distance of the pulley C and the particle D below the ceiling are Y and y respectively, show that for an inextensible rope these variables are connected by the constraint $2Y + y = \text{constant}$.

(b) How many degrees of freedom does this system have?

(c) Explain why the tension is the same along the length of the rope.

4.3.4 Kinetic and potential energy

In this section we introduce two concepts that are essential for the development of Lagrange's equation of motion: the kinetic energy and the potential energy for a system of interacting particles.

Kinetic energy

The kinetic energy of a single particle of mass m and with velocity $\mathbf{v}$ is a scalar and is defined to be $\frac{1}{2}mv^2$, where $v = |\mathbf{v}|$. For a set of N interacting particles the total kinetic energy, T, is defined to be the sum of the individual energies,

$$T = \tfrac{1}{2} \sum_{i=1}^{N} m_i v_i^2. \tag{4.48}$$

Note that the symbol T is also used to denote a tension: this should not cause confusion because these quantities normally appear in quite different contexts and have different units. Each component of this sum is proportional to the square of the speed of each particle, and depends on the first time-derivative of the associated position vector,

$$\mathbf{v}_i = \frac{d\mathbf{r}_i}{dt} = (\dot{x}_i, \dot{y}_i, \dot{z}_i), \quad \text{so} \quad v_i^2 = \dot{x}_i^2 + \dot{y}_i^2 + \dot{z}_i^2. \tag{4.49}$$

Each vector $\mathbf{r}_i$ will depend upon the n generalised coordinates, q_k, $k = 1, 2, \ldots, n$, so each of the $3N$ components $(\dot{x}_i, \dot{y}_i, \dot{z}_i)$ will depend on these

4.3 A discussion of Newton's laws

generalised components and their derivatives, $\dot{q}_k$, $k = 1, 2, \ldots, n$. For example, if

$$x_i = f_i(q_1, q_2, \ldots, q_n, t), \quad i = 1, 2, \ldots, N, \tag{4.50}$$

then using the chain rule to differentiate with respect to t gives

$$\frac{dx_i}{dt} = \sum_{k=1}^{n} \frac{\partial f_i}{\partial q_k} \frac{dq_k}{dt} + \frac{\partial f_i}{\partial t}, \quad i = 1, 2, \ldots, N, \tag{4.51}$$

with similar relations for the y- and z-components. The derivatives $\dot{q}_k$, $k = 1, 2, \ldots, n$, are called the *generalised velocities*. We see that the time derivatives $(\dot{x}_i, \dot{y}_i, \dot{z}_i)$ are all *linear* functions of $\dot{q}_k$, but not necessarily of the generalised coordinates. Because all the first time derivatives of $\mathbf{r}_i$ are linearly dependent on the generalised velocities, the kinetic energy depends quadratically on the generalised velocities.

Two examples should make the preceding analysis clear. First, consider a single particle moving in a plane with Cartesian coordinates (x, y): its velocity is $\mathbf{v} = (\dot{x}, \dot{y})$. In plane polar coordinates, (r, θ), we have

$$x = r\cos\theta, \quad y = r\sin\theta. \tag{4.52}$$

We may use either (x, y) or (r, θ) as generalised coordinates and the respective generalised velocities are $\dot{x}$, $\dot{y}$ and $\dot{r}$, $\dot{\theta}$. Differentiating the above two equations we obtain

$$\dot{x} = \dot{r}\cos\theta - r\dot{\theta}\sin\theta, \quad \dot{y} = \dot{r}\sin\theta + r\dot{\theta}\cos\theta. \tag{4.53}$$

These are the equivalent of equations (4.51), and we see that $\dot{x}$ and $\dot{y}$ depend linearly upon $\dot{r}$ and $\dot{\theta}$, because of the chain rule. The speed of the particle is given by

$$v^2 = \dot{x}^2 + \dot{y}^2 = \left(\dot{r}\cos\theta - r\dot{\theta}\sin\theta\right)^2 + \left(\dot{r}\sin\theta + r\dot{\theta}\cos\theta\right)^2$$
$$= \dot{r}^2 + r^2\dot{\theta}^2, \tag{4.54}$$

and this depends quadratically on the generalised velocities $\dot{r}$ and $\dot{\theta}$, as expected.

As a second example, we consider a single particle moving in three dimensions. The Cartesian coordinates are (x, y, z) and these are appropriate generalised coordinates, with the associated generalised velocities $\dot{x}$, $\dot{y}$ and $\dot{z}$.

Another set of coordinates are the spherical polar coordinates, (r, θ, ϕ), useful when there is a spherical symmetry. These can be defined by the equations

$$x = r\sin\theta\cos\phi, \quad y = r\sin\theta\sin\phi, \quad z = r\cos\theta. \tag{4.55}$$

The new generalised coordinates are r, θ and ϕ and the physical meaning of these is shown in Figure 4.11.

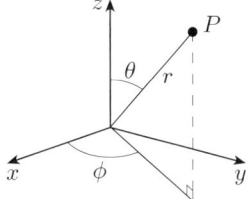

Figure 4.11 Diagram showing the physical meaning of the spherical polar coordinates (r, θ, ϕ).

Differentiation of these equations gives

$$\dot{x} = \dot{r}\sin\theta\cos\phi + r\dot{\theta}\cos\theta\cos\phi - r\dot{\phi}\sin\theta\sin\phi,$$
$$\dot{y} = \dot{r}\sin\theta\sin\phi + r\dot{\theta}\cos\theta\sin\phi + r\dot{\phi}\sin\theta\cos\phi, \qquad (4.56)$$
$$\dot{z} = \dot{r}\cos\theta - r\dot{\theta}\sin\theta.$$

The speed is $\dot{x}^2 + \dot{y}^2 + \dot{z}^2$; it is easiest to first find $\dot{x}^2 + \dot{y}^2$:

$$\begin{aligned}\dot{x}^2 + \dot{y}^2 &= \dot{r}^2\left(\sin^2\theta\cos^2\phi + \sin^2\theta\sin^2\phi\right) \\ &\quad + r^2\dot{\theta}^2\left(\cos^2\theta\cos^2\phi + \cos^2\theta\sin^2\phi\right) \\ &\quad + r^2\dot{\phi}^2\left(\sin^2\theta\sin^2\phi + \sin^2\theta\cos^2\phi\right) \\ &\quad + 2r\dot{r}\dot{\theta}\left(\sin\theta\cos\theta\cos^2\phi + \sin\theta\cos\theta\sin^2\phi\right) \\ &= \dot{r}^2\sin^2\theta + r^2\dot{\theta}^2\cos^2\theta + r^2\dot{\phi}^2\sin^2\theta + 2r\dot{r}\dot{\theta}\sin\theta\cos\theta. \qquad (4.57)\end{aligned}$$

Hence on adding $\dot{z}^2$ to this we obtain

$$\dot{x}^2 + \dot{y}^2 + \dot{z}^2 = \dot{r}^2 + r^2\dot{\theta}^2 + r^2\dot{\phi}^2\sin^2\theta. \qquad (4.58)$$

This is a sum of the squares of the generalised velocities, with coefficients depending upon the generalised coordinates.

Exercise 4.8

The generalised coordinates (x, y) and (u, v) are related by $x = u + v$ and $y = u - v$.

(a) Express the generalised velocities $(\dot{u}, \dot{v})$ in terms of $(\dot{x}, \dot{y})$.

(b) Show that $\dot{x}^2 + \dot{y}^2 = 2\left(\dot{u}^2 + \dot{v}^2\right)$.

(c) If the relations between (x, y) and (u, v) are replaced by $x = u + vt$ and $y = u - vt$ find the generalised velocities and show that $\dot{x}^2 + \dot{y}^2 = 2\left(\dot{u}^2 + \dot{v}^2 t^2\right) + 4v\dot{u}t + 2v^2$.

Exercise 4.9

The Cartesian coordinates (x, y) are related to new coordinates (u, v) by $x = uv$, $y = (u^2 - v^2)/2$. Show that the square of the speed is $\dot{x}^2 + \dot{y}^2 = (u^2 + v^2)(\dot{u}^2 + \dot{v}^2)$.

Exercise 4.10

A particle of mass m moves on a curve described by the equation $y = f(x)$. If s is the distance along this curve from some fixed point, show that the kinetic energy of the particle is

$$T = \tfrac{1}{2}m\dot{x}^2\left(1 + f'(x)^2\right) = \tfrac{1}{2}m\dot{s}^2.$$

[Hint: use the relation $\delta s^2 \simeq \delta x^2 + \delta y^2$, see Figure 1.5 (page 9).]

Exercise 4.11

A particle of mass m moves on an Archimedes spiral which has the equation $r = a\theta$, for some positive constant a and where (r, θ) are the plane polar coordinates. Show that the kinetic energy can be expressed in terms of $(r, \dot{r})$ or $(\theta, \dot{\theta})$ by the equations

$$T = \tfrac{1}{2}m\dot{r}^2\left(1 + \frac{r^2}{a^2}\right) \quad \text{or} \quad T = \tfrac{1}{2}ma^2\dot{\theta}^2\left(1 + \theta^2\right).$$

Exercise 4.12

Show that the kinetic energy of the pulley system illustrated in Figure 4.9 is given by $T = (M + 4m)\dot{y}^2/8$, where y is the variable defined in Exercise 4.7.

Potential energy

Next we assume that all forces are *conservative*, which means that the force on a particle is given by the gradient of a *potential function*, $V(\mathbf{r}, t)$, with respect to the position of the particle. For a single particle, with position $\mathbf{r} = (x, y, z)$, this means that the coordinates of the force on it are given by

$$\mathbf{F} = -\left(\frac{\partial V}{\partial x}, \frac{\partial V}{\partial y}, \frac{\partial V}{\partial z}\right) = -\operatorname{grad} V = -\frac{\partial V}{\partial \mathbf{r}}, \quad (4.59)$$

where $V = V(\mathbf{r}, t)$ is a scalar function of $\mathbf{r}$ and possibly the time t. The notation $\partial V/\partial \mathbf{r}$ for $\operatorname{grad} V$ may not be familiar, but it is very useful when there are several particles with positions $(\mathbf{r}_1, \mathbf{r}_2, \ldots, \mathbf{r}_n)$ and V depends on two or more of these. In such cases we often need to differentiate with respect to a particular $\mathbf{r}_k$ and this can be denoted by $\partial V/\partial \mathbf{r}_k$, whereas the alternative notation is not so convenient.

Note that in dynamics the adjective *conservative* has several different meanings. A discussion can be found on page 173.

The significance of equation (4.59) is that all three components of the force are defined by a single scalar function, V, which is variously named the *potential*, the *potential energy* or the *scalar potential*. Because the force is given by the gradient of V, potentials differing by either a constant or a function of only the time, give the same force, and are hence equivalent.

Conservative forces are so named because the work done when moving round a closed curve in configuration space is zero. Friction is not a conservative force because it always opposes the motion and the work done is always positive; gravity, where $\mathbf{F} = -mg\hat{\mathbf{z}}$ and $V = mgz$, on the other hand, is conservative because the work done going up is recovered on the way down.

Consider a particle in the three-dimensional configuration space, shown in Figure 4.12, with a force field $\mathbf{F}(\mathbf{r})$ and a curve C. The work done against the force when moving a distance $\delta \mathbf{r}$ along this curve is defined to be $\delta W = \mathbf{F} \cdot \delta \mathbf{r}$, that is force × distance travelled. Notice that the curve C is not necessarily a possible motion; it can be any smooth curve that does not violate any constraint.

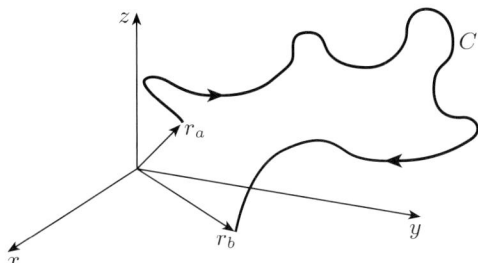

Figure 4.12 An arbitrary curve joining $\mathbf{r}_a$ to $\mathbf{r}_b$.

Hence the work done when moving along the curve C, from $\mathbf{r}_a$ to $\mathbf{r}_b$, is the line integral

$$W = \int_C d\mathbf{r} \cdot \mathbf{F}, \quad (4.60)$$

which will normally depend upon the path taken. If, however, $\mathbf{F}$ is conservative, with $\mathbf{F} = -\operatorname{grad} V$, for some potential V, then, for small changes

in $\mathbf{r}$,
$$\delta V = \frac{\partial V}{\partial x}\delta x + \frac{\partial V}{\partial y}\delta y + \frac{\partial V}{\partial z}\delta z = -\mathbf{F}\cdot\delta\mathbf{r}, \qquad (4.61)$$

so the expression for the work becomes
$$W = -\int_C dV = V(\mathbf{r}_a) - V(\mathbf{r}_b), \qquad (4.62)$$

which is independent of the path. For a closed curve, $\mathbf{r}_a = \mathbf{r}_b$, hence $W = 0$. If V depends upon the time, then this is held fixed in the integration round C, so it is possible to have conservative, time-dependent forces.

The interaction between two particles at $\mathbf{r}_i$ and $\mathbf{r}_j$, giving rise to the conservative force $\mathbf{F}_{ij}$, is described by a potential function $V_{ij}(r,t)$ that depends on these position vectors only through the distance $r = |\mathbf{r}_i - \mathbf{r}_j|$, and possibly (but not normally) the time, *but no other position vector*. Then
$$\mathbf{F}_{ij} = -\frac{\partial V_{ij}}{\partial \mathbf{r}_i} \quad \text{and} \quad \mathbf{F}_{ji} = -\frac{\partial V_{ij}}{\partial \mathbf{r}_j}. \qquad (4.63)$$

Note that for V_{ij} the suffix order is immaterial. These interactions automatically satisfy the third law $\mathbf{F}_{ij} = -\mathbf{F}_{ji}$. The proof of this is not central to this course and is therefore given in Exercise 4.42 at the end of the chapter.

The external forces, $\mathbf{F}_i^{(e)}(\mathbf{r}_i, t)$, are also assumed to be conservative, but these may depend upon the position vector $\mathbf{r}_i$, not just its length. A simple example of an external force is an electron in a uniform electric field. If $-e$ is the electron charge ($e > 0$) and $\mathbf{E}(t)$ the electric field, then the force on the electron is $\mathbf{F} = -e\mathbf{E}(t)$; this is independent of the position and the potential energy is just $V = e\mathbf{E}(t)\cdot\mathbf{r}$.

Conservative forces are important: gravitational forces are conservative and electric forces are also conservative, as is the force due to Hooke's law, described in Block I, Chapter 1; see also Exercise 4.14. Thus atoms, molecules, the solar system and galaxies are all held together by conservative forces. Important exceptions are frictional forces and air resistance.

Gravitational forces are conservative only for Newtonian gravitation; this is not true in general relativity.

Not all forces can be defined in terms of potentials. However, for one-dimensional systems, if the force $F(x,t)$ depends only upon x and t because F is a scalar it is always possible to find a potential $V(x,t)$, simply by integrating $-F(x,t)$ with respect to x, with t fixed: note that if the force depends upon $\dot{x}$ such a potential cannot be found.

Exercise 4.13

(a) The radial distance r from the origin is given by $r^2 = x^2 + y^2 + z^2$. Show that
$$\frac{\partial r}{\partial x} = \frac{x}{r}, \quad \frac{\partial r}{\partial y} = \frac{y}{r}, \quad \frac{\partial r}{\partial z} = \frac{z}{r}.$$

(b) Show that a potential energy function, $V(r)$, depending only upon the radial distance r, produces the force
$$\mathbf{F} = -\frac{V'(r)}{r}\mathbf{r} = -V'(r)\hat{\mathbf{r}}.$$

Potential energies add, so the potential energy of a set of N interacting particles is just the sum of the $\frac{1}{2}N(N-1)$ interactions between each pair,
$$V = \sum_{i=1}^{N-1}\sum_{j=i+1}^{N} V_{ij}(|\mathbf{r}_i - \mathbf{r}_j|, t), \qquad (4.64)$$

4.3 A discussion of Newton's laws

or in full

$$V = V_{12} + V_{13} + V_{14} \cdots + V_{1\,N-1} + V_{1N}$$
$$+ V_{23} + V_{24} \cdots + V_{2\,N-1} + V_{2N}$$
$$+ V_{34} \cdots + V_{3\,N-1} + V_{3N} \quad (4.65)$$
$$\vdots \quad \vdots$$
$$+ V_{N-2\,N-1} + V_{N-2\,N}$$
$$+ V_{N-1\,N}.$$

It follows that the interaction force on the ith particle is

$$\mathbf{F}_i = -\frac{\partial V}{\partial \mathbf{r}_i} = \sum_{\substack{j=1 \\ j \neq i}}^{N} \mathbf{F}_{ij}. \quad (4.66)$$

If there are many sources of potential energy the total potential energy is always taken as the sum of the values that each would contribute on its own.

Exercise 4.14

Consider a spring of natural length l_0 and stiffness k lying along the x-axis. Assuming Hooke's law (see Block I, Chapter 1) the force acting to shorten the spring when it is extended to the length $l_0 + x$, is $\mathbf{F} = -k\mathbf{x}$ show that this can be derived from the potential function $V = kx^2/2$.

Gravity

One ubiquitous force is that due to gravitational attraction. Two point particles of masses m_1 and m_2 attract each other with a force inversely proportional to the square of the distance, r, between them and proportional to the product of their masses, giving rise to the gravitational potential energy

$$V(r) = -\frac{Gm_1m_2}{r}, \quad r > 0, \quad (4.67)$$

where G is the universal gravitational constant. For a particle of mass m at a height z above the Earth's surface Newton showed, by assuming the Earth to be spherical and integrating over the Earth's volume, that the gravitational potential is

$$V(z) = -\frac{GMm}{R+z}, \quad z \geq 0, \quad (4.68)$$

where M and R are the Earth's mass and radius respectively.

Exercise 4.15

Show that if $R \gg z$ then the gravitational potential discussed above reduces to the approximation $V(z) = mgz$, where $g = GM/R^2$ is the acceleration due to gravity. Taking $M = 5.972 \times 10^{27}$ g, $R = 6.378 \times 10^8$ cm and $G = 6.673 \times 10^{-8}$ g^{-1} cm^3 s^{-2}, calculate the value of g.

Exercise 4.16

The formula for g found in Exercise 4.15 shows that the value of g decreases as the height above the mean sea level increases.

(a) Show that the relative change in g between its value at a height h above the mean sea level, $g(h)$ and its value at sea level, $g = g(0)$, is given approximately by
$$\frac{\Delta g}{g} = \frac{g(h) - g}{g} \simeq -\frac{2h}{R}.$$
For a hill 800 m high show that $\Delta g \simeq 2.5 \times 10^{-4} g$.

(b) The approximate period of a pendulum clock of length l is $T = 2\pi\sqrt{l/g}$. Show that the relative change in T and g are related by
$$\frac{\Delta T}{T} \simeq -\frac{\Delta g}{2g}.$$
Deduce that a pendulum clock accurate to 1 s/day could be used to measure the relative change in g between sea level and the top of an 800 m hill.

The significance of this calculation is that during his trip to St Helena (1677–1678), Halley was the first to measure the slowing of a pendulum clock due to altitude. Halley deduced that the force of gravity decreased as the distance from the centre of the Earth increased.

On the same trip he noted that at sea level the same clock ran slower in England than in St Helena. This is due to the flattening of the Earth, though Halley did not know this; see the solution on page 202 for more details.

Energy conservation

We end this section by showing that if the forces are conservative and independent of time, then the total energy, E, of the system is a constant. The total energy of N interacting particles is defined to be the sum of the kinetic and the potential energies:
$$E = \tfrac{1}{2} \sum_{i=1}^{N} m_i v_i^2 + V + \sum_{i=1}^{N} V_i^{(e)}, \tag{4.69}$$

where V is the potential energy due to the internal forces and defined by equation (4.64). The time derivative of V is, on using the chain rule,
$$\frac{dV}{dt} = \sum_{i=1}^{N} \frac{\partial V}{\partial \mathbf{r}_i} \cdot \frac{d\mathbf{r}_i}{dt}, \tag{4.70}$$

and since
$$\frac{\partial V}{\partial \mathbf{r}_i} = -\sum_{\substack{j=1 \\ j \neq i}}^{N} \mathbf{F}_{ij}, \tag{4.71}$$

we have
$$\frac{dE}{dt} = \sum_{i=1}^{N} \left(m_i \ddot{\mathbf{r}}_i \cdot \mathbf{v}_i + \frac{\partial V}{\partial \mathbf{r}_i} \cdot \mathbf{v}_i + \frac{\partial V_i^{(e)}}{\partial \mathbf{r}_i} \cdot \mathbf{v}_i \right)$$
$$= \sum_{i=1}^{N} \mathbf{v}_i \cdot \left(m_i \frac{d^2 \mathbf{r}_i}{dt^2} - \sum_{\substack{j=1 \\ j \neq i}}^{N} \mathbf{F}_{ij} - \mathbf{F}_i^{(e)} \right) = 0, \tag{4.72}$$

where the last equality follows from Newton's equations of motion, equation (4.42). Hence the total energy, E, is a constant on each solution, with a value that depends upon the initial conditions; such dynamical systems are named *conservative* systems. Note that the adjective *conservative* is used in two contexts: a conservative system involves conservative forces, but conservative forces that depend upon time do not normally produce conservative systems.

For a conservative system with one degree of freedom the expression for the total energy gives a first-order differential equation which replaces the second-order equation of Newton, which generally makes the analysis easier: an example of this process is found in Exercise 4.1. The energy E can be obtained by integrating the equations of motion once, so it is often referred to as an *integral of the motion* or, because it is constant, as a *constant of the motion*.

Exercise 4.17

(a) Consider a particle of mass m moving vertically upwards from the surface of the Earth, with height z above the surface. Use the potential energy defined in equation (4.68) to show that the total energy is

$$E = \tfrac{1}{2}m\dot{z}^2 - \frac{GMm}{R+z}.$$

(b) The escape velocity is defined to be the value of $\dot{z}$ when $z = 0$ such that as $z \to \infty$, $\dot{z} \to 0$, that is the particle can just escape the gravitational field: any smaller velocity results in it returning. Use the fact that $E = $ constant to show that the escape speed is

$$v_e = \sqrt{\frac{2GM}{R}}.$$

(c) Use the values given in the following table and the fact that $G = 6.673 \times 10^{-8}$ cm^{3}g^{-1}s^{-2} to find the escape speeds and the values of g using the formula found in Exercise 4.15, for the Earth, the Moon and Jupiter.

	Earth	Moon	Jupiter
M/g	5.972×10^{27}	7.350×10^{25}	1.899×10^{30}
R/cm	6.378×10^{8}	1.738×10^{8}	7.190×10^{9}

Exercise 4.18

Show that for a single particle with mass m, moving in a force $\mathbf{F}(\mathbf{r})$, Newton's equations of motion give the rate of change of the kinetic energy, T, to be

$$\frac{dT}{dt} = \mathbf{F} \cdot \mathbf{v}.$$

4.4 Lagrange's equations and Hamilton's principle

In this section we introduce Lagrange's equations of motion. For holonomic systems, these are equivalent to Newton's equations, but are formulated directly in terms of the generalised coordinates; they avoid use of constraints, except in the initial definition of the coordinates, so are far easier to derive. Lagrange's equations can be derived from Newton's equations, but the analysis is fairly complicated and distracts from the main story.

Holonomic systems are defined on page 164.

Lagrange's method can be applied to any holonomic system of particles that interact through conservative forces. The configuration of the system is defined by n generalised coordinates $\mathbf{q} = (q_1, q_2, \ldots, q_n)$, so the potential energy is a function, $V(\mathbf{q}, t)$, of these variables and possibly the time. The kinetic energy, T, is a quadratic function of the generalised velocities $\dot{\mathbf{q}} = (\dot{q}_1, \dot{q}_2, \ldots, \dot{q}_n)$. Then the *Lagrangian* is defined to be the *difference* between the kinetic and the potential energies,

$$L(\mathbf{q}, \dot{\mathbf{q}}, t) = T(\mathbf{q}, \dot{\mathbf{q}}, t) - V(\mathbf{q}, t). \tag{4.73}$$

The Lagrangian is a function of the generalised coordinates and velocities, and possibly the time; it is sometimes referred to as the Lagrangian function. Notice that the Lagrangian has the units of energy.

Lagrange's equations of motion depend upon the derivatives of the Lagrangian. If there are n generalised coordinates, there are n equations of motion, which are

$$\frac{d}{dt}\left(\frac{\partial L}{\partial \dot{q}_k}\right) - \frac{\partial L}{\partial q_k} = 0, \quad k = 1, 2, \ldots, n. \tag{4.74}$$

You will recognise these as the Euler–Lagrange equations for n dependent variables, derived in Chapter 3 – specifically, equation (3.71) (page 108) – but without the boundary conditions. This similarity suggests that Lagrange's equations of motion can be derived from a variational principle, and we shall describe how this is done later.

In simple cases it is easy to see that Lagrange's and Newton's equations are the same. For instance, a free particle (that is, a particle acted upon by no forces) with Cartesian coordinates $\mathbf{r} = (x, y, z)$ has no potential energy, so the Lagrangian is just the kinetic energy,

$$L = \tfrac{1}{2}m\left(\dot{x}^2 + \dot{y}^2 + \dot{z}^2\right). \tag{4.75}$$

In this example Cartesian coordinates are the most convenient generalised coordinates. Lagrange's equation of motion for the x-coordinate is then

$$\frac{d}{dt}\left(\frac{\partial L}{\partial \dot{x}}\right) = 0, \quad \text{giving} \quad m\ddot{x} = 0, \tag{4.76}$$

which is Newton's equation. The algebra for the remaining two coordinates is identical, and this leads to $m\ddot{\mathbf{r}} = 0$. It is important that you do the next exercise.

4.4 Lagrange's equations and Hamilton's principle

Exercise 4.19

For a single particle moving in a conservative force, with potential $V(\mathbf{r}, t)$, the Lagrangian is

$$L = \tfrac{1}{2}m\left(\dot{x}^2 + \dot{y}^2 + \dot{z}^2\right) - V(\mathbf{r}, t). \tag{4.77}$$

Use Lagrange's equations of motion to show that

$$m\ddot{x} + \frac{\partial V}{\partial x} = 0, \quad m\ddot{y} + \frac{\partial V}{\partial y} = 0, \quad m\ddot{z} + \frac{\partial V}{\partial z} = 0,$$

and show that these are the same as Newton's equations of motion.

Exercise 4.20

For a particle of mass m moving in the Oxy-plane under the influence of a force always directed towards the origin, the Lagrangian is

$$L = \tfrac{1}{2}m\left(\dot{x}^2 + \dot{y}^2\right) - V(r), \quad r^2 = x^2 + y^2.$$

(a) Use Lagrange's equations of motion and Cartesian coordinates to show that the equations of motion are

$$m\ddot{x} + V'(r)\frac{x}{r} = 0 \quad \text{and} \quad m\ddot{y} + V'(r)\frac{y}{r} = 0.$$

(b) Show that with polar coordinates, (r, θ), where $x = r\cos\theta$ and $y = r\sin\theta$, the Lagrangian becomes

$$L = \tfrac{1}{2}m\left(\dot{r}^2 + r^2\dot{\theta}^2\right) - V(r)$$

and that the equations of motion are

$$m\ddot{r} - \frac{A^2}{mr^3} + V'(r) = 0 \quad \text{and} \quad \dot{\theta} = \frac{A}{mr^2}, \quad \text{where } A \text{ is a constant.}$$

Lagrange's equations of motion, defined in equation (4.74), are more elegant than the equivalent Newtonian equations because their form is independent of the choice of generalised coordinates, recall the discussion in Section 3.2.2, in particular equations (3.28) and (3.29) (page 101). The underlying reason for this is that they can be derived from a variational principle, which, as we saw in the previous chapter, is invariant under coordinate changes: this variational principle can be used as the fundamental principle of Newtonian dynamics, in place of Newton's three laws. This change of emphasis, from Newton's laws to a variational principle, may seem an unnecessary complication but it is an essential step for later developments in advanced dynamics and theoretical physics. In addition, in the spirit of Ockham's razor, it provides a single, simple principle from which everything follows.

Hamilton's Principle

The trajectories between two points $\mathbf{q} = \mathbf{a}$ and $\mathbf{q} = \mathbf{b}$ in configuration space, starting and ending at times t_a and t_b, are given by the stationary paths of the functional

$$S[\mathbf{q}] = \int_{t_a}^{t_b} dt\, L(\mathbf{q}, \dot{\mathbf{q}}, t), \quad \mathbf{q}(t_a) = \mathbf{a}, \quad \mathbf{q}(t_b) = \mathbf{b}. \tag{4.78}$$

The functional S is important in advanced dynamics (and in quantum mechanics). Hamilton used it to draw important analogies between particle dynamics and optics, which were important in the development of quantum mechanics in the early part of the 20th century. Because S is important, it is called *the action*; it has the units of energy × time, that is, dimensions ML^2T^{-1}, which is the same as angular momentum.

It follows from the theory described in Chapter 3 that the path between the end points is given by the solution of the associated n Euler–Lagrange equations,

$$\frac{d}{dt}\left(\frac{\partial L}{\partial \dot{q}_i}\right) - \frac{\partial L}{\partial q_i} = 0, \quad q_i(t_a) = a_i, \ q_i(t_b) = b_i, \ i = 1, 2, \ldots, n. \quad (4.79)$$

These coupled, second-order equations are called *Lagrange's equations* of motion, because Newton's equations were first cast in this form by Lagrange; the full account of this theory was published in *Mécanique analytique* (1788). A short history of the development of these ideas is given at the end of this section.

If L does not depend upon a particular generalised coordinate, q_l say, the associated Lagrange's equation can be integrated directly to give

$$\frac{\partial L}{\partial \dot{q}_l} = \text{constant}. \quad (4.80)$$

That is, for this coordinate Lagrange's equation is a first-order equation and hence easier to solve. A generalised coordinate like q_l is named an *ignorable coordinate*. As an aside, we note that this is an illustration of Noether's theorem (page 120) in operation; see Exercise 4.54. One aim in choosing generalised coordinates is to obtain as many ignorable coordinates as possible, because this helps to find solutions. An example is given in Section 4.5.1.

Note that if the time interval $|t_b - t_a|$ is sufficiently small it can be shown that the stationary path is an actual minimum of the action. The reason for this is that for these times the kinetic energy term dominates the integral and the motion is close to that of a free particle; see Exercise 4.49.

Apart from the advantages discussed above, you will recall that part of the original motivation for developing Lagrange's equations is the ease with which constraints are included. For instance, consider, the bead constrained on a rigid wire in the shape $y = f(x)$: in Section 4.2.1 we dealt with the case $f(x) = ax^2/2$. The potential energy is $V = mgf(x)$, and the speed is

$$v^2 = \dot{x}^2 + \dot{y}^2 = \dot{x}^2\left(1 + f'(x)^2\right) \quad \text{since} \quad \dot{y} = f'(x)\dot{x}, \quad (4.81)$$

so, with x chosen to be the generalised coordinate, the Lagrangian is

$$L = \tfrac{1}{2}m\dot{x}^2\left(1 + f'(x)^2\right) - mgf(x). \quad (4.82)$$

Lagrange's equation of motion is then

$$\frac{d}{dt}\left(\dot{x}\left(1 + f'(x)^2\right)\right) - \dot{x}^2 f'(x)f''(x) + gf'(x) = 0. \quad (4.83)$$

You may like to compare the efforts needed to derive this and equation (4.5) (page 147), where $f(x) = ax^2/2$.

4.4 Lagrange's equations and Hamilton's principle

Exercise 4.21

Show that with $f(x) = ax^2/2$, equation (4.83) expands to equation (4.5) (page 147).

Exercise 4.22

Two particles, of masses m_1 and m_2, are joined by a light inextensible string passing over a smoothly turning pulley of radius R without slipping, as shown in Figure 4.13.

(a) Assuming that the mass of the pulley is sufficiently small for its energy of rotation to be neglected, show that the Lagrangian for the motion of the system is
$$L(z, \dot{z}) = \tfrac{1}{2}(m_1 + m_2)\dot{z}^2 + (m_1 - m_2)gz.$$

(b) If the rotational kinetic energy of the pulley is $T_p = \tfrac{1}{2}M_p R^2 \dot{\theta}^2$, where $\dot{\theta}$ is the angular velocity of the pulley and M_p is a constant, show that the Lagrangian becomes
$$L(z, \dot{z}) = \tfrac{1}{2}M\dot{z}^2 + (m_1 - m_2)gz, \quad \text{where} \ \ M = m_1 + m_2 + M_p.$$

(c) Solve the equations of motion for the case described in part (b).

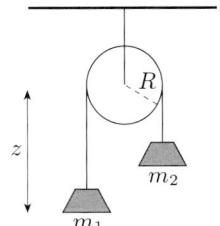

Figure 4.13

Exercise 4.23

(a) A particle of mass m moves under the influence of gravity along the smooth spiral wire defined, through the parameter ϕ, by
$$x = a\cos\phi, \quad y = a\sin\phi, \quad z = k\phi,$$
where a and k are positive constants and the z-axis is vertically upwards. Obtain the Lagrangian and the equations of motion, and solve these to find $\phi(t)$.

(b) If, instead of being fixed, the spiral rotates about the z-axis with angular velocity $\Omega = \dot{\theta}$, the parametric equations of the curve become
$$x = a\cos(\phi + \theta(t)), \quad y = a\sin(\phi + \theta(t)), \quad z = k\phi.$$
Find the Lagrangian and the equations of motion in this case.

Exercise 4.24

In this exercise we use Hamilton's principle to find an approximation to the solution of Lagrange's equations, rather than to derive the equations of motion.

Consider a particle of mass m moving along the x-axis in the potential $V(x) = ma^2 x^4/4$, where a is a positive constant. It can be shown that all solutions of the equation of motion are periodic.

(a) Show that the action for a T-periodic solution is
$$S[x] = m\int_0^T dt\, \left(\tfrac{1}{2}\dot{x}^2 - \tfrac{1}{4}a^2 x^4\right), \quad x(0) = A, \ x(T) = A,$$
for some A.

(b) By approximating a periodic solution with angular frequency ω and amplitude A with the function $x = A\sin\omega t$, show that the action is approximately
$$S(A) = \frac{\pi m A^2}{16\omega}\left(8\omega^2 - 3a^2 A^2\right).$$
You will need the integrals
$$\int_0^{2\pi} du\, \cos^{2n} u = \int_0^{2\pi} du\, \sin^{2n} u = 2\pi\, \frac{(2n-1)(2n-3)\ldots 3.1}{2n(2n-2)\ldots 4.2}.$$

(c) Hamilton's principle requires that $S[x]$ is stationary: hence an approximation to the motion is obtained by choosing A so that $S(A)$ is stationary. This will give an approximation to the amplitude of the motion with frequency ω.

By finding the solutions of $S'(A) = 0$, show that the amplitude and frequency are related approximately by

$$A = \frac{2\omega}{\sqrt{3}a} = 1.1547\frac{\omega}{a}.$$

Note that the exact relation between the amplitude and frequency is $A = 1.1803\omega/a$, so the approximation has a relative error of about 2%: the exact solution is, of course, a more complicated function of t than the approximation used here.

Conservation of energy

If the Lagrangian does not depend explicitly upon the time t, then it may be shown using Noether's theorem (page 120) that there is an integral of the motion, equivalent to the first integral derived in Exercise 2.6 (page 51) and used in Sections 2.5 and 2.7. This derivation is not an assessed part of the course, but the interested student should compare the functional (3.132) (page 120) and (4.73). We see that x and $\mathbf{y}$ corresponds to t and $\mathbf{q}$ and also, in the equivalent of equation (3.131) $\phi = 1$ and $\psi_k = 0$, $k = 1, 2, \ldots, n$. The first integral, equation (3.133), is

$$\sum_{k=1}^{N} \dot{q}_k \frac{\partial L}{\partial \dot{q}_k} - L = E, \tag{4.84}$$

where E is a constant, which is normally the energy. This expression is usually referred to as the *energy integral*.

Consider the simple Lagrangian defined in equation (4.77), for which $q_1 = x$, $q_2 = y$ and $q_3 = z$, in the case where the potential energy is independent of t. Then

$$\sum_{k=1}^{N} \dot{q}_k \frac{\partial L}{\partial \dot{q}_k} = m\left(\dot{x}^2 + \dot{y}^2 + \dot{z}^2\right) = 2T. \tag{4.85}$$

Hence $E = 2T - (T - V) = T + V$ so, in this case, E is the total energy.

Exercise 4.25

Show that the constant of motion associated with the Lagrangian (4.82) is

$$E = \tfrac{1}{2}m\dot{x}^2\left(1 + f'(x)^2\right) + mgf(x).$$

Use this to find an expression for $\dot{x}$ in terms of x and E.

Addition of a total derivative

If $\overline{L}$ and L are two Lagrangians related by

$$\overline{L}(\mathbf{q},\dot{\mathbf{q}},t) = L(\mathbf{q},\dot{\mathbf{q}},t) + \frac{d}{dt}f(\mathbf{q},t), \tag{4.86}$$

where $f(\mathbf{q},t)$ is any differentiable funtion of $\mathbf{q}$ and t, but not the generalised velocities, then Lagrange's equations of motion for $\overline{L}$ and L have the same solutions.

4.4 Lagrange's equations and Hamilton's principle

This result follows directly from the variational principle, because we have

$$\overline{S} = \int_{t_a}^{t_b} dt\, \overline{L} = \int_{t_a}^{t_b} dt \left(L + \frac{df}{dt} \right)$$
$$= S + f(\mathbf{b}, t_b) - f(\mathbf{a}, t_a). \quad (4.87)$$

The functionals $\overline{S}$ and S differ only by a constant, so a stationary path of S is also a stationary path of $\overline{S}$, which proves the result.

This result is often useful when the transformation between generalised and Cartesian coordinates involves the time: for example, in the case of a bead on a moving wire (page 148), we have $y = f(x) + \gamma(t)$ and the speed is given by

$$v^2 = \dot{x}^2 + \dot{y}^2 = \dot{x}^2 + \left(f'(x)\dot{x} + \dot{\gamma} \right)^2, \quad (4.88)$$

and the Lagrangian (4.82) becomes

$$L = \tfrac{1}{2} m\dot{x}^2 \left(1 + f'(x)^2\right) + m\dot{x} f'(x) \dot{\gamma} - mgf(x) - mg\gamma(t) + \tfrac{1}{2} m \dot{\gamma}^2. \quad (4.89)$$

Now observe that $\dot{x} f'(x) = df/dt$ and that $\dot{f}\dot{\gamma} = d(f\dot{\gamma})/dt - f(x)\ddot{\gamma}$. Hence, the Lagrangian may be written in the form

$$L = \tfrac{1}{2} m\dot{x}^2 \left(1 + f'(x)^2\right) - m\left(g + \ddot{\gamma}\right) f(x)$$
$$+ m \left[\frac{d}{dt} (f(x)\dot{\gamma}) - g\gamma(t) + \tfrac{1}{2} \dot{\gamma}^2 \right]. \quad (4.90)$$

The square brackets enclose three terms all of which are the derivatives of some function of the time so may be ignored. Hence the Lagrangian for this system is

$$\overline{L} = \tfrac{1}{2} m\dot{x}^2 \left(1 + f'(x)^2\right) - m\left(g + \ddot{\gamma}\right) f(x). \quad (4.91)$$

The only difference between this Lagrangian and that defined in equation (4.82) is that g is replaced by $g + \ddot{\gamma}$.

Exercise 4.26

If $f(x) = ax^2/2$, show that the equation of motion given by the Lagrangian (4.91) is the same as in equation (4.8) (page 148).

Exercise 4.27

In this exercise we derive the Lagrangian for a particle of mass m moving on a rigid wire in the vertical plane, with shape $y = f(x)$, where the y-axis is vertically upwards, as in the text, but using s, the distance along the wire, for the generalised coordinate.   *This is a hard exercise.*

(a) If s is the distance along the curve from some fixed point, show that the kinetic and potential energies are $T = m\dot{s}^2/2$ and $V = mgy(s)$, and hence that the equation of motion is $\ddot{s} + gy'(s) = 0$.

(b) Consider the cycloid discussed in Section 2.7.1 and shown in Figure 4.14.

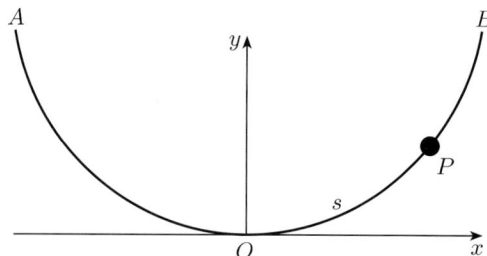

Figure 4.14 Diagram of a bead sliding on a wire in the shape of a cycloid. The distance along the curve from O to P is denoted by s.

The position of a particle at P, moving between the cusps at A and B, is defined uniquely by the distance, s, along the curve from O; alternatively, the coordinates of the point are defined by the parametric equations $x = a(\phi + \sin\phi)$, $y = a(1 - \cos\phi)$, $|\phi| \leq \pi$.

Show that $s = 4a\sin(\phi/2)$, where s is zero at the minimum, and $y = s^2/(8a)$ and hence that the equation of motion is

$$\frac{d^2 s}{dt^2} + \frac{g}{4a}s = 0, \quad |s| \leq 2a.$$

(c) Write down the general solution of this equation and show that the particle oscillates with a period that is independent of the amplitude of the motion. Find this period.

Deduce that the time to reach the bottom of the cycloid, starting from rest at any point between O and B is the same.

Exercise 4.28

Consider the pulley system illustrated in Figure 4.9 (page 163): show that the Lagrangian for the particle of mass m is

$$L = \tfrac{1}{8}(M + 4m)\dot{y}^2 + \tfrac{1}{2}(2m - M)gy,$$

where y is its distance below the ceiling. Use Lagrange's equations of motion to obtain the solution

$$y = A + Bt + \frac{2m - M}{4m + M}gt^2,$$

where A and B are constants.

A short history of Hamilton's principle

The development of the calculus of variations during the 18th century was complemented by contemporary work in physics, namely the *Principle of least action*. This, in turn was motivated by the observation, by Euclid, that light rays reflected in a mirror travel the shortest path, as described in Section 1.5.5 (page 23), which led to Snell's law and Fermat's principle (published in letters dated 1657 and 1662).

Euclid, ca. 325–265 BC.

Maupertius when working on the theory of light proposed, in 1744, his *Principle of least action* which he developed from Fermat's principle. The original formulation is rather vague, simply stating that the action is the product of the mass, speed and distance travelled and that this is a minimum along an actual path. Euler, also in 1744 after correspondence with Maupertius, formulated the same law a little more precisely by noting that the rate of change of the integral $\int dt\, v^2$ for a change in the path must be zero, though the precise meaning of this statement is not clear.

Pierre-Louis Moreau de Maupertius, 1698–1759.

Lagrange clarified the situation by defining the action to be the integral of the kinetic energy. Thus for a single particle the action was defined to be

$$W = \int_{t_1}^{t_2} dt\, T, \quad T = \tfrac{1}{2} m \left(\dot{x}^2 + \dot{y}^2 + \dot{z}^2 \right). \tag{4.92}$$

Lagrange assumed that energy was conserved, $T + V = E$, so dealt only with conservative systems. Lagrange's principle of least action states that W must be a minimum or a maximum along an actual path joining two given points in configuration space, at times t_1 and t_2. With this principle the admissible paths have the same energy as the actual path, so this variational principle is different from those dealt with in Chapter 2, because the admissible paths are constrained by an additional equation. Using this principle Lagrange was able to derive Newton's equations of motion.

Lagrange made the further step of introducing the generalised coordinates and showed that, for conservative systems with n degrees of freedom, Newton's equations of motion can be written in the form

$$\frac{d}{dt}\left(\frac{\partial T}{\partial \dot{q}_k}\right) - \left(\frac{\partial T}{\partial q_k} - \frac{\partial V}{\partial q_k}\right) = 0, \quad k = 1, 2, \ldots, n. \tag{4.93}$$

This analysis was described in *Mécanique analytique* (1788), but first suggested in *Miscellanea Taurinensia* II (1760).

The next major development was due to Poisson. In 1809 he defined the function $L = T - V$, that is, the Lagrangian function introduced in equation (4.73) (page 174), and noted that if V depends only upon the generalised coordinates (that is, neither the time nor the generalised velocities) the equations of motion of Lagrange can be written in the form given in equation (4.74). He also introduced the generalised momenta, defined by $p_k = \partial L/\partial \dot{q}_k$, $k = 1, 2, \ldots, n$, which play a major role in later developments of the theory, but do not occur again in our story.

Hamilton, in 1834, introduced the action integral defined in equation (4.78), but his theory was *not* limited to conservative systems. Hamilton's principle reduces to the original variational principle of Lagrange when (a) the system is conservative, and (b) the kinetic energy, T, is homogeneous of degree 2 in the generalised velocities. That is, for any constant λ, $T(\lambda \dot{\mathbf{q}}) = \lambda^2 T(\dot{\mathbf{q}})$, which is the case for many mechanical systems. Hamilton also demanded only that the action was *stationary* along the actual path. In his 1835 paper, Hamilton introduced the Hamiltonian function and set the scene for modern developments.

The kinetic energy was originally named, by Leibniz (1646–1716), the *visa viva*, or living force, and this name was in use until the last half of the 19th century.

Siméon-Denis Poisson, 1781–1840.

4.5 Applications of Lagrange's equations

In the following three sections we show how Lagrange's equations are derived for some typical systems with one or two degrees of freedom. By comparing this analysis with the equivalent analysis in Section 4.2, you will see how much simpler Lagrange's method can be. In the last example, Section 4.5.4, we show how the theory is extended to describe the partial differential equation for transverse vibrations of a taut string.

As you read through these applications you should be able to discern a simple pattern, easiest to describe for a particle moving in a Cartesian plane

Oxy, but constrained in some way, so there is only one generalised coordinate, q. In this case, the position of the particle is defined by two functions, $x = f(q,t)$ and $y = g(q,t)$, and everything follows from these functions. For instance, the potential energy will be some function of x and y, and hence q. The kinetic energy is proportional to the square of the speed, $v^2 = \dot{x}^2 + \dot{y}^2$, and since $\dot{x} = f_q \dot{q} + f_t$ and $\dot{y} = g_q \dot{q} + g_t$, the kinetic energy is given directly in terms of q, $\dot{q}$ and t. Hence, the Lagrangian is obtained directly from the functions $f(q,t)$ and $g(q,t)$. If there are more particles the method is exactly the same. As you read through the following sections, this pattern to the analysis should be noted.

4.5.1 The simple pendulum

We now return to a discussion of the simple pendulum, dealt with previously in Section 4.2.2, but first provide some general comments to explain why this is an important problem.

Some general comments

The simple pendulum is important for both historical and practical reasons and here we provide some background to explain why this is. First, it is worth noting that in 1889 a bibliography of pendulum papers was published by the Société Française de Physique: *Mémoires sur le Pendule* (Gauthier-Villars) contains a list of papers on the pendulum, published between 1629 and 1885 that runs to over 1300 items, many of which are concerned either with the construction of an accurate clock or with geophysical measurements.

Equations of motion similar to that of the pendulum occur in a variety of important problems. One is when a periodic driving force shakes a mechanical system at a frequency close to its frequency of natural vibrations and resonances occur. Another important, mathematically similar, system is found in the hindered rotations of certain types of molecules. An example is CH_3CF_3, the molecular structure of which is shown schematically in Figure 4.15.

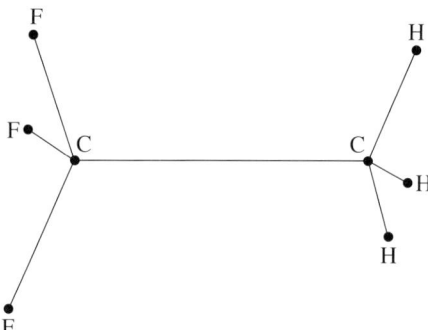

Figure 4.15 Diagram showing the structure of the CH_3CF_3 molecule.

In this molecule the end groups, CF_3 and CH_3, can each rotate rigidly and independently about the common C–C axis. The torque between the two groups clearly depends periodically upon the relative angle between the groups. It transpires that this relative rotation is dynamically similar to that of the simple pendulum considered here.

Historically the simple pendulum was of crucial importance in the development of Newtonian dynamics and it was studied and used extensively by Galileo and Newton, see Exercise 4.30. It is one of the simplest mechanical

4.5 Applications of Lagrange's equations

systems amenable to experiment and accurate measurement, so it is perhaps surprising that it was not studied by the Greeks. Moreover, it appears that there was little quantitative knowledge of the pendulum before the time of Galileo and his contemporaries.

It was Galileo, however, who discovered all the major properties of the pendulum. It is reported that in 1581, as a medical student in Pisa, his observations of a swinging lamp, in Pisa Cathedral, led him to believe in the isochrony of the pendulum. This was later confirmed experimentally and stated in a letter in 1602; here he noted that the period was independent of the mass, and also the veracity of the result found in Exercise 2.18 (page 66). Only much later, in 1638, did he show experimentally that the period was proportional to the square root of the length. This result was also stated by Mersenne in 1635 and the Italian physicist Baliani in 1638, although at this time the relevant length of the pendulum was not clearly defined. Of greater significance, however, Galileo used his observation of the pendulum to show that the, now unfamiliar, Aristotelian doctrine of 'mutually exclusive motion' was conceptually flawed and hence opened the way to the modern view of motion. Here is not the place to elaborate on this problem but a full account is given by Ariotti, which gives some idea of the intellectual effort required to overcome conventional wisdoms.

See *A short account of the history of mathematics* by W W Rouse Ball (Dover), 1960.

Marin Mersenne, 1588–1646.
Giovan Battista Baliani, 1582–1666.

P E Ariotti, 1972, *Aspects of the conception and development of the pendulum in the 17th century*, Arch Hist Exact Sci, Vol. 8.

After Galileo, the pendulum became a tool for regulating clocks and, through the simple formula for the period, $T = 2\pi\sqrt{l/g}$, an instrument for determining the local value of g and hence properties of the Earth. The first major step in clock design was due to Huygens, as described in Section 2.7.1. From 1656 onwards Huygens' invention brought significant improvements in the accuracy of time keeping. The first pendulum clock was accurate to about 30 seconds per day and the last, the Shortt pendulum clock (1921) was accurate to a one second per year; this was used in the Greenwich Observatory from 1921 until 1942, when it was replaced by a quartz crystal clock based at the Post Office research establishment in North London. The BBC 'pips', first started in 1924, were controlled by a pendulum clock until 1949.

Derivation of the equation of motion

We now return to our main pursuit, which is to derive, once more, the equation of motion for the pendulum. The idealised pendulum comprises a heavy particle, P, of mass m, firmly attached to a weightless, stiff, inextensible rod of length l. The other end of the rod is attached to a fixed point O, about which it can swing freely. Such a system is shown schematically in Figure 4.16.

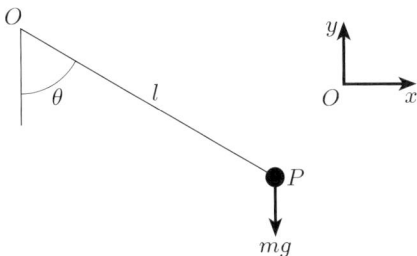

Figure 4.16 Diagram showing the idealised pendulum of length l, with smooth hinge at O and particle of mass m at P. The coordinate axes are shown with Oy vertically upwards.

As before we assume that there are no constraints on the value of θ, the angle between OP and the downward vertical.

The derivation starts by finding expressions for the position vector and speed of the particle P in terms of the generalised coordinate θ. The Cartesian coordinates of the position are given in equation (4.9) and for convenience we repeat them here:

$$x = l\sin\theta \quad \text{and} \quad y = -l\cos\theta, \tag{4.94}$$

where we use the same coordinate system and notation shown in Figure 4.16.

The potential energy is $V = mgy$, and in terms of the generalised coordinate this becomes

$$V(\theta) = -mgl\cos\theta. \tag{4.95}$$

The velocity of the mass is $(\dot{x}, \dot{y})$, which, on differentiating equations (4.94), becomes

$$\dot{x} = l\dot\theta\cos\theta \quad \text{and} \quad \dot{y} = l\dot\theta\sin\theta, \tag{4.96}$$

so its speed is given by $v^2 = \dot{x}^2 + \dot{y}^2 = l^2\dot\theta^2$, and hence the kinetic energy is

$$T = \tfrac{1}{2}mv^2 = \tfrac{1}{2}ml^2\dot\theta^2. \tag{4.97}$$

Thus the Lagrangian function is

$$L = T - V = \tfrac{1}{2}ml^2\dot\theta^2 + mgl\cos\theta. \tag{4.98}$$

Notice that all we need for the derivation of the Lagrangian is the position of the particle in terms of the generalised coordinate, equation (4.94), and its time derivative, equation (4.96).

Since $\partial L/\partial\dot\theta = ml^2\dot\theta$ and $\partial L/\partial\theta = -mgl\sin\theta$, the equation of motion is

$$\ddot\theta + \frac{g}{l}\sin\theta = 0, \tag{4.99}$$

which is the same as equation (4.13). You may like to compare the efforts needed to derive this and equation (4.13) (page 150).

Exercise 4.29

(a) Show that the first integral of the motion, equation (4.84), for the pendulum is just the total energy,

$$E = T + V = \tfrac{1}{2}ml^2\dot\theta^2 - mgl\cos\theta,$$

and that the minimum value of the energy is $E = -mgl$, corresponding to the pendulum hanging downwards, $\theta = 0$, and stationary, $\dot\theta = 0$.

(b) By putting $E = -mgl(1 - 2k^2)$, $k \geq 0$, show that the energy equation becomes

$$\dot\theta^2 = \frac{4g}{l}\left(k^2 - \sin^2(\theta/2)\right).$$

Deduce that if $0 \leq k < 1$, the value of θ is restricted to $|\theta| \leq \alpha < \pi$ where $\sin(\alpha/2) = k$.

If the pendulum starts at $\theta = -\alpha$ with $\dot\theta = 0$, show that the time, τ, to reach $\theta = \alpha$ is given by

$$\tau = \sqrt{\frac{l}{4g}} \int_{-\alpha}^{\alpha} d\theta \, \frac{1}{\sqrt{k^2 - \sin^2(\theta/2)}}, \quad k = \sin(\alpha/2).$$

Note that τ is half the period.

(c) Define a new variable ϕ by the equation $\sin(\theta/2) = k\sin\phi$ to show that the period, $T = 2\tau$ can be written in terms of the integral

$$T = 4\sqrt{\frac{l}{g}} \int_0^{\pi/2} d\phi \, \frac{1}{\sqrt{1 - k^2\sin^2\phi}}, \quad k = \sin(\alpha/2).$$

This is a slightly harder calculation which you should do only if time is available.

4.5 Applications of Lagrange's equations

Express the integrand as a power series in k using the binomial expansion and the integrals given in Exercise 4.24 (with appropriate adjustments to the limits), to obtain the approximation

$$T = 2\pi\sqrt{\frac{l}{g}}\left(1 + \tfrac{1}{4}k^2 + \tfrac{9}{64}k^4 + O(k^6)\right).$$

Exercise 4.30

Use the energy equation derived in the previous exercise to show that if the pendulum mass is released from rest at P_0, the speed of the mass at the bottom of the swing, $\theta = 0$, is proportional to the distance $d = PP_0$, shown in Figure 4.17.

The above result allows an indirect, accurate measurement of velocity, which was important in Newton's time because other methods were not available. Newton knew of this result, which he stated to be "a proposition well known to Geometers", and used it in his experiments involving the collision of two pendulum masses prior to formulating his laws of motion.

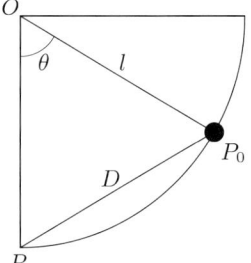

Figure 4.17

A moving point of support

In the second example considered in Section 4.2.2, when the pendulum hinge is attached to a freely moving particle of mass M, see Figure 4.3 (page 151), the Cartesian coordinates and velocity of P, with respect to the Oxy-axes, are

$$\mathbf{r} = (X + l\sin\theta, -l\cos\theta) \quad \text{and} \quad \dot{\mathbf{r}} = \left(\dot{X} + l\dot{\theta}\cos\theta, l\dot{\theta}\sin\theta\right), \quad (4.100)$$

so that the potential energy is $V = -mgl\cos\theta$ and the kinetic energy is

$$T = \tfrac{1}{2}M\dot{X}^2 + \tfrac{1}{2}m\left(\dot{X}^2 + l^2\dot{\theta}^2 + 2l\dot{X}\dot{\theta}\cos\theta\right). \quad (4.101)$$

Hence the Lagrangian is

$$L = \tfrac{1}{2}(M+m)\dot{X}^2 + \tfrac{1}{2}ml^2\dot{\theta}^2 + ml\dot{\theta}\dot{X}\cos\theta + mgl\cos\theta, \quad (4.102)$$

which depends on two generalised coordinates, X and θ. However, it depends upon $\dot{X}$ and not X, which is therefore an ignorable coordinate, so the X-equation of motion is simply

$$\frac{d}{dt}\left((M+m)\dot{X} + ml\dot{\theta}\cos\theta\right) = 0, \quad (4.103)$$

or

$$\dot{X} = A - \mu l\dot{\theta}\cos\theta, \quad \mu = \frac{m}{M+m}, \quad (4.104)$$

where A is a constant. On differentiating this with respect to t we obtain equation (4.20) (page 152).

On dividing by ml, the equation for θ is

$$\frac{d}{dt}\left(l\dot{\theta} + \dot{X}\cos\theta\right) + \dot{X}\dot{\theta}\sin\theta + g\sin\theta = 0 \quad (4.105)$$

which, on using the above expression for $\dot{X}$, becomes

$$\left(1 - \mu\cos^2\theta\right)l\ddot{\theta} + \mu l\dot{\theta}^2\sin\theta\cos\theta + g\sin\theta = 0, \quad (4.106)$$

which is just equation (4.21).

In this case the equations of motion simplify because X is ignorable. Using the expression for $\dot{X}$ in the energy integral (see equation (4.84) and

Exercise 4.31) we obtain

$$\tfrac{1}{2}\left(1-\mu\cos^2\theta\right)\dot\theta^2 - \frac{g}{l}\cos\theta = B, \quad \mu = \frac{m}{m+M}, \qquad (4.107)$$

where B is a constant. Since $0 < \mu < 1$, the coefficient of $\dot\theta^2$ is never zero.

Exercise 4.31

Show that the energy integral, equation (4.84), for the above system can be written in the form

$$\tfrac{1}{2}\dot X^2 + \tfrac{1}{2}\mu l^2 \dot\theta^2 + \mu l \dot X\dot\theta\cos\theta - \mu g l\cos\theta = \frac{E}{m+M}.$$

Using equation (4.104) for $\dot X$ show that this simplifies to equation (4.107).

Exercise 4.32

This is a hard exercise.

If the initial conditions at $t=0$ are $\theta = \theta_0$, $\dot\theta = 0$, $X = \dot X = 0$, use equations 4.104 and 4.107 to show that

$$X(t) = \mu l(\sin\theta_0 - \sin\theta) \quad \text{and} \quad \dot\theta^2 = \frac{4g}{l}\frac{\sin^2(\theta_0/2) - \sin^2(\theta/2)}{1-\mu\cos^2\theta}.$$

If $|\theta_0| \ll 1$, show that the approximate solution is

$$\theta(t) = \theta_0\cos\omega t, \quad X(t) = 2\mu l\theta_0 \sin^2(\omega t/2), \quad \text{where } \omega = \sqrt{\frac{g}{l(1-\mu)}}.$$

4.5.2 The double pendulum

For the double pendulum, we saw in Section 4.2.3 that the two generalised coordinates can be taken to be the angles θ_1 and θ_2 defined in Figure 4.4 (page 153). The derivation of the Lagrangian equations of motion for this system starts by deriving expressions for the Cartesian coordinates, (x_1, y_1) and (x_2, y_2) for each particle, in terms of the two generalised coordinates. This analysis is given in Section 4.2.3, but for convenience we repeat the relevant equations here:

$$\begin{aligned} x_1 &= l_1\sin\theta_1, & y_1 &= -l_1\cos\theta_1, \\ x_2 &= l_1\sin\theta_1 + l_2\sin\theta_2, & y_2 &= -l_1\cos\theta_1 - l_2\cos\theta_2. \end{aligned} \qquad (4.108)$$

The total potential energy is the sum of the potential energy of each particle, $V = m_1 g y_1 + m_2 g y_2$; in terms of the generalised coordinates, this is

$$V(\theta_1, \theta_2) = -m_1 g l_1 \cos\theta_1 - m_2 g \left(l_1\cos\theta_1 + l_2\cos\theta_2\right). \qquad (4.109)$$

The speed, v_1, of P_1 is given by the formula $v_1^2 = \dot x_1^2 + \dot y_1^2 = l_1^2\dot\theta_1^2$, where we have used equations (4.12) (page 150). The expression for the speed v_2 of the second particle is more complicated, though derived in the same manner:

$$\begin{aligned} v_2^2 &= \dot x_2^2 + \dot y_2^2 \\ &= \left(l_1\dot\theta_1\cos\theta_1 + l_2\dot\theta_2\cos\theta_2\right)^2 + \left(l_1\dot\theta_1\sin\theta_1 + l_2\dot\theta_2\sin\theta_2\right)^2 \\ &= l_1^2\dot\theta_1^2 + 2l_1 l_2 \dot\theta_1\dot\theta_2\cos(\theta_1-\theta_2) + l_2^2\dot\theta_2^2, \end{aligned} \qquad (4.110)$$

where we have used equations (4.26) (page 154) for $\dot x_2$ and $\dot y_2$. Thus the kinetic energy of the system is $T = (m_1 v_1^2 + m_2 v_2^2)/2$, which becomes

$$T = \tfrac{1}{2}l_1^2(m_1+m_2)\dot\theta_1^2 + \tfrac{1}{2}l_2^2 m_2 \dot\theta_2^2 + m_2 l_1 l_2 \dot\theta_1\dot\theta_2\cos(\theta_1-\theta_2). \qquad (4.111)$$

4.5 Applications of Lagrange's equations

Hence the Lagrangian function, $L = T - V$, is

$$L = \tfrac{1}{2}l_1^2(m_1 + m_2)\dot{\theta}_1^2 + m_2 l_1 l_2 \dot{\theta}_1 \dot{\theta}_2 \cos(\theta_1 - \theta_2) + \tfrac{1}{2}l_2^2 m_2 \dot{\theta}_2^2 \\ + (m_1 + m_2)g l_1 \cos\theta_1 + m_2 g l_2 \cos\theta_2. \quad (4.112)$$

The equations of motion are derived from this using Lagrange's equation of motion, equation (4.79). The simplest is that for θ_2, so we derive that first. Using the relations

$$\frac{\partial L}{\partial \theta_2} = -m_2 g l_2 \sin\theta_2 + m_2 l_1 l_2 \dot{\theta}_1 \dot{\theta}_2 \sin(\theta_1 - \theta_2), \quad (4.113)$$

$$\frac{\partial L}{\partial \dot{\theta}_2} = m_2 l_2^2 \dot{\theta}_2 + m_2 l_1 l_2 \dot{\theta}_1 \cos(\theta_1 - \theta_2). \quad (4.114)$$

Notice that the product $m_2 l_2$ is a factor of both these equations and hence does not appear in the equation of motion. Thus Lagrange's equation for θ_2 is

$$\frac{d}{dt}\left(l_2 \dot{\theta}_2 + l_1 \dot{\theta}_1 \cos(\theta_1 - \theta_2)\right) + g \sin\theta_2 - l_1 \dot{\theta}_1 \dot{\theta}_2 \sin(\theta_1 - \theta_2) = 0. \quad (4.115)$$

After expanding the time derivative, this simplifies to

$$l_1 \ddot{\theta}_1 \cos(\theta_1 - \theta_2) - l_1 \dot{\theta}_1^2 \sin(\theta_1 - \theta_2) + l_2 \ddot{\theta}_2 + g \sin\theta_2 = 0, \quad (4.116)$$

which is the same as equation (4.30).

The other equation of motion, for θ_1, is obtained in the same way, although it is more complicated. The basic partial derivatives are

$$\frac{\partial L}{\partial \theta_1} = -m_2 l_1 l_2 \dot{\theta}_1 \dot{\theta}_2 \sin(\theta_1 - \theta_2) - (m_1 + m_2)g l_1 \sin\theta_1, \quad (4.117)$$

$$\frac{\partial L}{\partial \dot{\theta}_1} = (m_1 + m_2)l_1^2 \dot{\theta}_1 + m_2 l_1 l_2 \dot{\theta}_2 \cos(\theta_1 - \theta_2), \quad (4.118)$$

and we see that Lagrange's equation for θ_1 is

$$\frac{d}{dt}\left((m_1 + m_2)l_1^2 \dot{\theta}_1 + m_2 l_1 l_2 \dot{\theta}_2 \cos(\theta_1 - \theta_2)\right) \\ + m_2 l_1 l_2 \dot{\theta}_1 \dot{\theta}_2 \sin(\theta_1 - \theta_2) + (m_1 + m_2) g l_1 \sin\theta_1 = 0, \quad (4.119)$$

which, on expanding the first bracket and dividing by $l_1(m_1 + m_2)$, simplifies to

$$l_1 \ddot{\theta}_1 + \frac{m_2 l_2}{(m_1 + m_2)}\left(\ddot{\theta}_2 \cos(\theta_1 - \theta_2) + \dot{\theta}_2^2 \sin(\theta_1 - \theta_2)\right) + g \sin\theta_1 = 0, \quad (4.120)$$

which is the same as equation (4.40).

The equations of motion, equations (4.116) and (4.120), are second-order nonlinear coupled differential equations which do not have any solutions that can be expressed as finite combinations of known functions. Indeed, under some circumstances the solutions display chaotic behaviour. The only limit in which simple solutions can be found is when both pendulums execute small oscillations: then the equations of motion can be approximated by coupled *linear* equations, which can be solved. These equations are derived in the next exercise.

Exercise 4.33

(a) Show that if $|\theta_i|$ and $|\dot{\theta}_i|$, $i = 1, 2$, are all small, so that second-order terms may be neglected, then the equations of motion (4.116) and (4.120) simplify to the pair of coupled linear equations

$$l_1\ddot{\theta}_1 + l_2\ddot{\theta}_2 + g\theta_2 = 0, \quad l_1\ddot{\theta}_1 + \mu l_2\ddot{\theta}_2 + g\theta_1 = 0, \quad \text{where } \mu = \frac{m_2}{m_1 + m_2}.$$

You will need the approximations $\sin x \simeq x$ and $\cos x \simeq 1$ for small $|x|$.

(b) Substitute the functions $\theta_1 = A_1 \cos\omega t$ and $\theta_2 = A_2 \cos\omega t$ into the above equations for the case $l_1 = l_2 = l$, and show that two solutions are

$$\theta_1 = a \cos\omega t, \quad \theta_2 = \pm\frac{a}{\sqrt{\mu}}\cos\omega t, \quad \text{where } \omega^2 = \frac{g}{l}\frac{1}{1 \pm \sqrt{\mu}}.$$

4.5.3 A simple pendulum with moving support

In this section we derive the equation of motion for a simple pendulum with a point of support that is oscillating vertically. This problem is interesting because it is one of the simplest examples showing the effects of vibration on a mechanical system. In addition, there are circumstances where the solutions behave counter-intuitively. Similar effects are seen in the interactions of powerful lasers and atoms.

The system is shown schematically in Figure 4.18.

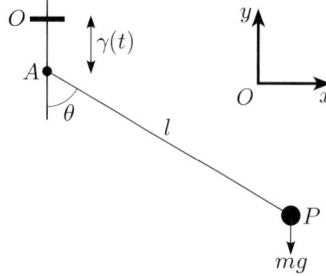

Figure 4.18 Diagram of a simple pendulum with a vertically moving hinge.

There is a particle P, of mass m, attached to the point A by a light inextensible rod of length l. The frictionless hinge A is moving vertically, so its distance below a fixed point O is $\gamma(t)$, a known function of time. The Lagrangian of this system is, as normal, determined from the position coordinates of P with respect to a fixed reference frame, expressed in terms of the generalised coordinate θ.

We use the axes Oxy, shown in the figure, with the y-axis vertically upwards. With respect to the moving point A, the coordinates of P are $(l\sin\theta, -l\cos\theta)$, so in the fixed reference frame

$$x = l\sin\theta, \quad y = -\gamma(t) - l\cos\theta. \tag{4.121}$$

The potential energy is $V = mgy$. The square of the speed is

$$v^2 = \left(l\dot{\theta}\cos\theta\right)^2 + \left(\dot{\gamma} - l\dot{\theta}\sin\theta\right)^2 = l^2\dot{\theta}^2 - 2l\dot{\gamma}\dot{\theta}\sin\theta + \dot{\gamma}^2. \tag{4.122}$$

Hence the Lagrangian is

$$L = \tfrac{1}{2}ml^2\dot{\theta}^2 + mgl\cos\theta - ml\dot{\gamma}\dot{\theta}\sin\theta + \left(\tfrac{1}{2}m\dot{\gamma}^2 + mg\gamma\right). \tag{4.123}$$

4.5 Applications of Lagrange's equations

The first two terms of this Lagrangian are identical to those in the Lagrangian (4.98), where the support is stationary. The third term is due to the motion of the support and may be simplified by noting that $\dot\theta \sin\theta = -d(\cos\theta)/dt$, and hence that

$$\frac{d}{dt}(\dot\gamma \cos\theta) = \ddot\gamma \cos\theta - \dot\gamma\dot\theta \sin\theta. \qquad (4.124)$$

On substituting for $\dot\gamma\dot\theta \sin\theta$ the Lagrangian becomes

$$L = \tfrac{1}{2} ml^2\dot\theta^2 + ml(g - \ddot\gamma)\cos\theta + m\left[\tfrac{1}{2}\dot\gamma^2 + g\gamma + l\frac{d}{dt}(\dot\gamma\cos\theta)\right]. \qquad (4.125)$$

Since the term in square brackets is the time derivative of a function of θ and t it may be ignored, to give the equivalent Lagrangian

$$\overline{L} = \tfrac{1}{2}ml^2\dot\theta^2 + ml(g - \ddot\gamma)\cos\theta. \qquad (4.126)$$

Thus the effect of moving the support vertically is to modify the value of g. For example, if the pendulum is in a lift uniformly accelerating upwards with acceleration a, we have $\ddot\gamma = -a$ and the effective value of g would be $g + a$, as might be expected.

The effect is more interesting if γ oscillates rapidly: for instance if $\gamma = a\sin\Omega t$, provided $a\Omega > \sqrt{2gl}$ (a result that can be proved using methods we cannot include in this course), the pendulum will oscillate stably in the upward vertical direction, that is, about $\theta = \pi$, which is impossible for the normal pendulum. The equation of motion for this system is

$$\frac{d^2\theta}{dt^2} + \frac{1}{l}\left(g + a\Omega^2\sin\Omega t\right)\sin\theta = 0, \qquad (4.127)$$

and in Figure 4.19 is shown the numerical solution of this equation with units chosen to give $g = l = 1$ and $a = 0.1$, $\Omega = 40$ and with the initial conditions $\theta(0) = 3.0$, $\dot\theta(0) = 0$. The value of $\theta(t)$ is shown for $0 \leq t \leq 2\pi$, that is 40 oscillations of the support: we note that the motion comprises a long period oscillation, with period about 2.4, on which is superimposed the high frequency oscillations of the support. The dashed line is at $\theta = \pi$, so we see that the pendulum is oscillating about the upward vertical.

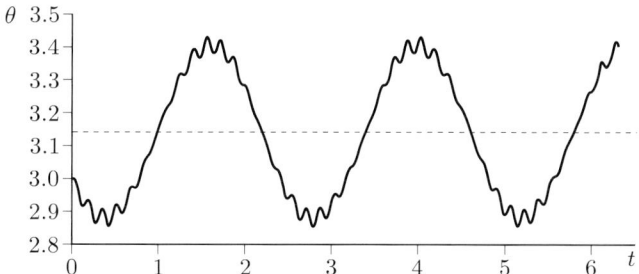

Figure 4.19 Graph of the solution $\theta(t)$, showing stable oscillations about the upward vertical. In this example $a = 0.1$, $\Omega = 40$, $g = l = 1$ and the initial conditions are $\theta(0) = 3.0$, $\dot\theta(0) = 0$.

The general theory mentioned above shows that the low frequency oscillations have the approximate frequency

$$\omega = \sqrt{\frac{g}{l}}\sqrt{\frac{a^2\Omega^2}{2gl} - 1}. \qquad (4.128)$$

For the parameters used in Figure 4.19, this gives the period as 2.37. It can also be shown that some initial conditions lead to chaotic motion.

Exercise 4.34

Consider the simple pendulum with its point of support moving horizontally, with known displacement $\gamma(t)$ from a fixed point. Using the notation defined in the previous text, show that the Lagrangian can be written in the form

$$L = \tfrac{1}{2}ml^2\dot{\theta}^2 + mgl\left(\cos\theta - \frac{\ddot{\gamma}}{g}\sin\theta\right),$$

and that the equation of motion is

$$\frac{d^2\theta}{dt^2} + \frac{g}{l}\left(\sin\theta + \frac{\ddot{\gamma}}{g}\cos\theta\right) = 0.$$

Note that this system will also oscillate about the upward vertical if $\gamma(t)$ is periodic, with sufficiently high frequency.

Exercise 4.35

Consider the simple pendulum with a fixed point of support, but with the length $l(t)$ a known function of time. Show that the Lagrangian is

$$L = \tfrac{1}{2}ml(t)^2\dot{\theta}^2 + mgl(t)\cos\theta$$

and that the equation of motion is

$$\frac{d}{dt}\left(l^2\frac{d\theta}{dt}\right) + gl\sin\theta = 0.$$

4.5.4 Transverse vibrations of a taut string (Optional)

In this final, optional, section we complete the circle of this course by showing how variational principles can be used to derive the wave equation for the transverse vibrations of a taut string, that is, the equation derived in Block I, Chapter 1. The ideas presented here are important because they form the basis of the modern descriptions of electromagnetic and gravitational fields. However, the actual method we describe was first used by Johann Bernoulli in 1727 and repeated by Lagrange in 1759, but his emphasis was on finding solutions that satisfied the boundary and initial conditions.

We start by approximating the string, initially lying along the x-axis, by a set of $N+2$ identical particles a distance h apart, where $(N+1)h = L$, L being the length of the string. Adjacent masses are connected by light, rigid rods. We assume that the ends of the string are fixed at $x = 0$ and L. In Figure 4.20 we show this approximation for $N = 5$, with the curved line depicting the disturbed string.

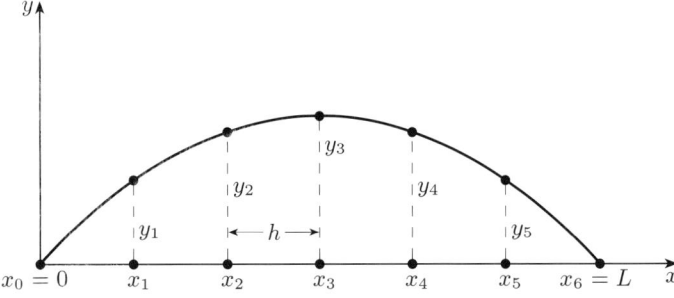

Figure 4.20 Diagram showing the discrete approximation to the string with seven particles, $N = 5$.

4.5 Applications of Lagrange's equations

The x-coordinate of the kth mass is $x_k = kh$, with the first ($k = 0$) and last ($k = N + 1$) particles stationary on the x-axis, so x_k is constant. If the string has a uniform line-density ρ, we approximate it by locating a mass $\mu = \rho h$ at each of the moving points.

In motion the y-coordinate of the kth particle is y_k and it is assumed to move perpendicularly to the x-axis: it therefore has kinetic energy $\frac{1}{2}\mu \dot{y}_k^2$ and the total kinetic energy of the system is

$$T = \tfrac{1}{2} h\rho \sum_{k=1}^{N} \dot{y}_k^2. \tag{4.129}$$

The potential energy, V is slightly more difficult to obtain. Consider the three particles centred at x_k, shown in Figure 4.21.

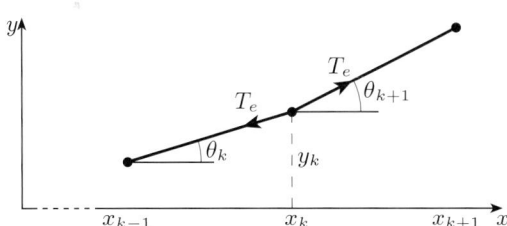

Figure 4.21 Diagram showing the forces on the particle at x_k.

The gradient of the kth element, between x_{k-1} and x_k, is $\tan\theta_k$ and, by definition $\tan\theta_k = (y_k - y_{k-1})/h$. We shall assume that $|\theta_k| \ll 1$, for all k, so it follows that $|y_k - y_{k-1}| \ll h$. The distance between the particles at x_{k-1} and x_k is, by Pythagoras' theorem,

$$s_k = \sqrt{h^2 + (y_k - y_{k-1})^2} = h\left(1 + O(\theta_k^2)\right) \simeq h, \tag{4.130}$$

since $\tan\theta_k = \theta_k + O(\theta_k^3)$. Thus, to first order, the separation between adjacent particles is constant; therefore the tensions in each segment are constant and the same, with value T_e. The force on the kth particle, in the y-direction, is

$$F_k = -\frac{\partial V_k}{\partial y_k} = T_e(\sin\theta_{k+1} - \sin\theta_k), \quad k = 1, 2, \ldots, N, \tag{4.131}$$

$$\simeq T_e(\theta_{k+1} - \theta_k) \simeq \frac{T_e}{h}(y_{k+1} - y_k) - \frac{T_e}{h}(y_k - y_{k-1}), \tag{4.132}$$

where V_k is the potential of the kth particle, all others being held fixed. Now integrate with respect to y_k (with y_{k-1} and y_{k+1} fixed) to obtain

$$V_k = \frac{T_e}{2h}(y_{k+1} - y_k)^2 + \frac{T_e}{2h}(y_k - y_{k-1})^2, \tag{4.133}$$

which is the potential producing the force F_k. Because V_k depends upon $y_{k\pm 1}$ we need to be careful not to include terms twice, hence the potential may be taken to be

$$V = \frac{T_e}{2h} \sum_{k=0}^{N} (y_{k+1} - y_k)^2, \quad y_0 = y_{N+1} = 0. \tag{4.134}$$

As a check we can differentiate this with respect to y_k, to obtain the above expression for F_k.

Thus the Lagrangian, $\mathcal{L}$, is

$$\mathcal{L} = T - V = \tfrac{1}{2} h\rho \sum_{k=0}^{N} \dot{y}_k^2 - \frac{T_e}{2h} \sum_{k=0}^{N} (y_{k+1} - y_k)^2. \tag{4.135}$$

The equation of motion for the kth particle is therefore
$$\rho h \ddot{y}_k - \frac{T_e}{h}(y_{k+1} - 2y_k + y_{k-1}) = 0, \quad k = 1, 2, \ldots, N. \tag{4.136}$$

If N is large, we assume that $y_k(t)$ is given in terms of a function $\eta(x,t)$, of x and t, by the relation
$$y_k = \eta(x_k, t), \quad \text{and that} \quad \dot{y}_k = \frac{\partial \eta}{\partial t} \quad \text{and} \quad \ddot{y}_k = \frac{\partial^2 \eta}{\partial t^2}, \tag{4.137}$$

where the derivatives are evaluated at $x = x_k$. Then we have
$$\frac{y_{k+1} - 2y_k + y_{k-1}}{h^2} = \frac{1}{h^2}\Big(\eta(x_k + h, t) - 2\eta(x_k, t) + \eta(x_k - h, t)\Big)$$
$$= \frac{\partial^2 \eta}{\partial x^2} + O(h), \tag{4.138}$$

as may be seen by substituting the Taylor expansions
$$\eta(x \pm h, t) = \eta(x, t) \pm h\frac{\partial \eta}{\partial x} + \tfrac{1}{2}h^2 \frac{\partial^2 \eta}{\partial x^2} + O(h^3). \tag{4.139}$$

Thus equation (4.136) can be written in the form
$$\frac{\partial^2 \eta}{\partial t^2} - c^2 \frac{\partial^2 \eta}{\partial x^2} = O(h), \quad c = \sqrt{\frac{T_e}{\rho}}. \tag{4.140}$$

In the limit as $h \to 0$ this gives the wave equation derived in Chapter 1 of Block I.

Our goal, however, is not the equations of motion, but the variational principle from which these equations can be derived, because these more general concepts are more useful in other circumstances. In particular, it is easier to derive the equations of motion in more complex conditions using a variational principle.

For the case of N particles, the action is
$$S[\mathbf{y}] = \frac{h}{2} \int_0^\tau dt \sum_{k=0}^N \left[\rho \dot{y}_k^2 - T_e \left(\frac{y_{k+1} - y_k}{h}\right)^2\right], \tag{4.141}$$

with
$$\mathbf{y}(0) = \mathbf{y}_0 \quad \text{and} \quad \mathbf{y}(\tau) = \mathbf{y}_1, \tag{4.142}$$

where $\mathbf{y}(0)$ and $\mathbf{y}(\tau)$ are the initial and final shapes of the string. We now approximate this using the function $\eta(x,t)$, introduced above, to obtain,
$$S = \tfrac{1}{2} \int_0^\tau dt\, h \sum_{k=0}^N \left[\rho\left(\frac{\partial \eta}{\partial t}\right)^2 - T_e \left(\frac{\partial \eta}{\partial x}\right)^2\right] + O(h^2), \tag{4.143}$$

where the kth term of the sum is evaluated at x_k. But an integral can be defined as the limit of a sum,
$$\int_0^L dx\, f(x) = \lim_{h \to 0} \sum_{k=0}^N h f(x_k), \quad x_k = kh, \quad (N+1)h = L. \tag{4.144}$$

Comparing this equation with the integrand in equation (4.143), we see that in the limit as $h \to 0$, that is, as $N \to \infty$, the functional becomes the double integral
$$S[\eta] = \tfrac{1}{2} \int_0^\tau dt \int_0^L dx \left[\rho\left(\frac{\partial \eta}{\partial t}\right)^2 - T_e \left(\frac{\partial \eta}{\partial x}\right)^2\right], \tag{4.145}$$

with the boundary conditions given by
$$\eta(0, t) = \eta(L, t) = 0 \quad \text{for all } t, \tag{4.146}$$

4.5 Applications of Lagrange's equations

and the initial and final conditions given respectively by

$$\eta(x,0) = \eta_0(x), \quad \eta(x,\tau) = \eta_1(x) \quad \text{for } 0 \le x \le L. \tag{4.147}$$

Notice that the first set of boundary conditions are the end conditions, corresponding to $y_0 = y_{N+1} = 0$ for all t. This integral is the difference between the kinetic and potential energies of the string,

$$T = \tfrac{1}{2}\int_0^L dx\, \rho \left(\frac{\partial \eta}{\partial t}\right)^2 \quad \text{and} \quad V = \tfrac{1}{2}T_e \int_0^L dx \left(\frac{\partial \eta}{\partial x}\right)^2. \tag{4.148}$$

Equation (4.145) gives the action for a string, and the functional has become a double integral: it is not, therefore, clear how to obtain the wave equation using the fact that S is stationary. In fact, we use exactly the same technique. That is, we evaluate the action on a neighbouring accessible path $\eta(x,t) + \epsilon g(x,t)$: because this a possible path it fits the boundary conditions, so $g(x,0) = g(x,\tau) = 0$ and $g(0,t) = g(L,t) = 0$. The Gâteaux differential, equation (2.24) (page 48), is

$$\Delta S[\eta,g] = \int_0^\tau dt \int_0^L dx \left(\rho \frac{\partial \eta}{\partial t}\frac{\partial g}{\partial t} - T_e \frac{\partial \eta}{\partial x}\frac{\partial g}{\partial x}\right). \tag{4.149}$$

We now proceed exactly as in Chapter 2, but a little bit more care is needed and it is convenient to treat each term separately.

For the potential energy term we have

$$\int_0^\tau dt \left(\int_0^L dx\, \frac{\partial \eta}{\partial x}\frac{\partial g}{\partial x}\right) = \int_0^\tau dt \left(\left[\frac{\partial \eta}{\partial x}g\right]_{x=0}^L - \int_0^L dx\, \frac{\partial^2 \eta}{\partial x^2} g\right), \tag{4.150}$$

where we have used integration by parts to rewrite the x-integral. But $g(0,t) = g(L,t) = 0$ for all t, so the first term on the right-hand side is zero and

$$\int_0^\tau dt \left(\int_0^L dx\, \frac{\partial \eta}{\partial x}\frac{\partial g}{\partial x}\right) = -\int_0^\tau dt \int_0^L dx\, \frac{\partial^2 \eta}{\partial x^2} g. \tag{4.151}$$

For the kinetic energy integral we change the order of integration to obtain, by the same method,

$$\int_0^L dx\, \rho \left(\int_0^\tau dt\, \frac{\partial \eta}{\partial t}\frac{\partial g}{\partial t}\right) = \int_0^L dx\, \rho \left(\left[\frac{\partial \eta}{\partial t}g\right]_{t=0}^\tau - \int_0^\tau dt\, \frac{\partial^2 \eta}{\partial t^2} g\right), \tag{4.152}$$

and since $g(x,0) = g(x,\tau) = 0$, this gives

$$\int_0^\tau dt \left(\int_0^L dx\, \rho \frac{\partial \eta}{\partial t}\frac{\partial g}{\partial t}\right) = -\int_0^\tau dt \int_0^L dx\, \rho \frac{\partial^2 \eta}{\partial t^2} g. \tag{4.153}$$

Thus the Gâteaux differential becomes

$$\Delta S[\eta,g] = -\int_0^\tau dt \int_0^L dx \left(\rho \frac{\partial^2 \eta}{\partial t^2} - T_e \frac{\partial^2 \eta}{\partial x^2}\right) g. \tag{4.154}$$

On a stationary path $\Delta S = 0$ for all $g(x,t)$ and using a version of the fundamental lemma of the calculus of variations (not proven here), we find that

$$\rho \frac{\partial^2 \eta}{\partial t^2} - T_e \frac{\partial^2 \eta}{\partial x^2} = 0, \tag{4.155}$$

which is just the wave equation. Notice that in this derivation we did not make the assumption that the density was the same along the length of the wire, so this equation is valid if the density, ρ, depends upon x.

The general form of a functional depending on two independent variables x and t will be

$$S[\eta] = \int_0^\tau dt \int_0^L dx\, F(x, t, \eta, \eta_x, \eta_t), \qquad (4.156)$$

Recall that $\eta_x = \partial \eta / \partial x$ and $\eta_t = \partial \eta / \partial t$.

where the integrand depends upon the unknown function, $\eta(x,t)$, and its first partial derivatives with respect to x and t. There will also be boundary conditions given by

$$\eta(0, t) = \eta(L, t) = 0 \quad \text{for all } t, \qquad (4.157)$$

and initial and final conditions given respectively by

$$\eta(x, 0) = \eta_0(x), \quad \eta(x, \tau) = \eta_1(x) \quad \text{for } 0 < x < L. \qquad (4.158)$$

The required function(s) make this functional stationary and it can be shown that these satisfy the partial differential equation

$$\frac{\partial}{\partial t}\left(\frac{\partial F}{\partial \eta_t}\right) + \frac{\partial}{\partial x}\left(\frac{\partial F}{\partial \eta_x}\right) - \frac{\partial F}{\partial \eta} = 0. \qquad (4.159)$$

This is the Euler–Lagrange equation for the functional (4.156).

Exercise 4.36

Use equation (4.159) to derive (4.155) from (4.145).

Exercise 4.37

A string with linear density $\rho(x)$ is embedded in a light, uniform rubber sheet such that in equilibrium it lies in a straight line, along the x-axis, with end points fixed at $x = 0$ and $x = L$, under tension T_e. It is free to make small oscillations in the y-direction, perpendicular to the x-axis, and in the plane of the sheet.

The rubber sheet acts on the element δx of the string to exert a force of magnitude, $\delta F = k\eta\,\delta x$, where k is a constant, proportional to the displacement, η, and towards the x-axis. Show that the action of the system is

$$S = \tfrac{1}{2}\int_0^\tau dt \int_0^L dx \left(\rho(x)\left(\frac{\partial \eta}{\partial t}\right)^2 - T_e\left(\frac{\partial \eta}{\partial x}\right)^2 - k\eta^2\right),$$

and find the equation of motion of the string. Recall that this system was also considered in Block I, Section 6.4.1.

Exercise 4.38

In previous derivations of the equation for transverse oscillations of a string it was assumed to be perfectly flexible, that is, no energy was needed to bend it. Here, we include this effect.

An approximation to the bending energy is obtained by considering the force needed to bend an element of the string. Provided the curvature is small the potential energy due to bending an element of length δx can be shown to be $\tfrac{1}{2}B(x)\eta_{xx}^2\,\delta x$, for some positive function $B(x)$, which reduces to a constant, B, for a uniform string.

In this case, show that the action for a finite string, fixed at $x = 0$ and $x = L$, is

$$S = \tfrac{1}{2}\int_0^\tau dt \int_0^L dx\,\left(\rho(x)\eta_t^2 - T_e\eta_x^2 - B\eta_{xx}^2\right).$$

Show that the Gâteaux derivative of this functional for admissible functions satisfying the conditions

$$\eta(0, t) = \eta(L, t) = 0 \quad \text{for all } t,$$

and

$$\eta(x, 0) = \eta_0(x), \quad \eta(x, \tau) = \eta_1(x) \quad \text{for } 0 \le x \le L,$$

is

$$\Delta S[\eta, g] = \int_0^\tau dt \int_0^L dx \left[-\rho \eta_{tt} + T_e \eta_{xx} - B \eta_{xxxx} \right] g - B \int_0^\tau dt \left[\eta_{xx} g_x \right]_{x=0}^L.$$

Deduce that if the admissible functions satisfy the additional boundary conditions $\eta_x(0, t) = \alpha(t)$ and $\eta_x(L, t) = \beta(t)$, where α and β are arbitrary functions of t, then the equation satisfied by η is

$$-B \frac{\partial^4 \eta}{\partial x^4} + T_e \frac{\partial^2 \eta}{\partial x^2} - \rho(x) \frac{\partial^2 \eta}{\partial t^2} = 0.$$

Exercise 4.39

Consider a string with linear density $\rho(x)$, fixed at the points $x = 0$ and $x = L$, and with each point moving perpendicular to the x-axis, in the y-direction. A force $F(x)$ per unit length acts in the y-direction and is independent of η.

(a) Show that the potential energy of the string is given by

$$V = \tfrac{1}{2} T_e \int_0^L dx \left(\frac{\partial \eta}{\partial x} \right)^2 - \int_0^L dx \, F(x) \, \eta.$$

(b) Write down the Lagrangian and the action functional for the *static* string. Use equation (4.159) to find the equation of motion and integrate this to find an expression for the displacement $\eta(x)$. Evaluate this expression in the case $F(x) = $ constant.

(c) Write down the Lagrangian and the action functional for the vibrating string and find the equation of motion.

4.6 Further Exercises

Exercise 4.40

For each of the following systems, define suitable generalised coordinate(s) and state the number of its degrees of freedom.

(a) A particle sliding on the interior surface of a hemispherical bowl.

(b) A particle swinging in a vertical plane, at one end of a stiff, elastic rod, whose other end is fixed.

(c) A particle swinging at one end of a stiff, elastic rod, whose other end is fixed.

(d) A particle at one end of a light rigid rod, free to move in a vertical plane, whose other end slides freely on a given smooth, rigid curve.

Exercise 4.41

(a) Show directly from Lagrange's equation of motion that the two Lagrangians

$$L_1(q, \dot{q}) = \tfrac{1}{2} (\dot{q} + q)^2, \quad L_2(q, \dot{q}) = \tfrac{1}{2} (\dot{q}^2 + q^2),$$

produce the same equations of motion.

(b) Show that these Lagrangians differ by a total time derivative.

Exercise 4.42

In this exercise we show that a potential function $V(r)$ that depends on the two vectors $\mathbf{r}_1$ and $\mathbf{r}_2$ through the distance $r = |\mathbf{r}_1 - \mathbf{r}_2|$ gives a force on each of the particles at $\mathbf{r}_1$ and $\mathbf{r}_2$ that satisfies Newton's third law.

(a) Using the definition $r^2 = (x_1 - x_2)^2 + (y_1 - y_2)^2 + (z_1 - z_2)^2$, show that
$$r\frac{\partial r}{\partial x_1} = -r\frac{\partial r}{\partial x_2} = x_1 - x_2,$$
and that similar relations hold for (y_k, z_k), $k = 1, 2$.

(b) Using this result and the chain rule, show that
$$\frac{\partial V}{\partial x_1} = -\frac{\partial V}{\partial x_2}, \quad \frac{\partial V}{\partial y_1} = -\frac{\partial V}{\partial y_2}, \quad \frac{\partial V}{\partial z_1} = -\frac{\partial V}{\partial z_2},$$
and hence that $\mathbf{F}_{12} = -\mathbf{F}_{21}$, where these forces are given by equations (4.63), that is $\mathbf{F}_{12} = -\partial V/\partial \mathbf{r}_1$ and $\mathbf{F}_{21} = -\partial V/\partial \mathbf{r}_2$.

Exercise 4.43

A set of coordinates, (u, v, ϕ), defined in terms of the Cartesian coordinates (x, y, z) by the equations
$$x = uv\cos\phi, \quad y = uv\sin\phi, \quad z = \tfrac{1}{2}\left(u^2 - v^2\right),$$
are sometimes useful when there is a symmetry along the z-axis; these are known as parabolic coordinates. Show that
$$\dot{x}^2 + \dot{y}^2 + \dot{z}^2 = \left(u^2 + v^2\right)\left(\dot{u}^2 + \dot{v}^2\right) + u^2 v^2 \dot{\phi}^2.$$

Exercise 4.44

Consider the motion of a single particle, P, in the Oxy-plane with respect to Cartesian coordinates (x, y) and (u, v), where the Ouv-axes are rotating with respect to Oxy with constant angular speed Ω, as shown in Figure 4.22.

The position of a point P can be defined by the Cartesian coordinates (x, y) and (u, v), with obvious notation. The relation between these two coordinates is
$$u = x\cos\Omega t + y\sin\Omega t, \quad v = -x\sin\Omega t + y\cos\Omega t.$$

(a) Show that the inverse transformation is
$$x = u\cos\Omega t - v\sin\Omega t, \quad y = u\sin\Omega t + v\cos\Omega t.$$

(b) If V is the speed of the particle in the Oxy coordinate system, so $V^2 = \dot{x}^2 + \dot{y}^2$, show that $V^2 = \dot{u}^2 + \dot{v}^2 + \Omega^2\left(u^2 + v^2\right) - 2\Omega(\dot{u}v - u\dot{v})$.

(c) If (ρ, ϕ) are the plane polar coordinates in Ouv, so $u = \rho\cos\phi$, $v = \rho\sin\phi$, show that $V^2 = \dot{\rho}^2 + \rho^2(\dot{\phi} + \Omega)$.

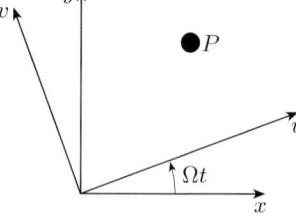

Figure 4.22

Exercise 4.45

Two blocks, each of mass M, are connected by an inextensible string of length l. One block is placed on a smooth horizontal surface, and the other block hangs over the side, the string passing over a frictionless pulley, as shown in Figure 4.23.

Determine the Lagrangian in the following two cases:

(a) the mass of the string is negligible;

(b) the string has uniform mass m.

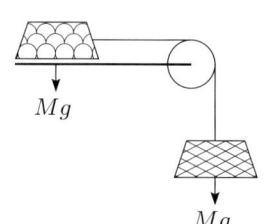

Figure 4.23

Exercise 4.46

A particle of mass m slides on a smooth straight wire which is pivoted about a point O and rotates in a prescribed manner, in a vertical plane, with variable angular speed so that at time t the angle between the wire and the horizontal is $\alpha(t)$. Initially the wire is horizontal, so $\alpha(0) = 0$, and the particle is at rest, a distance A from O.

Show that the Lagrangian of the system is

$$L = \tfrac{1}{2}m\left(\dot{q}^2 + q^2\dot{\alpha}^2\right) - mgq\sin\alpha,$$

where q is the distance of the particle from the pivot.

In the case where $\alpha(t) = \beta t$, show that the general solution of the equation of motion is

$$q(t) = A\cosh\beta t + B\sinh\beta t + \frac{g}{2\beta^2}\sin\beta t.$$

Exercise 4.47

A rigid wire in the shape of a parabola moves horizontally so that the coordinates, (x, y), of a point on it are related by the equation

$$y = \tfrac{1}{2}a\left(x - \gamma(t)\right)^2,$$

where the y-axis is vertically upward and γ is a known function of time. Using the generalised coordinate $q = x - \gamma(t)$ show that the Lagrangian for a particle of mass m sliding smoothly on this wire is

$$L = \tfrac{1}{2}m\dot{q}^2\left(1 + a^2q^2\right) - \tfrac{1}{2}magq^2 - mq\ddot{\gamma}.$$

Exercise 4.48

Find the Lagrangian and the equation of motion for a particle of mass m sliding smoothly on the parabolic wire $y = a(t)x^2/2$, where $a(t)$ is a positive, known function of the time and the y-axis is vertically upwards.

Exercise 4.49

Show that for a free particle the stationary path of the action gives a minimum of the action.

4.7 Harder Exercises

Exercise 4.50

A governor is a device for controlling variations in the speed of rotation of an axle. An elementary mechanical governor consists of a mass m_2 moving on a vertical axle AB, and two masses m_1 attached to it and to a fixed point O on the axle by rods of length a, as shown in the diagram. As the whole system rotates about AB with constant angular velocity Ω, the angle θ changes; see Figure 4.24.

Show that the Lagrangian of the system is

$$L = a^2\dot{\theta}^2\left(m_1 + 2m_2\sin^2\theta\right) + m_1a^2\Omega^2\sin^2\theta + 2ag(m_1 + m_2)\cos\theta,$$

and that the energy is

$$E = a^2\dot{\theta}^2\left(m_1 + 2m_2\sin^2\theta\right) - m_1a^2\Omega^2\sin^2\theta - 2ag(m_1 + m_2)\cos\theta.$$

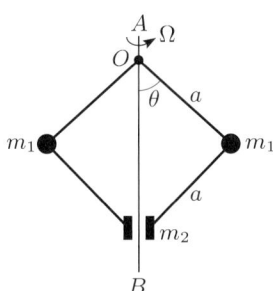

Figure 4.24

Exercise 4.51

A particle of mass m is attached to a vertical cylindrical post of radius a by a long horizontal string, which can wrap itself round the post. If $R \gg a$ is the length of the unwrapped part of the string, show that the Lagrangian is

$$L(R, \dot{R}) = \frac{m}{2a^2} R^2 \dot{R}^2.$$

Also show that the first integral of the motion is $mR^2 \dot{R}^2 = 2a^2 E$, where E is a constant.

Exercise 4.52

A particle is constrained to slide under gravity on a smooth wire in the shape of a vertical circle with radius R. The wire rotates about its vertical diameter with constant angular velocity Ω. Show that the Lagrangian is

$$L = \tfrac{1}{2} m R^2 \dot{\theta}^2 + \tfrac{1}{2} m \Omega^2 R^2 \sin^2 \theta + mgR \cos \theta,$$

where θ is the angular displacement of the particle from the downward vertical, with the centre of the circle as origin.

Exercise 4.53

Kepler's third law

In this exercise we derive general properties of solutions when the potential energy is homogeneous of degree λ in q, that is $V(kq) = k^\lambda V(q)$; for instance the potential $V = q^2/2$ is homogeneous of degree 2. This exercise is interesting because, when extended to three dimensions, the same idea can be used to derive Kepler's third law without solving the equations of motion.

Consider the Lagrangian

$$L(q, \dot{q}) = \tfrac{1}{2} m \dot{q}^2 - V(q), \quad \text{where } V(kq) = k^\lambda V(q).$$

(a) Show that the arbitrary Lagrangians $L(q, \dot{q}, t)$ and $aL(q, \dot{q}, t)$, a being a non-zero constant, produce the same equations of motion.

(b) Introduce the scaled variables Q and T by the relations $Q = aq$ and $T = bt$, a and b being constants, and show that if $b = a^{1-\lambda/2}$, the expression for the Lagrangian in terms of Q and $Q' = dQ/dT$ is

$$\overline{L}(Q, Q') = a^{-\lambda} \left(\tfrac{1}{2} m Q'^2 - V(Q) \right).$$

(c) Deduce that if $q = f(t)$ is a solution of the equations of motion for L, then $q = af(a^{\lambda/2-1}t)$ is also a solution. Show, in particular, that for periodic motion under such a potential, the period varies as $A^{1-\lambda/2}$, where A is the amplitude.

(d) Deduce that in the case of a linear oscillator, where $V(q) \propto q^2$, the period is independent of the amplitude.

(e) For the one-dimensional Coulomb potential, $V(q) \propto q^{-1}$, $q > 0$, deduce that the period varies as $A^{3/2}$, which is just Kepler's third law; see page 156.

Exercise 4.54

A Lagrangian $L(q_1, \dot{q}_1, \dot{q}_2)$ is independent of q_2, so q_2 is an ignorable coordinate. Use the appropriate form of Noether's theorem (given on page 120) to show that $\partial L/\partial \dot{q}_2 = $ constant.

Solutions to Exercises in Chapter 4

Solution 4.1

Since m, g and a are all positive, $E \geq 0$. Differentiation of E with respect to t gives

$$\frac{dE}{dt} = m\dot{x}\ddot{x}\left(1 + a^2x^2\right) + \tfrac{1}{2}m\dot{x}^2\left(2a^2x\dot{x}\right) + mga x\dot{x}$$
$$= m\dot{x}\left[\ddot{x}\left(1 + a^2x^2\right) + a^2x\dot{x}^2 + agx\right] = 0,$$

the last equality following from the equation of motion (4.5). Rearranging equation (4.6) gives

$$m\dot{x}^2\left(1 + a^2x^2\right) = 2E - mgax^2 \quad \text{or} \quad \frac{dx}{dt} = \pm\sqrt{\frac{2E - mgax^2}{m\left(1 + a^2x^2\right)}}.$$

Solution 4.2

The expression for $\dot{E}$ is derived in the solution of Exercise 4.1. Substituting equation (4.8) into this gives

$$\frac{dE}{dt} = -max\dot{x}\ddot{\gamma} = -\frac{1}{2}ma\ddot{\gamma}\frac{d}{dt}\left(x^2\right).$$

Note that $x(t) = 0$ is the only solution of the differential equation for which $\dot{E} = 0$.

Solution 4.3

If $\mu = 0$, equations (4.21) and (4.20) reduce to $\ddot{\theta} + (g/l)\sin\theta = 0$ and $\ddot{X} = 0$, respectively. The first of these is just equation (4.13). The second gives $X = \alpha + \beta t$, where α and β are constants, so the particle A moves with uniform speed along the rail and is unaffected by the swinging pendulum below it, as would be expected for a heavy particle. Also, the pendulum motion is unaffected by the uniform motion of the support, A: this is an example of Galilean invariance, discussed in Section 4.3.

Solution 4.4

Differentiation with respect to t gives

$$\frac{dE}{dt} = (M+m)\dot{X}\ddot{X} + ml^2\dot{\theta}\ddot{\theta} + ml\left(\ddot{\theta}\dot{X}\cos\theta + \dot{\theta}\ddot{X}\cos\theta - \dot{\theta}^2\dot{X}\sin\theta + g\dot{\theta}\sin\theta\right).$$

Substituting for $(M+m)\ddot{X}$ using equation (4.20) reduces this to

$$\frac{dE}{dt} = ml\dot{\theta}\left(l\ddot{\theta} + \ddot{X}\cos\theta + g\sin\theta\right),$$

and using equation (4.19) we see that $\dot{E} = 0$, so E is a constant.

Solution 4.5

(a) Figure 4.25 shows the two vectors $\mathbf{r}_1$ and $\mathbf{r}_2$, the vector $\overline{\mathbf{R}}$ of the centre of mass at C, all relative to an origin O. By definition, $(m_1 + m_2)\overline{\mathbf{R}} = m_1\mathbf{r}_1 + m_2\mathbf{r}_2$. Also, addition of the vectors gives $\mathbf{r}_1 + \mathbf{r}_{12} - \mathbf{r}_2 = 0$, $\mathbf{r}_1 + \overline{\mathbf{r}}_1 - \overline{\mathbf{R}} = 0$ and $\overline{\mathbf{R}} - \overline{\mathbf{r}}_2 - \mathbf{r}_2 = 0$. Thus $\overline{\mathbf{r}}_1 = \overline{\mathbf{R}} - \mathbf{r}_1$ and $\overline{\mathbf{r}}_2 = \overline{\mathbf{R}} - \mathbf{r}_2$, and using the centre of mass equation to eliminate $\overline{\mathbf{R}}$ gives

$$(m_1 + m_2)\overline{\mathbf{r}}_1 = m_1\mathbf{r}_1 + m_2\mathbf{r}_2 - (m_1 + m_2)\mathbf{r}_1 = m_2(\mathbf{r}_2 - \mathbf{r}_1),$$

or

$$\overline{\mathbf{r}}_1 = \frac{m_2}{m_1 + m_2}\mathbf{r}_{12}.$$

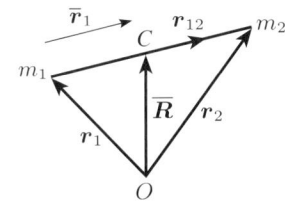

Figure 4.25

Similarly,

$$(m_1 + m_2)\overline{\mathbf{r}}_2 = m_1\mathbf{r}_1 + m_2\mathbf{r}_2 - (m_1 + m_2)\mathbf{r}_2 = m_1(\mathbf{r}_1 - \mathbf{r}_2),$$

or
$$\bar{\mathbf{r}}_2 = -\frac{m_1}{m_1 + m_2}\mathbf{r}_{12}.$$

(b) In the case where $m_1 = m_S$ and $m_2 = m_E$, we have
$$\frac{m_2}{m_1 + m_2} = \frac{1}{328901}, \quad \text{so} \quad |\bar{\mathbf{r}}_1| = \frac{149.6 \times 10^6}{3.289 \times 10^5} = 454.8 \, \text{km}.$$

(c) In the case where $m_1 = m_S$ and $m_2 = m_J$, we have
$$\frac{m_2}{m_1 + m_2} = \frac{1}{1048.4}, \quad \text{so} \quad |\bar{\mathbf{r}}_1| = \frac{778.3 \times 10^6}{1.048 \times 10^3} = 7.427 \times 10^5 \, \text{km}.$$

Solution 4.6

(a) The angle from any fixed point on the circle. One degree of freedom.

(b) The perpendicular distance from the plane. One degree of freedom.

(c) Two coordinates are needed to specify the positions on a two-dimensional surface. Suitable coordinates for a sphere are the latitude and longitude. Two degrees of freedom.

(d) Since the axis of rotation is known, only one further coordinate is needed to specify completely the orientation of the body. This may be taken as the angle of rotation about the axis. One degree of freedom.

(e) The motion of the circle is given, so only one variable is needed to fix the position of the particle on the circle. The same variable as in part (a) may be used. One degree of freedom.

(f) If the circle is free to rotate, its motion is affected by that of the particle moving on it, so two variables are needed. One of these may be the same as in part (a) and the other may be taken to be the angle of rotation of the circle. Two degrees of freedom.

(g) For any body moving in three-dimensional space, whether it is a particle or an extended solid body, three coordinates are needed to specify its position. For a solid body one would normally take these to be the Cartesian coordinates of the centre of mass, with respect to some given, fixed coordinate system. For a rigid body a further three variables are needed because it can rotate; these can be defined in a variety of ways, for example by using two to defined an instantaneous axis of rotation and the third to specify the angle of rotation about this axis. Thus the system has six degrees of freedom, or three if the centre of mass is fixed.

(h) Much depends upon how we approximate the system. If we regard the planet as a particle, then three coordinates are required to fix its position, which could be the three spherical polar coordinates with the sun at the origin; this system has three degrees of freedom. If the planet is regarded as a rigid sphere rotating about a given axis one extra degree of freedom, the angle of rotation, is required. If the structure of the planet is to be taken into account then its orientation must be specified as well as its position in space; this requires a further three coordinates, as in part (g), giving six degrees of freedom.

Solution 4.7

(a) The length of the rope is $2Y + y + \text{constant}$, hence the result.

(b) There is one degree of freedom.

(c) There are two reasons why the tension along the rope does not change. First, the rope is light, so any change in tension in a vertical section would result in an infinite acceleration. Second, the pulleys are smooth and light so the tension on either side must be the same, otherwise the angular acceleration would be infinite.

Solution 4.8

(a) In this example $\dot{x} = \dot{u} + \dot{v}$ and $\dot{y} = \dot{u} - \dot{v}$, hence $2\dot{u} = \dot{x} + \dot{y}$, $2\dot{v} = \dot{x} - \dot{y}$.

(b) Squaring and adding these expressions gives
$$\dot{x}^2 + \dot{y}^2 = (\dot{u} + \dot{v})^2 + (\dot{u} - \dot{v})^2 = 2\left(\dot{u}^2 + \dot{v}^2\right).$$

(c) Differentiation gives $\dot{x} = \dot{u} + \dot{v}t + v$ and $\dot{y} = \dot{u} - \dot{v}t - v$. Squaring and adding these expressions gives
$$\dot{x}^2 + \dot{y}^2 = \dot{u}^2 + \dot{v}^2 t^2 + v^2 + 2\dot{u}\dot{v}t + 2v\dot{v}t + 2v\dot{u}$$
$$+ \dot{u}^2 + \dot{v}^2 t^2 + v^2 - 2\dot{u}\dot{v}t + 2v\dot{v}t - 2v\dot{u}$$
$$= 2\left(\dot{u}^2 + \dot{v}^2 t^2\right) + 4v\dot{v}t + 2v^2.$$

Solution 4.9

Differentiation gives $\dot{x} = \dot{u}v + u\dot{v}$ and $\dot{y} = u\dot{u} - v\dot{v}$, so that squaring and adding gives the quoted result.

Solution 4.10

Since $\delta s^2 = \delta x^2 + \delta y^2 +$ higher-order terms, on dividing by δt and taking the limit as δt tends to zero, we see that $v^2 = \dot{x}^2 + \dot{y}^2 = \dot{s}^2$.

The kinetic energy can be expressed in a variety of ways. If $y = f(x)$, so $\dot{y} = f'(x)\dot{x}$, we have
$$T = \tfrac{1}{2}mv^2 = \tfrac{1}{2}m\left(\dot{x}^2 + \dot{y}^2\right) = \tfrac{1}{2}m\dot{x}^2\left(1 + f'(x)^2\right) = \tfrac{1}{2}m\dot{s}^2.$$

Solution 4.11

The kinetic energy is $T = m\left(\dot{r}^2 + r^2\dot{\theta}^2\right)/2$, where we have used the result derived in the text, equation (4.54). But since $r = a\theta$, $\dot{r} = a\dot{\theta}$, so we may eliminate either $\dot{\theta}$ or $\dot{r}$ to give, respectively,
$$T = \tfrac{1}{2}m\dot{r}^2\left(1 + \frac{r^2}{a^2}\right) \quad \text{or} \quad T = \tfrac{1}{2}ma^2\dot{\theta}^2\left(1 + \theta^2\right).$$

Solution 4.12

The total kinetic energy is $T = M\dot{Y}^2/2 + m\dot{y}^2/2$. But $2Y + y =$ constant, so $2\dot{Y} = -\dot{y}$ and $T = M\dot{y}^2/8 + m\dot{y}^2/2$, which gives the required result.

Solution 4.13

(a) Partial differentiation of r^2 with respect to the independent variables x, y and z gives
$$2r\frac{\partial r}{\partial x} = 2x, \quad 2r\frac{\partial r}{\partial y} = 2y, \quad 2r\frac{\partial r}{\partial z} = 2z.$$

(b) The chain rule gives $\partial V/\partial x = \partial V/\partial r \times \partial r/\partial x$, with similar relations for differentiation with respect to y and z. Hence
$$\frac{\partial V}{\partial \mathbf{r}} = \left(\frac{\partial V}{\partial x}, \frac{\partial V}{\partial y}, \frac{\partial V}{\partial z}\right) = \frac{\partial V}{\partial r}\left(\frac{x}{r}, \frac{y}{r}, \frac{z}{r}\right) = \frac{1}{r}V'(r)\mathbf{r}$$
$$= V'(r)\hat{\mathbf{r}}.$$

Since $\mathbf{F} = -\partial V/\partial \mathbf{r}$, the result follows.

Solution 4.14

The force is $\mathbf{F} = -k\mathbf{x} = -\partial V/\partial \mathbf{r}$, hence $\partial V/\partial x = kx$ and $V = kx^2/2$.

Solution 4.15

If $z \ll R$ we may expand in powers of z/R by writing the potential energy as

$$V = -\frac{GMm}{R}\left(1+\frac{z}{R}\right)^{-1}$$
$$= -\frac{GMm}{R}\left[1-\frac{z}{R}+O\left(\frac{z^2}{R^2}\right)\right]$$
$$= -\frac{GMm}{R}+\frac{GMm}{R^2}z+O\left(\frac{z^2}{R^2}\right).$$

The first term is a constant and can be ignored so, to the lowest order, $V = mgz$ where $g = GM/R^2$. Using the given values $g = 979.7\,\text{cm}\,\text{s}^{-2}$.

The actual value on the surface of the Earth varies with latitude because the Earth is not perfectly spherical. At the equator $g = 983.221$ and at the poles $g = 978.049$.

Solution 4.16

(a) The variation of the acceleration due to gravity, $g(h)$, with the height, h, above the mean sea level is derived in the previous exercise, and is $g(h) = GM/(R+h)^2$. A first-order Taylor expansion about $h = 0$ (that is, sea level) gives $g(h) = g(0) + hg'(0)$; that is, if we denote $g(0)$ by g,

$$g(h) - g = -h\frac{2GM}{R^3}, \quad \text{so} \quad \frac{\Delta g}{g} = \frac{g(h)-g}{g} \simeq -\frac{2h}{R}.$$

Putting $R = 6.378 \times 10^6\,\text{m}$ and $h = 800\,\text{m}$ gives $\Delta g = -2.5 \times 10^{-4}g$.

(b) The rate of change of T with g can be written in the form $T'(g) = -T/(2g)$ and to first order $T(g+\Delta g) - T(g) = \Delta g T'(g)$, which gives

$$\frac{\Delta T}{T} \simeq -\frac{\Delta g}{2g}.$$

There are $24 \times 60 \times 60 = 86\,400$ seconds in a day, so the relative accuracy of the clock is $\Delta T/T = 1.2 \times 10^{-5}$. But the relative change in T due to the change in height is, from the above formula, $\Delta T/T = h/R \simeq 1.2 \times 10^{-4}$ so this change can be measured by the clock in less than a day.

Solution 4.17

(a) The kinetic energy is $T = m\dot{z}^2/2$, and since $E = T + V$ the result follows.

If the initial speed is v, at $z = 0$, the energy is

$$E = \tfrac{1}{2}mv^2 - \frac{GMm}{R} = \tfrac{1}{2}m\dot{z}^2 - \frac{GMm}{R+z},$$

since E is a constant.

(b) As $z \to \infty$ and $\dot{z} \to 0$, the right-hand side tends to zero to give the escape speed,

$$\tfrac{1}{2}mv_e^2 = \frac{GMm}{R} \quad \text{or} \quad v_e = \sqrt{\frac{2GM}{R}}.$$

(c) Using the data given for the Earth, the Moon and Jupiter we obtain the following.

	Earth	Moon	Jupiter
v_e	$11.18\,\text{km}\,\text{s}^{-1}$	$2.37\,\text{km}\,\text{s}^{-1}$	$59.37\,\text{km}\,\text{s}^{-1}$
g	$979.6\,\text{cm}\,\text{s}^{-2}$	$162.4\,\text{cm}\,\text{s}^{-2}$	$2451\,\text{cm}\,\text{s}^{-2}$

Solution 4.18

Since $T = mv^2/2$ and $m\dot{\mathbf{v}} = \mathbf{F}$, we have $\dot{T} = mv\dot{v} = m\mathbf{v}\cdot\dot{\mathbf{v}} = \mathbf{v}\cdot\mathbf{F}$, since

$$v^2 = \mathbf{v}\cdot\mathbf{v} \quad \text{giving} \quad 2v\frac{dv}{dt} = \frac{d}{dt}(\mathbf{v}\cdot\mathbf{v}) = 2\mathbf{v}\cdot\frac{d\mathbf{v}}{dt}.$$

Solution 4.19

In this example we have
$$\frac{\partial L}{\partial \dot{x}} = m\dot{x}, \quad \frac{\partial L}{\partial \dot{y}} = m\dot{y}, \quad \frac{\partial L}{\partial \dot{z}} = m\dot{z},$$

and
$$\frac{\partial L}{\partial x} = -\frac{\partial V}{\partial x}, \quad \frac{\partial L}{\partial y} = -\frac{\partial V}{\partial y}, \quad \frac{\partial L}{\partial z} = -\frac{\partial V}{\partial z}.$$

Hence Lagrange's equation for the x-coordinate is
$$\frac{d}{dt}\left(\frac{\partial L}{\partial \dot{x}}\right) - \frac{\partial L}{\partial x} = m\frac{d\dot{x}}{dt} + \frac{\partial V}{\partial x} = 0,$$

which reduces to the quoted equation: since $-\partial V/\partial x$ is the force in the x-direction, this is just Newton's equation of motion for this coordinate. The results for the y- and z-directions follow similarly.

Solution 4.20

(a) First note that
$$\frac{\partial V}{\partial x} = \frac{\partial V}{\partial r}\frac{\partial r}{\partial x} \quad \text{and} \quad \frac{\partial V}{\partial y} = \frac{\partial V}{\partial r}\frac{\partial r}{\partial y},$$

and since $\partial r/\partial x = x/r$ and $\partial r/\partial y = y/r$ (see Exercise 4.13(a)), Lagrange's equations of motion in Cartesian coordinates are
$$\frac{d}{dt}(m\dot{x}) + V'(r)\frac{x}{r} = 0 \quad \text{and} \quad \frac{d}{dt}(m\dot{y}) + V'(r)\frac{y}{r} = 0.$$

(b) Since $\dot{x}^2 + \dot{y}^2 = \dot{r}^2 + r^2\dot{\theta}^2$ (see equation (4.54)), the Lagrangian becomes
$$L = \tfrac{1}{2}m\left(\dot{r}^2 + r^2\dot{\theta}^2\right) - V(r).$$

This is independent of θ, so this equation of motion becomes
$$\frac{d}{dt}\left(\frac{\partial L}{\partial \dot{\theta}}\right) = \frac{d}{dt}\left(mr^2\dot{\theta}\right) = 0, \quad \text{hence} \quad \dot{\theta} = \frac{A}{mr^2},$$

where A is a constant. The equation for r is
$$\frac{d}{dt}\left(\frac{\partial L}{\partial \dot{r}}\right) - \frac{\partial L}{\partial r} = m\ddot{r} - mr\dot{\theta}^2 + V'(r) = 0.$$

Using the equation for $\dot{\theta}$ gives the quoted result.

Solution 4.21

Since $f(x) = ax^2/2$, $f'(x) = ax$ and $f''(x) = a$, equation (4.83) becomes
$$\frac{d}{dt}\left(\dot{x}(1 + a^2x^2)\right) - a^2x\dot{x}^2 + agx = 0,$$

which expands to $\ddot{x}\left(1 + a^2x^2\right) + 2a^2x\dot{x}^2 - a^2x\dot{x}^2 + agx = 0$ and is the same as equation (4.5).

Solution 4.22

(a) If the length of the string between the particle of mass m_1 and the pulley axis is z, and the total length of the string is l, then the length of the string between the particle of mass m_2 and the pulley is $l - \pi R - z$. Thus the potential energies of the two particles are $V_1 = -m_1 g z$, and $V_2 = -m_2 g (l - \pi R - z)$. Notice that the signs are negative because distance is measured downwards. The total potential energy is

$$V(z) = V_1 + V_2 = -(m_1 - m_2)gz - m_2 g l + \pi m_2 g R,$$

and the last two terms may be ignored because they are constants.

When the mass of the pulley is neglected, the kinetic energy is just the sum of the kinetic energy of the particles $T = m_1 \dot{z}^2/2 + m_2 \dot{z}^2/2 = (m_1 + m_2)\dot{z}^2/2$, so the Lagrangian is

$$L = \tfrac{1}{2}(m_1 + m_2)\dot{z}^2 + (m_1 - m_2)gz.$$

(b) If the mass of the pulley wheel cannot be neglected then its rotation will add a component to the kinetic energy, but not to the potential energy. The extra kinetic energy is given in the question in terms of $\dot{\theta}$; since the rope does not slip we have the relation $R\dot{\theta} = \dot{z}$, so the new kinetic energy is $T = (m_1 + m_2 + M_p)\dot{z}^2/2$ and the Lagrangian is

$$L = \tfrac{1}{2}(m_1 + m_2 + M_p)\dot{z}^2 + (m_1 - m_2)gz.$$

(c) With $M = m_1 + m_2 + M_p$ we have $\partial L/\partial z = (m_1 - m_2)g$ and $\partial L/\partial \dot{z} = M\dot{z}$, so Lagrange's equation of motion is

$$M\ddot{z} = (m_1 - m_2)g,$$

with general solution

$$z = z(0) + \dot{z}(0)t + \frac{1}{2}\frac{m_1 - m_2}{M}gt^2.$$

As would be expected, if $m_1 > m_2$ the particle with mass m_2 moves up, and if $m_1 < m_2$ it moves down. Also, the heavier the pulley wheel the less the acceleration.

Solution 4.23

(a) In this example the motion is along a line in three dimensions so the speed v is obtained from $v^2 = \dot{x}^2 + \dot{y}^2 + \dot{z}^2 = (k^2 + a^2)\dot{\phi}^2$. The kinetic energy is thus $T = m(k^2 + a^2)\dot{\phi}^2/2$. The potential energy is $V(\phi) = mgz = mgk\phi$, giving the Lagrangian

$$L = T - V = \tfrac{1}{2}m(k^2 + a^2)\dot{\phi}^2 - mgk\phi,$$

which is similar to the Lagrangian of a particle falling in a gravitational field. Lagrange's equations of motion are $(k^2 + a^2)\ddot{\phi} + gk = 0$. Direct integration of this equation gives $\phi(t) = \phi(0) + \dot{\phi}(0)t - \dfrac{gkt^2}{2(k^2 + a^2)}$, which means that the particle slides down the spiral with ever increasing speed.

(b) When the spiral is rotating the coordinates of the particle are

$$x = a\cos(q + \theta(t)), \quad y = a\sin(\phi + \theta(t)), \quad z = k\phi.$$

For a rotating wire the potential energy is unaffected by the rotation, but the speed is now given by $v^2 = a^2\left(\dot{\phi} + \dot{\theta}\right)^2 + k^2\dot{\phi}^2$ giving the Lagrangian

$$L = \tfrac{1}{2}m\left((a^2 + k^2)\dot{\phi}^2 + 2a^2\dot{\phi}\dot{\theta}\right) - mgk\phi,$$

where we have, as usual, ignored the term $m\dot{\theta}^2/2$ because it is a known function of time.

This Lagrangian can be cast in a more convenient form by removing the dependence on $\dot{\phi}$, using a trick discussed later, on page 178. Observe that

$$\dot{\phi}\ddot{\theta} = \frac{d}{dt}(\phi\dot{\theta}) - \phi\ddot{\theta},$$

so the Lagrangian becomes

$$L = \tfrac{1}{2}m\left(a^2 + k^2\right)\dot{\phi}^2 - m\left(gk + a^2\ddot{\theta}\right)\phi + ma^2\frac{d}{dt}(\phi\dot{\theta}).$$

On page 178 we show that the last term, being a derivative with respect to time, may be ignored to give the simpler, equivalent Lagrangian,

$$L = \tfrac{1}{2}m\left(a^2 + k^2\right)\dot{\phi}^2 - m\left(gk + a^2\ddot{\theta}\right)\phi.$$

The equations of motion are obtained using the derivatives of the first Lagrangian,

$$\frac{\partial L}{\partial \dot{\phi}} = m\left(a^2 + k^2\right)\dot{\phi} + ma^2\dot{\theta} \quad \text{and} \quad \frac{\partial L}{\partial \phi} = -mgk,$$

which give the Lagrangian equations of motion

$$m\left(a^2 + k^2\right)\ddot{\phi} + ma^2\ddot{\theta} + mgk = 0 \quad \text{and} \quad \ddot{\phi} = -\frac{1}{a^2 + k^2}\left(a^2\ddot{\theta} + gk\right).$$

Solution 4.24

(a) For a T-periodic solution $x(t + T) = x(t)$, for all t. If $x(0) = A$, then $x(T) = A$ and the action is

$$S = \int_0^T dt\, \left(\tfrac{1}{2}m\dot{x}^2 - \tfrac{1}{4}ma^2x^4\right), \quad x(0) = x(T) = A.$$

(b) If $x = A\sin\omega t$, $\omega = 2\pi/T$, the functional becomes

$$S = \tfrac{1}{2}mA^2\omega^2 \int_0^{2\pi/\omega} dt\, \cos^2\omega t - \tfrac{1}{4}mA^4a^2 \int_0^{2\pi/\omega} dt\, \sin^4\omega t$$

$$= \tfrac{1}{2}m\omega A^2 \int_0^{2\pi} d\theta\, \cos^2\theta - \tfrac{1}{4}\frac{mA^4 a^2}{\omega} \int_0^{2\pi} d\theta\, \sin^4\theta$$

$$= \frac{m\pi}{16\omega}\left(8\omega^2 A^2 - 3a^2 A^4\right).$$

(c) The functional must be stationary, so $S'(A) = 0$, that is, $16\omega^2 A - 12a^2 A^3 = 0$. So

$$A^2 = \frac{4\omega^2}{3a^2} \quad \text{or} \quad A = \frac{2}{\sqrt{3}}\frac{\omega}{a}.$$

Solution 4.25

The constant is

$$E = \dot{x}\frac{\partial L}{\partial \dot{x}} - L = m\dot{x}^2\left(1 + f'(x)^2\right) - \tfrac{1}{2}m\dot{x}^2\left(1 + f'(x)^2\right) + mgf(x),$$

which reduces to the given expression. Rearranging this gives

$$m\left(\frac{dx}{dt}\right)^2 = \frac{2E - 2mgf(x)}{1 + f'(x)^2}.$$

Solution 4.26

The equation of motion is

$$\ddot{x}\left(1 + f'(x)^2\right) + \dot{x}^2 f'(x) f''(x) + (g + \ddot{\gamma}) f'(x) = 0$$

and if $f = ax^2/2$ this becomes $\ddot{x}\left(1 + (ax)^2\right) + a^2 x \dot{x}^2 + (g + \ddot{\gamma})ax = 0$, as required.

Solution 4.27

(a) It is shown in Exercise 4.10 that $v^2 = \dot{s}^2$, thus $T = mv^2/2 = m\dot{s}^2/2$. The potential energy is $V = mgy(s)$, so the Lagrangian is $L = m\dot{s}^2/2 - mgy(s)$ and the equation of motion is $m\ddot{s} + mgy'(s) = 0$, hence $\ddot{s} + gy'(s) = 0$.

(b) On the cycloid

$$s(\phi) = \int_0^\phi d\phi \sqrt{\left(\frac{dx}{d\phi}\right)^2 + \left(\frac{dy}{d\phi}\right)^2}$$

$$= a \int_0^\phi d\phi \sqrt{(1 + \cos\phi)^2 + \sin^2\phi}$$

$$= a \int_0^\phi d\phi \sqrt{2(1 + \cos\phi)}.$$

But $\cos\phi = 2\cos^2(\phi/2) - 1$, hence

$$s = 2a \int_0^\phi d\phi \cos(\phi/2) = 4a\sin(\phi/2).$$

Using the related trigonometric identity, $\cos\phi = 1 - 2\sin^2(\phi/2)$, we obtain $y = 2a\sin^2(\phi/2) = s^2/(8a)$. Hence, from part (a), the equation of motion is as given.

(c) The general solution of the equation of motion is

$$s(t) = s(0)\cos\omega t + \frac{\dot{s}(0)}{\omega}\sin\omega t, \quad \omega = \sqrt{\frac{g}{4a}}.$$

Thus $s(t) = s(t+T)$, $T = 2\pi/\omega$, for all valid initial conditions. That is, the period of the motion is independent of the initial conditions: no matter where the particle starts, it reaches the bottom in the same time.

Solution 4.28

The kinetic energy is found in Exercise 4.12 (page 169). The potential energy is $V = -(MY + my)g$ and, since $2Y + y$ is constant, the potential energy becomes, apart from a constant, $V = (M - 2m)yg/2$, giving the quoted Lagrangian.

The equation of motion is $(M + 4m)\ddot{y} = 2(2m - M)g$; integrating twice gives the general solution quoted.

Solution 4.29

(a) Using the expressions given for T and V we obtain the quoted expression,

$$E = \frac{\partial L}{\partial \dot{\theta}}\dot{\theta} - L = ml^2\dot{\theta}^2 - \left(\tfrac{1}{2}ml^2\dot{\theta}^2 + mgl\cos\theta\right) = \tfrac{1}{2}ml^2\dot{\theta}^2 - mgl\cos\theta.$$

The minimum of E is at $\dot{\theta} = 0$ and the maximum of $mgl\cos\theta$, that is $\theta = 0$; thus $E \geq -mgl$.

(b) Using the half-angle formula, the energy integral becomes

$$-mgl(1 - 2k^2) = \tfrac{1}{2}ml^2\dot{\theta}^2 - mgl\left[1 - 2\sin^2(\theta/2)\right],$$

which can be rearranged to give $\dot{\theta}^2 = 4g\left(k^2 - \sin^2(\theta/2)\right)/l$. Since $\dot{\theta}$ is real, we need $|\sin(\theta/2)| < k$; if $0 \leq k < 1$, this means that $|\theta| \leq \alpha < \pi$ where $\sin(\alpha/2) = k$.

Since θ increases from $-\alpha$ to α, $\dot{\theta} > 0$, and hence

$$\frac{d\theta}{dt} = \sqrt{\frac{4g}{l}}\sqrt{k^2 - \sin^2(\theta/2)},$$

and integration gives

$$\int_{-\alpha}^{\alpha} d\theta \frac{1}{\sqrt{k^2 - \sin^2(\theta/2)}} = \sqrt{\frac{4g}{l}}\tau.$$

(c) First note that the values of the contributions from the intervals $(-\alpha, 0)$ and $(0, \alpha)$ are identical, so

$$T = 2\sqrt{\frac{l}{g}} \int_0^\alpha d\theta \, \frac{1}{\sqrt{k^2 - \sin^2(\theta/2)}} = 2\sqrt{\frac{l}{g}} \int_0^{\pi/2} d\phi \, \frac{d\theta}{d\phi} \frac{1}{\sqrt{k^2 - \sin^2(\theta/2)}}.$$

But

$$\frac{1}{2}\cos(\theta/2)\frac{d\theta}{d\phi} = k\cos\phi, \quad \text{so} \quad \frac{d\theta}{d\phi} = \frac{2k\cos\phi}{\sqrt{1 - k^2\sin^2\phi}},$$

hence

$$T = 4\sqrt{\frac{l}{g}} \int_0^{\pi/2} d\phi \, \frac{1}{\sqrt{1 - k^2\sin^2\phi}}, \quad k = \sin(\alpha/2).$$

Now use the binomial expansion $(1-z)^{-1/2} = 1 + z/2 + 3z^2/8 + O(z^3)$ to write the expression in the form

$$T = 4\sqrt{\frac{l}{g}} \int_0^{\pi/2} d\phi \left(1 + \frac{k^2}{2}\sin^2\phi + \frac{3k^4}{8}\sin^4\phi + O(k^6)\right)$$

and use the integrals, given in Exercise 4.24, but with modified integration limits,

$$\int_0^{\pi/2} d\phi \, \sin^2\phi = \frac{\pi}{4}, \quad \int_0^{\pi/2} d\phi \, \sin^4\phi = \frac{3\pi}{16},$$

to give the quoted result.

Note that for small k the relative difference between this expression and the small amplitude limit, $k = 0$, is 1% when $k^2/4 = 1/100$, that is $k = 2/10$ or $\alpha \simeq 4/10$, which is a swing of about $23°$.

Solution 4.30

At P_0 we have $(\theta, \dot\theta) = (\theta_0, 0)$ and at P, $(\theta, \dot\theta) = (0, \dot\theta_1)$, so the energy equation gives

$$-mgl\cos\theta_0 = \tfrac{1}{2}ml^2\dot\theta_1^2 - mgl.$$

Thus the speed at P, $v = l\dot\theta_1$, is given by $\tfrac{1}{2}v^2 = gl(1 - \cos\theta_0)$. But $\cos u = 1 - 2\sin^2(u/2)$ and elementary trigonometry gives $d/(2l) = \sin(\theta_0/2)$, so

$$\tfrac{1}{2}v^2 = 2gl\sin^2(\theta_0/2) \quad \text{or} \quad v = d\sqrt{\frac{g}{l}}.$$

Solution 4.31

The total energy is the sum of the kinetic and potential energies,

$$E = \tfrac{1}{2}(M+m)\dot X^2 + \tfrac{1}{2}ml^2\dot\theta^2 + ml\dot\theta\dot X\cos\theta - mgl\cos\theta.$$

This result can also be obtained from equation (4.84). Dividing by $M+m$, putting $E = C(M+m)$ and using the relation $\dot X = A - \mu l\dot\theta\cos\theta$ gives

$$C = \tfrac{1}{2}\left(A - \mu l\dot\theta\cos\theta\right)^2 + \tfrac{1}{2}\mu l^2\dot\theta^2 + \mu l\dot\theta\cos\theta\left(A - \mu l\dot\theta\cos\theta\right) - \mu gl\cos\theta$$

$$= \tfrac{1}{2}A^2 + \tfrac{1}{2}\mu l^2\dot\theta^2 - \tfrac{1}{2}\mu^2 l^2\dot\theta^2\cos^2\theta - \mu gl\cos\theta,$$

which gives

$$\tfrac{1}{2}\dot\theta^2\left(1 - \mu\cos^2\theta\right) - \frac{g}{l}\cos\theta = \frac{2C - A^2}{2\mu l^2} = B.$$

Solution 4.32

The equation (4.104) for $\dot{X}$ can be written as

$$\dot{X} = A - \mu l \frac{d}{dt}(\sin\theta) \quad \text{which integrates to} \quad X = C + At - \mu l \sin\theta.$$

The initial conditions give $A = 0$ and $C = \mu l \sin\theta_0$, hence $X = \mu l(\sin\theta_0 - \sin\theta)$.

Using the initial conditions in equation (4.107) gives $B = -(g/l)\cos\theta_0$, and on using the identity $\cos u = 1 - 2\sin^2(u/2)$ this becomes

$$\dot{\theta}^2 = \frac{4g}{l}\frac{\sin^2(\theta_0/2) - \sin^2(\theta/2)}{1 - \mu\cos^2\theta}.$$

If $|\theta_0| \ll 1$, then using the approximations $\sin u \simeq u$ this becomes $\dot{\theta}^2 + \omega^2\theta^2 = \omega^2\theta_0^2$ where $\omega = \sqrt{g/(l(1-\mu))}$. The solution of this equation is that quoted. For $X(t)$ we have, for small θ,

$$X(t) = \mu l(\theta_0 - \theta) = \mu l\theta_0(1 - \cos\omega t),$$

which gives the quoted result after using the identity $\cos\phi = 1 - 2\sin^2(\phi/2)$.

Solution 4.33

(a) If both angles are small, on using the approximations given, we see that equation (4.120) becomes $l_1\ddot{\theta}_1 + \mu l_2\ddot{\theta}_2 + g\theta_1 = 0$ and that equation (4.116) becomes $l_1\ddot{\theta}_1 + l_2\ddot{\theta}_2 + g\theta_2 = 0$, with μ as quoted.

(b) Using the substitution given, these equations become

$$A_1\left(\omega^2 l_1 - g\right) + \mu A_2 l_2\omega^2 = 0, \quad A_1 l_1\omega^2 + \left(\omega^2 l_2 - g\right) A_2 = 0.$$

These equations, for A_1 and A_2, have non-trivial solutions only if

$$\begin{vmatrix} l_1\omega^2 & \omega^2 l_2 - g \\ \omega^2 l_1 - g & \mu l_2\omega^2 \end{vmatrix} = 0,$$

that is, if

$$\omega^4 l_1 l_2 (1-\mu) - \omega^2(l_1 + l_2)g + g^2 = 0.$$

Putting $l_1 = l_2 = l$ we see that the solutions of this quadratic for ω are

$$\omega^2 = \frac{g}{l}\frac{1 \pm \sqrt{\mu}}{1 - \mu}, \quad \text{that is,} \quad \omega^2 = \frac{g}{l}\frac{1}{1 \pm \sqrt{\mu}}.$$

If ω is taken to be either of these values, then

$$A_2 = -A_1\frac{\omega^2 l_1 - g}{\mu l_2\omega^2} = -\frac{A_1}{\mu}\left(1 - \frac{g}{l\omega^2}\right) = \pm\frac{A_1}{\sqrt{\mu}}, \quad l_1 = l_2 = l.$$

Hence, putting $A_1 = a$ we have

$$\theta_1 = a\cos\omega t, \quad \theta_2 = \pm\frac{a}{\sqrt{\mu}}\cos\omega t \quad \text{where} \quad \omega^2 = \frac{g}{l}\frac{1}{1 \pm \sqrt{\mu}}.$$

Solution 4.34

The coordinates of the pendulum mass with respect to the point of support are $(l\sin\theta, -l\cos\theta)$, so with respect to a fixed point the coordinates are $(l\sin\theta + \gamma(t), -l\cos\theta)$. The kinetic energy is

$$T = \tfrac{1}{2}m\left[\left(l\dot\theta\cos\theta + \dot\gamma\right)^2 + \left(l\dot\theta\sin\theta\right)^2\right]$$

$$= \tfrac{1}{2}ml^2\dot\theta^2 + ml\dot\gamma\frac{d}{dt}(\sin\theta) + \tfrac{1}{2}m\dot\gamma^2$$

$$= \tfrac{1}{2}ml^2\dot\theta^2 + ml\left(-\ddot\gamma\sin\theta + \frac{d}{dt}(\dot\gamma\sin\theta)\right) + \tfrac{1}{2}m\dot\gamma^2$$

The potential energy is $V = -mgl\cos\theta$, so the Lagrangian can be taken to be

$$L = \tfrac{1}{2}ml^2\dot\theta^2 - ml\ddot\gamma\sin\theta + mgl\cos\theta,$$

and the equation of motion is as given.

Solution 4.35

Consider Figure 4.16 (page 183): the coordinates of the point P are as in equation (4.94), that is $x = l(t)\sin\theta$ and $y = -l(t)\cos\theta$. Differentiating with respect to t gives
$$\dot{x} = \dot{l}\sin\theta + l\dot{\theta}\cos\theta, \quad \dot{y} = -\dot{l}\cos\theta + l\dot{\theta}\sin\theta,$$
hence the square of the speed is $v^2 = \dot{l}^2 + l^2\dot{\theta}^2$. Since the kinetic energy is $T = mv^2/2$ and the potential energy is $V = mgy = -mgl(t)\cos\theta$, the Lagrangian is
$$L = \tfrac{1}{2}ml(t)^2\dot{\theta}^2 + mgl(t)\cos\theta + \tfrac{1}{2}m\dot{l}^2.$$
The final is a function of time only, so may be ignored. Differentiation gives $\partial L/\partial\dot{\theta} = ml^2\dot{\theta}$ and $\partial L/\partial\theta = -mgl\sin\theta$, which gives the quoted equation of motion.

Solution 4.36

From equation (4.145) we see that $F = \dfrac{\rho}{2}\eta_t^2 - \dfrac{T_e}{2}\eta_x^2$, so
$$\frac{\partial F}{\partial \eta_t} = \rho\eta_t, \quad \frac{\partial F}{\partial \eta_x} = -T_e\eta_x, \quad \frac{\partial F}{\partial \eta} = 0.$$
Hence, equation (4.159) gives $\rho\eta_{tt} - T_e\eta_{xx} = 0$, which is the wave equation.

Solution 4.37

Each element of the string, of length δx, has an extra potential energy $k\eta^2\delta x/2$ because of the energy stored in the stretched rubber sheet. Hence the action is that given in equation (4.145) with the extra potential energy subtracted from it, that is
$$S = \tfrac{1}{2}\int_0^\tau dt \int_0^L dx\,\left(\rho\eta_t^2 - T_e\eta_x^2 - k\eta^2\right)$$
Using the general result (4.159) this gives the equation of motion $\rho\eta_{tt} - T_e\eta_{xx} + k\eta = 0$: this equation also occurs in quantum mechanics where it is known as the *Klein–Gordon equation*.

Solution 4.38

Using the potential energy given in the question we see that the total potential energy of bending is
$$V_B = \tfrac{1}{2}B \int_0^L dx\,\left(\frac{\partial^2\eta}{\partial x^2}\right)^2.$$
Including this contribution, the action (4.145) gives the result quoted.

Consider the Gâteaux differential of the bending contribution, S_B, to the action:
$$\Delta S_B[\eta, g] = -B\int_0^\tau dt \int_0^L dx\,\eta_{xx}g_{xx}.$$
Integrating by parts with respect to x twice and using the conditions $g(0,t) = g(L,t) = 0$ gives
$$\Delta S_B[\eta, g] = -B\int_0^\tau dt\,\left[\eta_{xx}g_x\right]_{x=0}^L - B\int_0^\tau dt \int_0^L dx\,\frac{\partial^4\eta}{\partial x^4}g.$$
If there are the additional boundary conditions then $g_x(0,t) = g_x(L,t) = 0$ and the boundary term vanishes and the stationary path is determined by the given equation.

Solution 4.39

(a) The potential energy comprises the sum of the component due to work being done against the tension, found in the text,

$$V_1 = \tfrac{1}{2} T_e \int_0^L dx \left(\frac{\partial \eta}{\partial x}\right)^2$$

and the work done against the force $F(x)$. For an element of the string, length δx, this potential is $\delta V_2 = -F(x)\eta\, \delta x$. Compare this with the potential gained by lifting a mass m a distance η in a gravitational field, where the force is mg, so the potential energy gained is $-mg\eta$.

Hence the potential energy is

$$V = \tfrac{1}{2} T_e \int_0^L dx \left(\frac{\partial \eta}{\partial x}\right)^2 - \int_0^L dx\, F(x)\, \eta(x).$$

(b) The corresponding action is

$$S = \tfrac{1}{2} \int_0^\tau dt \int_0^L dx \left[\rho(x)\left(\frac{\partial \eta}{\partial t}\right)^2 - T_e \left(\frac{\partial \eta}{\partial x}\right)^2 + 2F(x)\eta(x)\right].$$

For a static string, $\eta_t = 0$ and all functions are independent of t, so the action simplifies to

$$S = \frac{1}{2} \int_0^\tau dt \int_0^L dx \left[-T_e \left(\frac{\partial \eta}{\partial x}\right)^2 + 2F(x)\eta(x)\right].$$

Using equation (4.159) we obtain the equation of motion $T_e \eta_{xx} + F(x) = 0$. Integrating this once gives

$$\frac{\partial \eta}{\partial x} = B - \frac{1}{T_e}\int_0^x du\, F(u),$$

where B is a constant. Integrating again gives

$$\eta(x) = A + Bx - \frac{1}{T_e}\int_0^x dv \int_0^v du\, F(u) = A + Bx - \frac{1}{T_e}\int_0^x du\,(x-u)F(u),$$

where A is another constant. The boundary conditions, $\eta(0) = \eta(L) = 0$ give $A = 0$ and $BL - \frac{1}{T_e}\int_0^L du\,(L-u)F(u) = 0$, respectively, so that

$$\eta(x) = \frac{1}{T_e}\left[\frac{x}{L}\int_0^L du\,(L-u)F(u) - \int_0^x du\,(x-u)F(u)\right].$$

If $F = k$, for some constant k these integrals give $\eta = kx(L-x)/(2T_e)$.

(c) Using equation (4.159) we obtain the equation of motion

$$\rho(x)\frac{\partial^2 \eta}{\partial t^2} - T_e \frac{\partial^2 \eta}{\partial x^2} - F(x) = 0.$$

Solution 4.40

(a) On any surface two coordinates are needed to specify a position, so this system has two degrees of freedom. Possible coordinates are the latitude and longitude.

(b) If the rod were rigid the system would be a simple pendulum, with one degree of freedom. Because the length of the rod can change another coordinate is required to specify the position of the particle, so the system has two degrees of freedom. Two possible coordinates are the angle between the rod and the downward vertical and the length of the rod.

(c) The only difference between this and the previous example is that the end of the rigid rod moves on a sphere, not a vertical plane, and requires two coordinates to define its position. Thus for the elastic rod, three coordinates

Solutions to Exercises in Chapter 4

are needed and the system has three degrees of freedom: possible generalised coordinates are the two spherical polar angles of the rod.

(d) If the end of the rod on the curve were fixed we should have a conventional vertical pendulum, with one degree of freedom. Because its hinge can move this adds a second degree of freedom. Coordinates could be the angle between the rod and the downward vertical and the distance of the hinge along the curve.

Solution 4.41

(a) The equation of motion for L_1 is $d(\dot{q}+q)/dt - (\dot{q}+q)$ or $\ddot{q}-q=0$. The equation of motion for L_2 is $d(\dot{q})/dt - q$ or $\ddot{q}-q=0$.

(b) We have
$$L_1 = \tfrac{1}{2}(\dot{q}^2+q^2) + q\dot{q} = L_2 + \tfrac{1}{2}\frac{d}{dt}(q^2).$$

Solution 4.42

(a) The independent variables are the coordinates (x_1, y_1, z_1) and (x_2, y_2, z_2). Differentiate r^2 with respect to x_1 and x_2 to obtain
$$2r\frac{\partial r}{\partial x_1} = 2(x_1-x_2) \quad \text{and} \quad 2r\frac{\partial r}{\partial x_2} = -2(x_1-x_2),$$
which gives the quoted results. Differentiation with respect to (y_1, y_2) and (z_1, z_2) gives similar relations.

(b) The chain rule gives
$$\frac{\partial V}{\partial x_k} = \frac{\partial V}{\partial r}\frac{\partial r}{\partial x_k}, \quad \frac{\partial V}{\partial y_k} = \frac{\partial V}{\partial r}\frac{\partial r}{\partial y_k}, \quad \frac{\partial V}{\partial z_k} = \frac{\partial V}{\partial r}\frac{\partial r}{\partial z_k}, \quad k=1,2.$$
Since $\partial r/\partial x_1 = -\partial r/\partial x_2$, etc., we see that $\partial V/\partial x_1 = -\partial V/\partial x_2$, etc. Hence
$$\mathbf{F}_{12} = -\frac{\partial V}{\partial \mathbf{r}_1} = -\frac{\partial V}{\partial r}\left(\frac{x_1-x_2}{r}, \frac{y_1-y_2}{r}, \frac{z_1-z_2}{r}\right) = -\frac{\mathbf{r}_1-\mathbf{r}_2}{r}\frac{\partial V}{\partial r}$$
and $\mathbf{F}_{21} = -\partial V/\partial \mathbf{r}_2 = -\mathbf{F}_{12}$.

Solution 4.43

The generalised coordinates are u, v and ϕ, and differentiation gives
$$\dot{x} = (\dot{u}v+u\dot{v})\cos\phi - uv\dot{\phi}\sin\phi,$$
$$\dot{y} = (\dot{u}v+u\dot{v})\sin\phi + uv\dot{\phi}\cos\phi,$$
$$\dot{z} = u\dot{u} - v\dot{v}.$$

The speed is $v^2 = \dot{x}^2+\dot{y}^2+\dot{z}^2$; it is easiest to first evaluate $\dot{x}^2+\dot{y}^2$:
$$\dot{x}^2+\dot{y}^2 = (\dot{u}v+u\dot{v})^2\cos^2\phi - 2(\dot{u}v+u\dot{v})uv\dot{\phi}\sin\phi\cos\phi + u^2v^2\dot{\phi}^2\sin^2\phi$$
$$+ (\dot{u}v+u\dot{v})^2\sin^2\phi + 2(\dot{u}v+u\dot{v})uv\dot{\phi}\sin\phi\cos\phi + u^2v^2\dot{\phi}^2\cos^2\phi$$
$$= (\dot{u}v+u\dot{v})^2 + (uv\dot{\phi})^2.$$

Hence, on adding $\dot{z}^2$ to this, we obtain
$$\dot{x}^2+\dot{y}^2+\dot{z}^2 = (\dot{u}v+u\dot{v})^2 + (uv\dot{\phi})^2 + (u\dot{u}-v\dot{v})^2$$
$$= (u^2+v^2)(\dot{u}^2+\dot{v}^2) + u^2v^2\dot{\phi}^2.$$

This is a sum of the squares of the generalised velocities, with coefficients depending upon the generalised coordinates.

Solution 4.44

(a) A neat way of obtaining the inverse transformation is to use matrices. The transformation is linear and can be represented as follows

$$\begin{pmatrix} u \\ v \end{pmatrix} = R(\Omega t) \begin{pmatrix} x \\ y \end{pmatrix}, \quad R(\theta) = \begin{pmatrix} \cos\theta & \sin\theta \\ -\sin\theta & \cos\theta \end{pmatrix}.$$

So

$$\begin{pmatrix} x \\ y \end{pmatrix} = R(\Omega t)^{-1} \begin{pmatrix} u \\ v \end{pmatrix}, \quad R(\theta)^{-1} = \begin{pmatrix} \cos\theta & -\sin\theta \\ \sin\theta & \cos\theta \end{pmatrix}.$$

Hence

$$x = u\cos\Omega t - v\sin\Omega t, \quad y = u\sin\Omega t + v\cos\Omega t.$$

(b) Differentiating these expressions and putting $c = \cos\Omega t$ and $s = \sin\Omega t$ gives

$$\dot{x} = \dot{u}c - \dot{v}s + \Omega(-us - vc) \quad \text{and} \quad \dot{y} = \dot{u}s + \dot{v}c + \Omega(uc - vs).$$

Squaring and adding these gives

$$\dot{x}^2 + \dot{y}^2 = (\dot{u}c - \dot{v}s)^2 + (\dot{u}s + \dot{v}c)^2$$
$$+ \Omega^2 \left[(-us - vc)^2 + (uc - vs)^2\right]$$
$$+ 2\Omega\left[(\dot{u}c - \dot{v}s)(-us - vc) + (\dot{u}s + \dot{v}c)(uc - vs)\right].$$

Expanding all these terms gives

$$V^2 = \dot{u}^2 + \dot{v}^2 + \Omega^2 \left(u^2 + v^2\right) - 2\Omega(\dot{u}v - u\dot{v}).$$

(c) We have $\dot{u} = \dot{\rho}\cos\phi - \rho\dot{\phi}\sin\phi$, $\dot{v} = \dot{\rho}\sin\phi + \rho\dot{\phi}\cos\phi$, giving

$$u^2 + v^2 = \rho^2, \quad \dot{u}^2 + \dot{v}^2 = \dot{\rho}^2 + (\rho\dot{\phi})^2, \quad \dot{u}v - u\dot{v} = -\rho^2\dot{\phi},$$

hence

$$V^2 = \dot{\rho}^2 + (\rho\dot{\phi})^2 + 2\Omega\rho^2\dot{\phi} + \Omega^2\rho^2 = \dot{\rho}^2 + \rho^2\left(\dot{\phi} + \Omega\right)^2.$$

Solution 4.45

We consider both situations together. If l is the length of the string and z the distance of the hanging block below the pulley then, because all components are moving with speed $\dot{z}$ the kinetic energy is

$$T = \tfrac{1}{2}M\dot{z}^2 + \tfrac{1}{2}M\dot{z}^2 + \tfrac{1}{2}m\dot{z}^2.$$

The potential energy of the hanging block is $-Mgz$ (because the distance is measured downwards). The mass of string above the pulley has potential energy that is independent of z. The mass of string below the pulley is mz/l, and the centre of mass of this is a distance $z/2$ below the pulley, so the potential energy of this portion of the string is $g \times (mz/l) \times (-z/2)$. Hence the total potential energy is

$$V = -Mgz - \frac{mg}{2l}z^2$$

and the Lagrangian is

$$L = \tfrac{1}{2}(2M + m)\dot{z}^2 + Mgz + \frac{mg}{2l}z^2.$$

In the first case the string has zero mass so $L = M\dot{z}^2 + Mgz$.

Solution 4.46

Let q be the distance between the particle and the origin, so its coordinates are $x = q\cos\alpha$ and $y = q\sin\alpha$. The square of the speed is
$$v^2 = (\dot{q}\cos\alpha - q\dot{\alpha}\sin\alpha)^2 + (\dot{q}\sin\alpha + q\dot{\alpha}\cos\alpha)^2 = \dot{q}^2 + q^2\dot{\alpha}^2.$$
The potential energy is $V = mgy$, so the Lagrangian is
$$L = \tfrac{1}{2}m\left(\dot{q}^2 + q^2\dot{\alpha}^2\right) - mgq\sin\alpha.$$
The equation of motion is $\ddot{q} - q\dot{\alpha}^2 + g\sin\alpha = 0$ and with $\alpha = \beta t$ this becomes
$$\ddot{q} - \beta^2 q = -g\sin\beta t, \quad q \geq 0.$$
The solution of this linear equation comprises two parts; first there is the general solution of the homogeneous equation $\ddot{q} - q\beta^2 = 0$, $q_1(t) = A\cosh(\beta t) + B\sinh(\beta t)$. Then there is a particular solution of the inhomogeneous equation, which is clearly of the form $q_p(t) = C\sin(\beta t)$ for some constant C. This is found by substituting into the equation of motion:
$$C\left(-\beta^2 - \beta^2\right)\sin(\beta t) = -g\sin(\beta t), \quad \text{so} \quad C = g/2\beta^2,$$
giving the solution satisfying $q(0) = A$,
$$q(t) = A\cosh(\beta t) + B\sinh(\beta t) + \frac{g}{2\beta^2}\sin(\beta t).$$

Solution 4.47

The speed of the particle is given by
$$\begin{aligned}v^2 &= \dot{x}^2 + \dot{y}^2 = (\dot{q} + \dot{\gamma})^2 + (aq\dot{q})^2 \\ &= \dot{q}^2\left(1 + a^2q^2\right) + 2\dot{q}\dot{\gamma} + \dot{\gamma}^2 \\ &= \dot{q}^2\left(1 + a^2q^2\right) - 2q\ddot{\gamma} + \left[\dot{\gamma}^2 + 2\frac{d}{dt}(q\dot{\gamma})\right],\end{aligned}$$
where the terms in the square brackes can be ignored because they are time derivatives. The potential energy is $V = mgy$, hence the Lagrangian is
$$L = \tfrac{1}{2}m\dot{q}^2\left(1 + a^2q^2\right) - \tfrac{1}{2}magq^2 - mq\ddot{\gamma}.$$

Solution 4.48

The square of the speed is $\dot{x}^2 + \dot{y}^2$ and since $\dot{y} = ax\dot{x} + \dot{a}x^2/2$ we have
$$v^2 = \dot{x}^2 + \left(ax\dot{x} + \tfrac{1}{2}\dot{a}x^2\right)^2 = \dot{x}^2\left(1 + a^2x^2\right) + a\dot{a}x^3\dot{x} + \tfrac{1}{4}\dot{a}^2x^4,$$
so the Lagrangian is
$$L = \tfrac{1}{2}m\dot{x}^2\left(1 + a^2x^2\right) + \tfrac{1}{2}ma\dot{a}x^3\dot{x} + \tfrac{1}{8}m\dot{a}^2x^4 - \tfrac{1}{2}mgax^2.$$
But,
$$a\dot{a}x^3\dot{x} = \tfrac{1}{4}a\dot{a}\frac{dx^4}{dt} = \tfrac{1}{4}\frac{d}{dt}\left(a\dot{a}x^4\right) - \tfrac{1}{4}x^4\frac{d}{dt}(a\dot{a})$$
so that, on ignoring the first term, the Lagrangian becomes
$$L = \tfrac{1}{2}m\dot{x}^2\left(1 + a^2x^2\right) - \tfrac{1}{2}mgax^2 - \tfrac{1}{8}mx^4\gamma(t), \quad \text{where} \quad \gamma = \tfrac{1}{2}\frac{d^2a^2}{dt^2} - \left(\frac{da}{dt}\right)^2.$$
The partial derivatives required for the equation of motion are:
$$\frac{\partial L}{\partial \dot{x}} = m\dot{x}\left(1 + a^2x^2\right), \quad \frac{\partial L}{\partial x} = ma^2x\dot{x}^2 - mgax - \tfrac{1}{2}m\gamma(t)x^3,$$
so the equation of motion is
$$\frac{d}{dt}\left(\dot{x}\left(1 + a^2x^2\right)\right) - a^2x\dot{x}^2 + gax + \tfrac{1}{2}m\gamma(t)x^3 = 0.$$
Expanding this gives
$$\ddot{x}\left(1 + a^2x^2\right) + 2a\dot{a}x^2\dot{x} + a^2x\dot{x}^2 + gax + \tfrac{1}{2}\gamma(t)x^3 = 0.$$

If $\dot{a} = 0$ this reduces to equation (4.5) (page 147).

Solution 4.49

The action is
$$S = \tfrac{1}{2}m \int_{t_1}^{t_2} dt\, \left(\dot{x}^2 + \dot{y}^2 + \dot{z}^2\right).$$

If $(x(t), y(t), z(t))$ is the actual path and $(x(t) + \epsilon\,\xi(t), y(t) + \epsilon\,\eta(t), z(t) + \epsilon\,\zeta(t))$ is an admissible varied path, then (ξ, η, ζ) are zero at $t = t_1$ and t_2 and the change in the action is
$$\delta S = \frac{m}{2} \int_{t_1}^{t_2} dt\, \left[\left(\dot{x} + \epsilon\dot{\xi}\right)^2 - \dot{x}^2 + \left(\dot{y} + \epsilon\dot{\eta}\right)^2 - \dot{y}^2 + \left(\dot{z} + \epsilon\dot{\zeta}\right)^2 - \dot{z}^2\right]$$
$$= m\epsilon \int_{t_1}^{t_2} dt\, \left(\dot{x}\dot{\xi} + \dot{y}\dot{\eta} + \dot{z}\dot{\zeta}\right) + \tfrac{1}{2}m\epsilon^2 \int_{t_1}^{t_2} dt\, \left(\dot{\xi}^2 + \dot{\eta}^2 + \dot{\zeta}^2\right).$$

But integration by parts gives
$$\int_{t_1}^{t_2} dt\, \dot{x}\dot{\xi} = [\xi\dot{x}]_{t_1}^{t_2} - \int_{t_1}^{t_2} dt\, \xi\ddot{x}.$$

Since $\xi(t_1) = \xi(t_2) = 0$ and $\ddot{x} = 0$, because there are no forces, this integral is zero. Similarly, the integrals over $\dot{y}\dot{\eta}$ and $\dot{z}\dot{\zeta}$ are zero. Hence
$$\delta S = \tfrac{1}{2}m\epsilon^2 \int_{t_1}^{t_2} dt\, \left(\dot{\xi}^2 + \dot{\eta}^2 + \dot{\zeta}^2\right).$$

Thus $\delta S > 0$ (unless $|\xi| + |\eta| + |\zeta| = 0$), so the action is a minimum on the stationary path.

Solution 4.50

Use the notation defined in the figure and put the z-axis along the governor axis BA. The heights of the masses m_1 and m_2 are $z_1 = -a\cos\theta$, $z_2 = -2a\cos\theta$, so that the total potential energy is
$$V(\theta) = 2m_1 g z_1 + m_2 g z_2 = -2ag\,(m_1 + m_2)\cos\theta.$$

Since the particle of mass m_2 moves only along the z-axis its kinetic energy is $T_2 = \tfrac{1}{2}m_2\dot{z}_2^2 = 2m_2 a^2 \dot{\theta}^2 \sin^2\theta$. The motion of either one of the particles of mass m_1 is more complicated as it moves in three dimensions. Consider the one on the right in the figure. In the plane of the paper it has the vertical velocity component $\dot{z}_1$ and also the horizontal velocity component $\dot{x}_1$ where $x_1 = a\sin\theta$; there is also a component of velocity into the paper due to the rotation of the system about the axis, which is $x_1\Omega$. Thus the speed v_1 of this particle is given by
$$v_1^2 = a^2\dot{\theta}^2\sin^2\theta + a^2\dot{\theta}^2\cos^2\theta + a^2\Omega^2\sin^2\theta = a^2\left(\dot{\theta}^2 + \Omega^2\sin^2\theta\right).$$

The speed of the other particle of mass m_1 is clearly the same. It follows that the combined kinetic energy of these particles is $T_1 = m_1 a^2\left(\dot{\theta}^2 + \Omega^2\sin^2\theta\right)$ and hence the Lagrangian is
$$L(\theta, \dot{\theta}) = m_1 a^2\left(\dot{\theta}^2 + \Omega^2\sin^2\theta\right) + 2m_2 a^2\dot{\theta}^2\sin^2\theta + 2ag\,(m_1 + m_2)\cos\theta$$
$$= a^2\dot{\theta}^2\left(m_1 + 2m_2\sin^2\theta\right) + m_1 a^2\Omega^2\sin^2\theta + 2ag\,(m_1 + m_2)\cos\theta.$$

The energy, E, is given by the first integral
$$E = \dot{\theta}\frac{\partial L}{\partial\dot{\theta}} - L$$
$$= 2a^2\dot{\theta}^2\left(m_1 + 2m_2\sin^2\theta\right) - L$$
$$= a^2\dot{\theta}^2\left(m_1 + 2m_2\sin^2\theta\right) - m_1 a^2\Omega^2\sin^2\theta - 2ag(m_1 + m_2)\cos\theta.$$

Solution 4.51

The string is always tangential to the surface of the perpendicular, cylindrical post; let θ be the angle between the radius to this tangential point and some fixed radius, as shown in the diagram, and $R(t)$ the length of the string from this point to the particle.

The distance R depends upon θ: the total length of the rope is, to within an additive constant, $R + a\theta$, so $\dot{R} = -a\dot{\theta}$. We now determine the coordinates of the mass m with respect to the axes Oxy, shown in Figure 4.26.

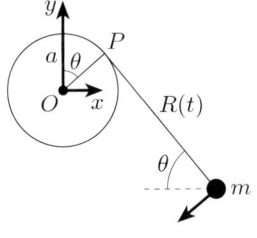

Figure 4.26

Point P, where the rope leaves the cylinder, has coordinates $(a\sin\theta, a\cos\theta)$. The x and y distances of the mass from P are, respectively, $R\cos\theta$ and $-R\sin\theta$, so the coordinates of the mass are $(a\sin\theta + R\cos\theta, a\cos\theta - R\sin\theta)$. The velocity components are

$$\dot{x} = (a\cos\theta - R\sin\theta)\dot{\theta} + \dot{R}\cos\theta \quad \text{and} \quad \dot{y} = -(a\sin\theta + R\cos\theta)\dot{\theta} - \dot{R}\sin\theta,$$

so that the square of the speed of the mass is

$$v^2 = (a^2 + R^2)\dot{\theta}^2 + \dot{R}^2 + 2a\dot{R}\dot{\theta}.$$

But $\dot{\theta} = -\dot{R}/a$ so this reduces to $v^2 = (R\dot{R}/a)^2$.

The kinetic energy is therefore $T = m(R\dot{R})^2/(2a^2)$. Since the gravitational forces are to be neglected the potential energy is constant and the Lagrangian is $L = T$. The first integral, equation (4.84), is

$$E = \dot{R}\frac{\partial L}{\partial \dot{R}} - L = T = \frac{m}{2a^2}R^2\dot{R}^2.$$

Solution 4.52

Take the origin O to be the centre of the circle and the generalised coordinate to be θ, as shown in Figure 4.27. The potential energy is $V(\theta) = -mgR\cos\theta$ and the speed of the particle is $v^2 = R^2\dot{\theta}^2 + (\Omega R\sin\theta)^2$, the last term being the component of the velocity due to the rotation. The Lagrangian is

$$L = \tfrac{1}{2}mR^2\dot{\theta}^2 + \tfrac{1}{2}m\Omega^2 R^2\sin^2\theta + mgR\cos\theta.$$

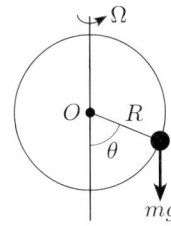

Figure 4.27

Solution 4.53

(a) Lagrange's equation for the Lagrangian aL is

$$\frac{d}{dt}\left(a\frac{\partial L}{\partial \dot{q}}\right) - a\frac{\partial L}{\partial q} = a\left[\frac{d}{dt}\left(\frac{\partial L}{\partial \dot{q}}\right) - \frac{\partial L}{\partial q}\right] = 0.$$

Thus a solution $q(t)$ of Lagrange's equation for L is also a solution of Lagrange's equation for aL, and vice versa.

(b) If $Q = aq$ and $T = bt$, then

$$V(q) = V(Q/a) = a^{-\lambda}V(Q) \quad \text{and} \quad \frac{dq}{dt} = \frac{d(Q/a)}{d(T/b)} = \frac{b}{a}\frac{dQ}{dT},$$

and the Lagrangian is

$$\overline{L}\left(Q, \frac{dQ}{dT}\right) = L(q, \dot{q}) = \tfrac{1}{2}m\left(\frac{b}{a}\right)^2\left(\frac{dQ}{dT}\right)^2 - a^{-\lambda}V(Q)$$

$$= a^{-\lambda}\left(\tfrac{1}{2}m\left(\frac{dQ}{dT}\right)^2 - V(Q)\right) \quad \text{if } b = a^{1-\lambda/2}.$$

(c) If $q = f(t)$ is a solution of the original equation of motion for $L(q, \dot{q})$ then $Q = af(T/b)$ is a solution of the equation of motion for $\overline{L}(Q, dQ/dT)$. Using the result of part (a) we see that when $b = a^{1-\lambda/2}$, $Q = af(T/b)$ is a solution of the equation of motion for $a^\lambda \overline{L}(Q, dQ/dT)$, and since this is formally the same as the original Lagrangian we conclude that $q = af(t/b)$ will be a solution of the equation of motion for $L(q, \dot{q})$. But, in this case

$$af(t/b) = af\left(a^{\lambda/2-1}t\right).$$

In particular, if $f(t)$ is a periodic solution with unit amplitude and period τ then $Af(A^{\lambda/2-1}t)$ is a periodic solution with amplitude A, and its period is $A^{1-\lambda/2}\tau$.

(d) If $V(q) \propto q^2$ then $\lambda = 2$ and $A^{1-\lambda/2} = 1$. Therefore the period is independent of the amplitude in this case.

(e) If $V(q) \propto q^{-1}$ then $\lambda = -1$ and $A^{1-\lambda/2} = A^{3/2}$, so the period increases as the 3/2 power of the amplitude.

Kepler's third law of planetary motion states that the square of the period of each planet in its elliptical orbit is proportional to the cube of the orbit's semi-major axis. This is analogous to the 3/2 power relationship between period and amplitude obtained above for one-dimensional motion; of course the gravitational potential varies as r^{-1}, where r is the planet's distance from the Sun. The following table of the mean distances a of each planet from the Sun and their periods confirms Kepler's third law.

Planet	Mean distance a		Period τ	$a\tau^{-2/3}$
	million miles	AU	(years)	
Mercury	35.98	0.3871	0.2409	92.94
Venus	67.23	0.7233	0.6152	92.91
Earth	92.96	1.000	1.0000	92.96
Mars	141.7	1.524	1.881	93.00
Jupiter	483.6	5.203	11.86	93.00
Saturn	886.7	9.539	29.46	92.93
Uranus	1783.1	19.18	84.01	92.95
Neptune	2794.1	30.06	164.8	92.96
Pluto	3666.1	39.44	247.7	92.93

Solution 4.54

In this example the independent variable, t, is unchanged so $\phi = 0$, in equation (3.131) (page 119), $\psi_1 = 0$ and $\psi_2 =$ constant: here we have used the correspondence $y_1 = q_1$, $y_2 = q_2$ and $x = t$. Hence, equation (3.133) becomes $\partial L/\partial \dot{q}_2 =$ constant, which is the integrated form of Lagrange's equation of motion for q_2, that is

$$0 = \frac{d}{dt}\left(\frac{\partial L}{\partial \dot{q}_2}\right) - \frac{\partial L}{\partial q_2} = \frac{d}{dt}\left(\frac{\partial L}{\partial \dot{q}_2}\right).$$

INDEX

action   176
admissible functions   44
Aristotle   24, 156
astronomical unit   159

Baliani, G B   183
Bernoulli
    Daniel   152
    James   143
    Johann   21, 62, 152, 190
brachistochrone   21, 62
    functional for   65
bubbles   60

Cam, bridge over   63
catenary equation   26
centre of mass   160
Clairaut, A C   156
Closter, S   63
configuration space   165
conservation of energy   178
conservative force   169
conservative system   173
constant of the motion   114, 173
constraining force   27
constraint   163
coupled differential equations   105
cycloid   21, 62, 179
    area and length   64
cycloidal pendulum   64

d'Alembert, J R   143
degrees of freedom   165
Descartes, R   63
distance along a curve   9
dot notation   145
double pendulum   152, 186
du Bois-Reymond, E H   50

Eddington, A S   13
Einstein, A   13, 157
energy integral   178
equations of motion   145
escape velocity   173
Essex, J   63
Euclid   7, 24
Euler, L   42, 152
Euler–Lagrange equation   41, 48
external force   159

Fermat, P de   24, 63
Fermat's principle   23
first integral   49, 114, 120
force of constraint   27, 163
free particle   174
frustum   52
functional   5
    differentiation of   43
    stationary value of   45
fundamental lemma   47

g
    variation with height   172
    variation with latitude   202
Gâteaux differential   45, 105, 193
Galilean relativity   157
Galileo   21, 62, 149, 155, 183
General theory of relativity   13
generalised coordinates   165
generalised velocities   167
geodesic   13
Goldschmidt solution   57, 60
gravitational lensing   13
gravity   171
great circle   13

Halley, E   156, 172
Hamilton's Principle   175
Hamilton, W R   143
hanging chain   25
Hero of Alexandria   24
hindered rotations   182
holonomic   164
homogeneous functions   198
Hooke's law   170
Huygens, C   63, 155, 183

ignorable coordinate   176
integral of the motion   114, 173
internal force   159, 161
invariant functional   118
isochronous   150
isoperimetric problem   22

Kepler's laws   156
Kepler, J   156
kinetic energy   166
Klein–Gordon equation   209

Lagrange, J-L   42, 143, 190
Lagrange's equations   176
Lagrangian   174
Lalande, J-J L de   156
Lalouvère, A de   63
Leibniz, G W   21, 181
Lepute, N-R   156

Mach, E   157
Maupertius, P-L M de   180
Mersenne, M   183

Newton, I   21, 62, 155
Newton's equations of motion   146, 158
Newton's laws   146
Newton's second law   26
Noether, P de   119
Noether's theorem   120
nonholonomic   164
norm   44

Pappus of Alexandria   23

parabolic coordinates   196
particle   144
Pascal, B   63
pendulum
    cycloidal   64
    double   152, 186
    Shortt   183
    simple   26, 149, 182
pendulum clock   63
Poincaré, J H   143, 157
Poisson, S D   181
potential energy   169
potential function   169
principle of least action   180

rigid body   145
Roberval, G. P. de   63

scalar potential   169
Shortt pendulum   183
simple pendulum   149, 182
Smith, R   63
Snell's law   25
soap films   60
speed of light   24, 157

spinning top   164
St Paul's Cathedral   63
stationary curve   10
stationary function   10
stationary path   10
stationary point   10, 45
    classification   14
surface of revolution
    minimum area   22, 51

Three-body problem   159
Trinity College, Cambridge   62
tropical year   159
Tycho Brahe   156

uncoupled equations   107

variational principle   5
visa viva   181

Wallis, J   63, 156
work   169
Wren, C   63, 156

Zenodorus   6, 23